Algorithms for Optimization

Algorithms for Optimization

Mykel J. Kochenderfer
Tim A. Wheeler

The MIT Press
Cambridge, Massachusetts
London, England

This book was set in TEX Gyre Pagella by the authors in LATEX.
Printed and bound in the United States of America.

Library of Congress Cataloging-in-Publication Data

Names: Kochenderfer, Mykel J., 1980– author. | Wheeler, Tim A. (Tim Allan), author.
Title: Algorithms for optimization / Mykel J. Kochenderfer and Tim A. Wheeler.
Description: Cambridge, MA : The MIT Press, [2019] | Includes bibliographical references and index.
Identifiers: LCCN 2018023091 | ISBN 9780262039420 (hardcover : alk. paper)
Subjects: LCSH: Algorithms. | Algorithms—Problems, exercises, etc.
Classification: LCC QA9.58 .K65425 2019 | DDC 518/.1—dc23
LC record available at https://lccn.loc.gov/2018023091

10 9 8 7 6 5 4

To our families.

Contents

Preface

This book provides a broad introduction to optimization with a focus on practical algorithms for the design of engineering systems. We cover a wide variety of optimization topics, introducing the underlying mathematical problem formulations and the algorithms for solving them. Figures, examples, and exercises are provided to convey the intuition behind the various approaches.

This text is intended for advanced undergraduates and graduate students as well as professionals. The book requires some mathematical maturity and assumes prior exposure to multivariable calculus, linear algebra, and probability concepts. Some review material is provided in the appendix. Disciplines where the book would be especially useful include mathematics, statistics, computer science, aerospace, electrical engineering, and operations research.

Fundamental to this textbook are the algorithms, which are all implemented in the Julia programming language. We have found the language to be ideal for specifying algorithms in human readable form. Permission is granted, free of charge, to use the code snippets associated with this book, subject to the condition that the source of the code is acknowledged. We anticipate that others may want to contribute translations of these algorithms to other programming languages. As translations become available, we will link to them from the book's webpage.

Mykel J. Kochenderfer
Tim A. Wheeler
Stanford, Calif.
October 30, 2018

Ancillary material is available on the book's webpage:
`http://mitpress.mit.edu/algorithms-for-optimization`

Acknowledgments

This textbook has grown from a course on engineering design optimization taught at Stanford. We are grateful to the students and teaching assistants who have helped shape the course over the past five years. We are also indebted to the faculty who have taught variations of the course before in our department based on lecture notes from Joaquim Martins, Juan Alonso, Ilan Kroo, Dev Rajnarayan, and Jason Hicken. Many of the topics discussed in this textbook were inspired by their lecture notes.

The authors wish to thank the many individuals who have provided valuable feedback on early drafts of our manuscript, including Mohamed Abdelaty, Atish Agarwala, Ross Alexander, Piergiorgio Alotto, David Ata, Rishi Bedi, Felix Berkenkamp, Raunak Bhattacharyya, Hans Borchers, Maxime Bouton, Ellis Brown, Abhishek Cauligi, Mo Chen, Zhengyu Chen, Vince Chiu, Anthony Corso, Holly Dinkel, Jonathan Cox, Katherine Driggs-Campbell, Thai Duong, Hamza El-Saawy, Sofiane Ennadir, Kaijun Feng, Tamas Gal, Christopher Lazarus Garcia, Wouter Van Gijseghem, Michael Gobble, Robert Goedman, Jayesh Gupta, Aaron Havens, Richard Hsieh, Zdeněk Hurák, Masha Itkina, Arec Jamgochian, Bogumił Kamiński, Walker Kehoe, Mindaugas Kepalas, Veronika Korneyeva, Erez Krimsky, Petr Krysl, Jessie Lauzon, Ruilin Li, Iblis Lin, Edward Londner, Charles Lu, Miles Lubin, Marcus Luebke, Jacqueline Machesky, Ashe Magalhaes, Zouhair Mahboubi, Pranav Maheshwari, Travis McGuire, Jeremy Morton, Robert Moss, Santiago Padrón, Jimin Park, Harsh Patel, Christian Peel, Derek Phillips, Brad Rafferty, Sidd Rao, Andreas Reschka, Alex Reynell, Stuart Rogers, Per Rutquist, Ryan Samuels, Orson Sandoval, Jeffrey Sarnoff, Chelsea Sidrane, Sumeet Singh, Nathan Stacey, Ethan Strijbosch, Alex Toews, Pamela Toman, Rachael Tompa, Zacharia Tuten, Raman Vilkhu, Yuri Vishnevsky, Julie Walker, Zijian Wang, Patrick Washington, Jacob West, Adam Wiktor, Brandon Yeung, Anil Yildiz, Robert Young,

Javier Yu, Andrea Zanette, and Remy Zawislak. In addition, it has been a pleasure working with Marie Lufkin Lee and Christine Bridget Savage from the MIT Press in preparing this manuscript for publication.

The style of this book was inspired by Edward Tufte. Among other stylistic elements, we adopted his wide margins and use of small multiples. In fact, the typesetting of this book is heavily based on the Tufte-LaTeX package by Kevin Godby, Bil Kleb, and Bill Wood. We were also inspired by the clarity of the textbooks by Donald Knuth and Stephen Boyd.

Over the past few years, we have benefited from discussions with the core Julia developers, including Jeff Bezanson, Stefan Karpinski, and Viral Shah. We have also benefited from the various open source packages on which this textbook depends (see appendix A.4). The typesetting of the code is done with the help of pythontex, which is maintained by Geoffrey Poore. Plotting is handled by pgfplots, which is maintained by Christian Feuersänger. The book's color scheme was adapted from the Monokai theme by Jon Skinner of Sublime Text. For plots, we use the viridis colormap defined by Stéfan van der Walt and Nathaniel Smith.

1   *Introduction*

Many disciplines involve optimization at their core. In physics, systems are driven to their lowest energy state subject to physical laws. In business, corporations aim to maximize shareholder value. In biology, fitter organisms are more likely to survive. This book will focus on optimization from an engineering perspective, where the objective is to design a system that optimizes a set of metrics subject to constraints. The system could be a complex physical system like an aircraft, or it could be a simple structure such as a bicycle frame. The system might not even be physical; for example, we might be interested in designing a control system for an automated vehicle or a computer vision system that detects whether an image of a tumor biopsy is cancerous. We want these systems to perform as well as possible. Depending on the application, relevant metrics might include efficiency, safety, and accuracy. Constraints on the design might include cost, weight, and structural soundness.

This book is about the *algorithms*, or computational processes, for optimization. Given some representation of the system design, such as a set of numbers encoding the geometry of an airfoil, these algorithms will tell us how to search the space of possible designs with the aim of finding the best one. Depending on the application, this search may involve running physical experiments, such as wind tunnel tests, or it might involve evaluating an analytical expression or running computer simulations. We will discuss computational approaches for addressing a variety of challenges, such as how to search high-dimensional spaces, handling problems where there are multiple competing objectives, and accommodating uncertainty in the metrics.

1.1 A History

We will begin our discussion of the history of algorithms for optimization[1] with the ancient Greek philosophers. Pythagoras of Samos (569–475 BCE), the developer of the Pythagorean theorem, claimed that "the principles of mathematics were the principles of all things,"[2] popularizing the idea that mathematics could model the world. Both Plato (427–347 BCE) and Aristotle (384–322 BCE) used reasoning for the purpose of societal optimization.[3] They contemplated the best style of human life, which involves the optimization of both individual lifestyle and functioning of the state. Aristotelian logic was an early formal process—an algorithm—by which deductions can be made.

Optimization of mathematical abstractions also dates back millennia. Euclid of Alexandria (325–265 BCE) solved early optimization problems in geometry, including how to find the shortest and longest lines from a point to the circumference of a circle. He also showed that a square is the rectangle with the maximum area for a fixed perimeter.[4] The Greek mathematician Zenodorus (200–140 BCE) studied Dido's problem, shown in figure 1.1.

Others demonstrated that nature seems to optimize. Heron of Alexandria (10–75 CE) showed that light travels between points through the path of shortest length. Pappus of Alexandria (290–350 CE), among his many contributions to optimization, argued that the hexagon repeated in honeycomb is the optimal regular polygon for storing honey; its hexagonal structure uses the least material to create a lattice of cells over a plane.[5]

Central to the study of optimization is the use of *algebra*, which is the study of the rules for manipulating mathematical symbols. Algebra is credited to the Persian mathematician al-Khwārizmī (790–850 CE) with the treatise "al-Kitāb al-jabr wal-muqābala," or "The Compendious Book on Calculation by Completion and Balancing." Algebra had the advantage of using Hindu-Arabic numerals, including the use of zero in base notation. The word *al'jabr* is Persian for restoration and is the source for the Western word *algebra*. The term *algorithm* comes from *algoritmi*, the Latin translation and pronunciation of al-Khwārizmī's name.

Optimization problems are often posed as a search in a space defined by a set of coordinates. Use of coordinates comes from René Descartes (1596–1650), who used two numbers to describe a point on a two-dimensional plane. His insight linked algebra, with its analytic equations, to the descriptive and visual field of geometry.[6] His work also included a method for finding the tangent to any curve

[1] This discussion is not meant to be comprehensive. A more detailed history is provided by X.-S. Yang, "A Brief History of Optimization," in *Engineering Optimization*. Wiley, 2010, pp. 1–13.

[2] Aristotle, *Metaphysics*, trans. by W. D. Ross. 350 BCE, Book I, Part 5.

[3] See discussion by S. Kiranyaz, T. Ince, and M. Gabbouj, *Multidimensional Particle Swarm Optimization for Machine Learning and Pattern Recognition*. Springer, 2014, Section 2.1.

[4] See books III and VI of Euclid, *The Elements*, trans. by D. E. Joyce. 300 BCE.

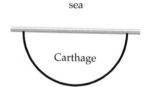

Figure 1.1. Queen Dido, founder of Carthage, was granted as much land as she could enclose with a bullhide thong. She made a semicircle with each end of the thong against the Mediterranean Sea, thus enclosing the maximum possible area. This problem is mentioned in Virgil's *Aeneid* (19 BCE).

[5] T. C. Hales, "The Honeycomb Conjecture," *Discrete & Computational Geometry*, vol. 25, pp. 1–22, 2001.

[6] R. Descartes, "La Géométrie," in *Discours de la Méthode*. 1637.

whose equation is known. Tangents are useful in identifying the minima and maxima of functions. Pierre de Fermat (1601–1665) began solving for where the derivative is zero to identify potentially optimal points.

The concept of *calculus*, or the study of continuous change, plays an important role in our discussion of optimization. Modern calculus stems from the developments of Gottfried Wilhelm Leibniz (1646–1716) and Sir Isaac Newton (1642–1727). Both differential and integral calculus make use of the notion of convergence of infinite series to a well-defined limit.

The mid-twentieth century saw the rise of the electronic computer, spurring interest in numerical algorithms for optimization. The ease of calculations allowed optimization to be applied to much larger problems in a variety of domains. One of the major breakthroughs came with the introduction of linear programming, which is an optimization problem with a linear objective function and linear constraints. Leonid Kantorovich (1912–1986) presented a formulation for linear programming and an algorithm to solve it.[7] It was applied to optimal resource allocation problems during World War II. George Dantzig (1914–2005) developed the simplex algorithm, which represented a significant advance in solving linear programs efficiently.[8] Richard Bellman (1920–1984) developed the notion of dynamic programming, which is a commonly used method for optimally solving complex problems by breaking them down into simpler problems.[9] Dynamic programming has been used extensively for optimal control. This textbook outlines many of the key algorithms developed for digital computers that have been used for various engineering design optimization problems.

Decades of advances in large scale computation have resulted in innovative physical engineering designs as well as the design of artificially intelligent systems. The intelligence of these systems have been demonstrated in games such as chess, Jeopardy!, and Go. IBM's Deep Blue defeated the world chess champion Garry Kasparov in 1996 by optimizing moves by evaluating millions of positions. In 2011, IBM's Watson played Jeopardy! against former winners Brad Futter and Ken Jennings. Watson won the first place prize of $1 million by optimizing its response with respect to probabilistic inferences about 200 million pages of structured and unstructured data. Since the competition, the system has evolved to assist in healthcare decisions and weather forecasting. In 2017, Google's AlphaGo defeated Ke Jie, the number one ranked Go player in the world. The system used neural networks with millions of parameters that were optimized from self-play and

[7] L. V. Kantorovich, "A New Method of Solving Some Classes of Extremal Problems," in *Proceedings of the USSR Academy of Sciences*, vol. 28, 1940.

[8] The simplex algorithm will be covered in chapter 11.

[9] R. Bellman, "On the Theory of Dynamic Programming," *Proceedings of the National Academy of Sciences of the United States of America*, vol. 38, no. 8, pp. 716–719, 1952.

data from human games. The optimization of deep neural networks is fueling a major revolution in artificial intelligence that will likely continue.[10]

[10] I. Goodfellow, Y. Bengio, and A. Courville, *Deep Learning*. MIT Press, 2016.

1.2   *Optimization Process*

A typical engineering design optimization process is shown in figure 1.2.[11] The role of the *designer* is to provide a problem *specification* that details the parameters, constants, objectives, and constraints that are to be achieved. The designer is responsible for crafting the problem and quantifying the merits of potential designs. The designer also typically supplies a baseline design or initial design point to the optimization algorithm.

[11] Further discussion of the design process in engineering is provided in J. Arora, *Introduction to Optimum Design*, 4th ed. Academic Press, 2016.

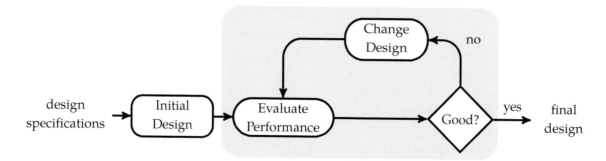

This book is about automating the process of refining the design to improve performance. An optimization algorithm is used to incrementally improve the design until it can no longer be improved or until the budgeted time or cost has been reached. The designer is responsible for analyzing the result of the optimization process to ensure its suitability for the final application. Misspecifications in the problem, poor baseline designs, and improperly implemented or unsuitable optimization algorithms can all lead to suboptimal or dangerous designs.

Figure 1.2. The design optimization process. We seek to automate the optimization procedure highlighted in blue.

There are several advantages of an optimization approach to engineering design. First of all, the optimization process provides a systematic, logical design procedure. If properly followed, optimization algorithms can help reduce the chance of human error in design. Sometimes intuition in engineering design can be misleading; it can be much better to optimize with respect to data. Optimization can speed the process of design, especially when a procedure can be written once and then be reapplied to other problems. Traditional engineering techniques are often visualized and reasoned about by humans in two or three dimensions. Modern optimization techniques, however, can be applied to problems with millions of variables and constraints.

There are also challenges associated with using optimization for design. We are generally limited in our computational resources and time, and so our algorithms have to be selective in how they explore the design space. Fundamentally, the optimization algorithms are limited by the designer's ability to specify the problem. In some cases, the optimization algorithm may exploit modeling errors or provide a solution that does not adequately solve the intended problem. When an algorithm results in an apparently optimal design that is counterintuitive, it can be difficult to interpret. Another limitation is that many optimization algorithms are not guaranteed to produce optimal designs.

1.3 Basic Optimization Problem

The basic optimization problem is:

$$\begin{aligned} \underset{\mathbf{x}}{\text{minimize}} \quad & f(\mathbf{x}) \\ \text{subject to} \quad & \mathbf{x} \in \mathcal{X} \end{aligned} \tag{1.1}$$

Here, $\mathbf{x}$ is a *design point*. A design point can be represented as a vector of values corresponding to different *design variables*. An n-dimensional design point is written:[12]

$$[x_1, x_2, \cdots, x_n] \tag{1.2}$$

where the ith design variable is denoted x_i. The elements in this vector can be adjusted to minimize the *objective function* f. Any value of $\mathbf{x}$ from among all points in the *feasible set* $\mathcal{X}$ that minimizes the objective function is called a *solution* or *minimizer*. A particular solution is written $\mathbf{x}^*$. Figure 1.3 shows an example of a one-dimensional optimization problem.

This formulation is general, meaning that any optimization problem can be rewritten according to equation (1.1). In particular, a problem

$$\underset{\mathbf{x}}{\text{maximize}} \ f(\mathbf{x}) \ \text{subject to} \ \mathbf{x} \in \mathcal{X} \tag{1.3}$$

can be replaced by

$$\underset{\mathbf{x}}{\text{minimize}} - f(\mathbf{x}) \ \text{subject to} \ \mathbf{x} \in \mathcal{X} \tag{1.4}$$

[12] As in Julia, square brackets with comma-separated entries are used to represent column vectors. Design points are column vectors.

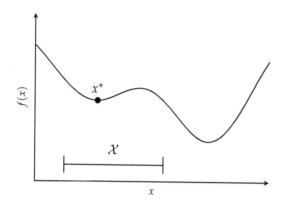

Figure 1.3. A one-dimensional optimization problem. Note that the minimum is merely the best in the feasible set—lower points may exist outside the feasible region.

The new form is the same problem in that it has the same set of solutions.

Modeling engineering problems within this mathematical formulation can be challenging. The way in which we formulate an optimization problem can make the solution process either easy or hard.[13] We will focus on the algorithmic aspects of optimization that arise after the problem has been properly formulated.[14]

Since this book discusses a wide variety of different optimization algorithms, one may wonder which algorithm is best. As elaborated by the *no free lunch theorems* of Wolpert and Macready, there is no reason to prefer one algorithm over another unless we make assumptions about the probability distribution over the space of possible objective functions. If one algorithm performs better than another algorithm on one class of problems, then it will perform worse on another class of problems.[15] For many optimization algorithms to work effectively, there needs to be some regularity in the objective function, such as Lipschitz continuity or convexity, both topics that we will cover later. As we discuss different algorithms, we will outline their assumptions, the motivation for their mechanism, and their advantages and disadvantages.

1.4 Constraints

Many problems have constraints. Each *constraint* limits the set of possible solutions, and together the constraints define the feasible set $\mathcal{X}$. Feasible design points do not violate any constraints. For example, consider the following optimization

[13] See discussion in S. Boyd and L. Vandenberghe, *Convex Optimization*. Cambridge University Press, 2004.

[14] Many texts provide examples of how to translate real-world optimization problems into optimization problems. See, for example, the following: R. K. Arora, *Optimization: Algorithms and Applications*. Chapman and Hall/CRC, 2015. A. D. Belegundu and T. R. Chandrupatla, *Optimization Concepts and Applications in Engineering*, 2nd ed. Cambridge University Press, 2011. A. Keane and P. Nair, *Computational Approaches for Aerospace Design*. Wiley, 2005. P. Y. Papalambros and D. J. Wilde, *Principles of Optimal Design*. Cambridge University Press, 2017.

[15] The assumptions and results of the no free lunch theorems are provided by D. H. Wolpert and W. G. Macready, "No Free Lunch Theorems for Optimization," *IEEE Transactions on Evolutionary Computation*, vol. 1, no. 1, pp. 67–82, 1997.

problem:

$$\underset{x_1, x_2}{\text{minimize}} \quad f(x_1, x_2)$$
$$\text{subject to} \quad x_1 \geq 0 \tag{1.5}$$
$$x_2 \geq 0$$
$$x_1 + x_2 \leq 1$$

The feasible set is plotted in figure 1.4.

Constraints are typically written with $\leq$, $\geq$, or $=$. If constraints involve $<$ or $>$ (i.e., strict inequalities), then the feasible set does not include the constraint boundary. A potential issue with not including the boundary is illustrated by this problem:

$$\underset{x}{\text{minimize}} \quad x$$
$$\text{subject to} \quad x > 1 \tag{1.6}$$

The feasible set is shown in figure 1.5. The point $x = 1$ produces values smaller than any x greater than 1, but $x = 1$ is not feasible. We can pick any x arbitrarily close to, but greater than, 1, but no matter what we pick, we can always find an infinite number of values even closer to 1. We must conclude that the problem has no solution. To avoid such issues, it is often best to include the constraint boundary in the feasible set.

1.5 Critical Points

Figure 1.6 shows a *univariate function*[16] $f(x)$ with several labeled *critical points*, where the derivative is zero, that are of interest when discussing optimization problems. When minimizing f, we wish to find a *global minimizer*, a value of x for which $f(x)$ is minimized. A function may have at most one global minimum, but it may have multiple global minimizers.

Unfortunately, it is generally difficult to prove that a given candidate point is at a global minimum. Often, the best we can do is check whether it is at a *local minimum*. A point x^* is at a local minimum (or is a local minimizer) if there exists a $\delta > 0$ such that $f(x^*) \leq f(x)$ for all x with $|x - x^*| < \delta$. In the multivariate context, this definition generalizes to there being a $\delta > 0$ such that $f(\mathbf{x}^*) \leq f(\mathbf{x})$ whenever $\|\mathbf{x} - \mathbf{x}^*\| < \delta$.

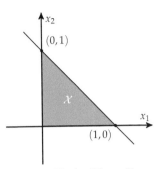

Figure 1.4. The feasible set $\mathcal{X}$ associated with equation (1.5).

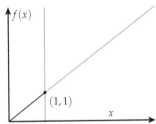

Figure 1.5. The problem in equation (1.6) has no solution because the constraint boundary is not feasible.

[16] A univariate function is a function of a single scalar. The term *univariate* describes objects involving one variable.

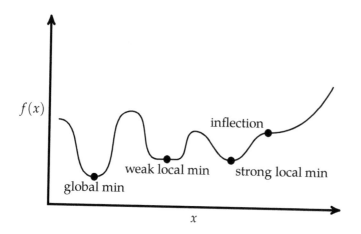

Figure 1.6. Examples of critical points of interest to optimization algorithms (where the derivative is zero) on a univariate function.

Figure 1.6 shows two types of local minima: *strong local minima* and *weak local minima*. A *strong local minimizer*, also known as a *strict local minimizer*, is a point that uniquely minimizes f within a neighborhood. In other words, x^* is a strict local minimizer if there exists a $\delta > 0$ such that $f(x^*) < f(x)$ whenever $x^* \neq x$ and $|x - x^*| < \delta$. In the multivariate context, this generalizes to there being a $\delta > 0$ such that $f(\mathbf{x}^*) < f(\mathbf{x})$ whenever $\mathbf{x}^* \neq \mathbf{x}$ and $\|\mathbf{x} - \mathbf{x}^*\| < \delta$. A weak local minimizer is a local minimizer that is not a strong local minimizer.

The derivative is zero at all local and global minima of continuous, unbounded objective functions. While having a zero derivative is a *necessary condition* for a local minimum,[17] it is not a *sufficient condition*.

Figure 1.6 also has an *inflection point* where the derivative is zero but the point does not locally minimize f. An inflection point is where the sign of the second derivative of f changes, which corresponds to a local minimum or maximum of f'.

[17] Points with nonzero derivatives are never minima.

1.6   Conditions for Local Minima

Many numerical optimization methods seek local minima. Local minima are locally optimal, but we do not generally know whether a local minimum is a global minimum. The conditions we discuss in this section assume that the objective function is differentiable. Derivatives, gradients, and Hessians are reviewed in the next chapter. We also assume in this section that the problem is unconstrained. Conditions for optimality in constrained problems are introduced in chapter 10.

1.6.1 Univariate

A design point is guaranteed to be at a strong local minimum if the local derivative is zero and the second derivative is positive:

1. $f'(x^*) = 0$

2. $f''(x^*) > 0$

A zero derivative ensures that shifting the point by small values does not significantly affect the function value. A positive second derivative ensures that the zero first derivative occurs at the bottom of a *bowl*.[18]

A point can also be at a local minimum if it has a zero derivative and the second derivative is merely nonnegative:

1. $f'(x^*) = 0$, the *first-order necessary condition* (FONC)[19]

2. $f''(x^*) \geq 0$, the *second-order necessary condition* (SONC)

These conditions are referred to as *necessary* because all local minima obey these two rules. Unfortunately, not all points with a zero derivative and a zero second derivative are local minima, as demonstrated in figure 1.7.

The first necessary condition can be derived using the Taylor expansion[20] about our candidate point x^*:

$$f(x^* + h) = f(x^*) + hf'(x^*) + O(h^2) \tag{1.7}$$
$$f(x^* - h) = f(x^*) - hf'(x^*) + O(h^2) \tag{1.8}$$
$$f(x^* + h) \geq f(x^*) \implies hf'(x^*) \geq 0 \tag{1.9}$$
$$f(x^* - h) \geq f(x^*) \implies hf'(x^*) \leq 0 \tag{1.10}$$
$$\implies f'(x^*) = 0 \tag{1.11}$$

where the asymptotic notation $O(h^2)$ is covered in appendix C.

The second-order necessary condition can also be obtained from the Taylor expansion:

$$f(x^* + h) = f(x^*) + \underbrace{hf'(x^*)}_{=0} + \frac{h^2}{2} f''(x^*) + O(h^3) \tag{1.12}$$

[18] If $f'(x) = 0$ and $f''(x) < 0$, then x is a local maximum.

[19] A point that satisfies the first-order necessary condition is sometimes called a *stationary point*.

[20] The Taylor expansion is derived in appendix C.

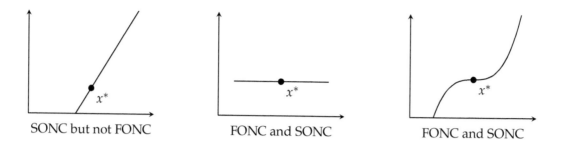

SONC but not FONC   FONC and SONC   FONC and SONC

Figure 1.7. Examples of the necessary but insufficient conditions for strong local minima.

We know that the first-order necessary condition must apply:

$$f(x^* + h) \geq f(x^*) \implies \frac{h^2}{2} f''(x^*) \geq 0 \qquad (1.13)$$

since $h > 0$. It follows that $f''(x^*) \geq 0$ must hold for x^* to be at a local minimum.

1.6.2   Multivariate

The following conditions are necessary for $\mathbf{x}$ to be at a local minimum of f:

1. $\nabla f(\mathbf{x}) = 0$, the first-order necessary condition (FONC)

2. $\nabla^2 f(\mathbf{x})$ is positive semidefinite (for a review of this definition, see appendix C.6), the second-order necessary condition (SONC)

The FONC and SONC are generalizations of the univariate case. The FONC tells us that the function is not changing at $\mathbf{x}$. Figure 1.8 shows examples of multivariate functions where the FONC is satisfied. The SONC tells us that $\mathbf{x}$ is in a bowl.

The FONC and SONC can be obtained from a simple analysis. In order for $\mathbf{x}^*$ to be at a local minimum, it must be smaller than those values around it:

$$f(\mathbf{x}^*) \leq f(\mathbf{x}^* + h\mathbf{y}) \quad \Leftrightarrow \quad f(\mathbf{x}^* + h\mathbf{y}) - f(\mathbf{x}^*) \geq 0 \qquad (1.14)$$

If we write the second-order approximation for $f(\mathbf{x}^*)$, we get:

$$f(\mathbf{x}^* + h\mathbf{y}) = f(\mathbf{x}^*) + h\nabla f(\mathbf{x}^*)^\top \mathbf{y} + \frac{1}{2}h^2 \mathbf{y}^\top \nabla^2 f(\mathbf{x}^*)\mathbf{y} + O(h^3) \qquad (1.15)$$

We know that at a minimum, the first derivative must be zero, and we neglect the higher order terms. Rearranging, we get:

$$\frac{1}{2}h^2 \mathbf{y}^\top \nabla^2 f(\mathbf{x}^*)\mathbf{y} = f(\mathbf{x}^* + h\mathbf{y}) - f(\mathbf{x}^*) \geq 0 \qquad (1.16)$$

This is the definition of a positive semidefinite matrix, and we recover the SONC. Example 1.1 illustrates how these conditions can be applied to the Rosenbrock banana function.

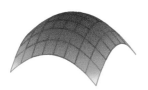

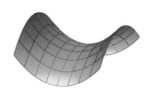

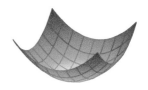

A *local maximum*. The gradient at the center is zero, but the Hessian is negative definite.

A *saddle*. The gradient at the center is zero, but it is not a local minimum.

A *bowl*. The gradient at the center is zero and the Hessian is positive definite. It is a local minimum.

Figure 1.8. The three local regions where the gradient is zero.

While necessary for optimality, the FONC and SONC are not sufficient for optimality. For unconstrained optimization of a twice-differentiable function, a point is guaranteed to be at a strong local minimum if the FONC is satisfied and $\nabla^2 f(\mathbf{x})$ is positive definite. These conditions are collectively known as the *second-order sufficient condition* (SOSC).

1.7 Contour Plots

This book will include problems with a variety of numbers of dimensions, and will need to display information over one, two, or three dimensions. Functions of the form $f(x_1, x_2) = y$ can be rendered in three-dimensional space, but not all orientations provide a complete view over the domain. A *contour plot* is a visual representation of a three-dimensional surface obtained by plotting regions with constant y values, known as *contours*, on a two-dimensional plot with axes indexed by x_1 and x_2. Example 1.2 illustrates how a contour plot can be interpreted.

1.8 Overview

This section provides a brief overview of the chapters of this book. The conceptual dependencies between the chapters are outlined in figure 1.9.

Consider the Rosenbrock banana function,

$$f(\mathbf{x}) = (1 - x_1)^2 + 5(x_2 - x_1^2)^2$$

Does the point $(1, 1)$ satisfy the FONC and SONC?

The gradient is:

$$\nabla f(\mathbf{x}) = \begin{bmatrix} \frac{\partial f}{\partial x_1} \\ \frac{\partial f}{\partial x_2} \end{bmatrix} = \begin{bmatrix} 2(10x_1^3 - 10x_1x_2 + x_1 - 1) \\ 10(x_2 - x_1^2) \end{bmatrix}$$

and the Hessian is:

$$\nabla^2 f(\mathbf{x}) = \begin{bmatrix} \frac{\partial^2 f}{\partial x_1 \partial x_1} & \frac{\partial^2 f}{\partial x_1 \partial x_2} \\ \frac{\partial^2 f}{\partial x_2 \partial x_1} & \frac{\partial^2 f}{\partial x_2 \partial x_2} \end{bmatrix} = \begin{bmatrix} -20(x_2 - x_1^2) + 40x_1^2 + 2 & -20x_1 \\ -20x_1 & 10 \end{bmatrix}$$

We compute $\nabla(f)([1,1]) = 0$, so the FONC is satisfied. The Hessian at $[1,1]$ is:

$$\begin{bmatrix} 42 & -20 \\ -20 & 10 \end{bmatrix}$$

which is positive definite, so the SONC is satisfied.

Example 1.1. Checking the first- and second-order necessary conditions of a point on the Rosenbrock function. The minimizer is indicated by the dot in the figure below.

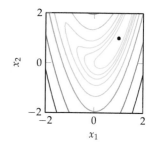

The function $f(x_1, x_2) = x_1^2 - x_2^2$. This function can be visualized in a three-dimensional space based on its two inputs and one output. It can also be visualized using a contour plot, which shows lines of constant y value. A three-dimensional visualization and a contour plot are shown below.

Example 1.2. An example three-dimensional visualization and the associated contour plot.

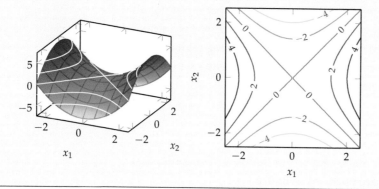

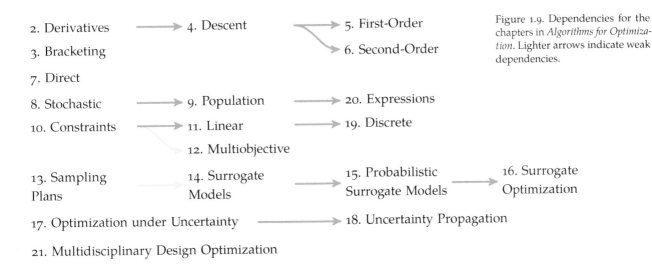

Figure 1.9. Dependencies for the chapters in *Algorithms for Optimization*. Lighter arrows indicate weak dependencies.

Chapter 2 begins by discussing *derivatives* and their generalization to multiple dimensions. Derivatives are used in many algorithms to inform the choice of direction of the search for an optimum. Often derivatives are not known analytically, and so we discuss how to estimate them numerically and using automatic differentiation techniques.

Chapter 3 discusses *bracketing*, which involves identifying an interval in which a local minimum lies for a univariate function. Different bracketing algorithms use different schemes for successively shrinking the interval based on function evaluations. One of the approaches we discuss uses knowledge of the Lipschitz constant of the function to guide the bracketing process. These bracketing algorithms are often used as subroutines within the optimization algorithms discussed later in the text.

Chapter 4 introduces *local descent* as a general approach to optimizing multivariate functions. Local descent involves iteratively choosing a descent direction and then taking a step in that direction and repeating that process until convergence or some termination condition is met. There are different schemes for choosing the step length. We will also discuss methods that adaptively restrict the step size to a region where there is confidence in the local model.

Chapter 5 builds upon the previous chapter, explaining how to use *first-order* information obtained through the gradient estimate as a local model to inform the descent direction. Simply stepping in the direction of steepest descent is often not the best strategy for finding a minimum. This chapter discusses a wide variety of different methods for using the past sequence of gradient estimates to better inform the search.

Chapter 6 shows how to use local models based on *second-order* approximations to inform local descent. These models are based on estimates of the Hessian of the objective function. The advantage of second-order approximations is that it can inform both the direction and step size.

Chapter 7 presents a collection of *direct methods* for finding optima that avoid using gradient information for informing the direction of search. We begin by discussing methods that iteratively perform line search along a set of directions. We then discuss pattern search methods that do not perform line search but rather perform evaluations some step size away from the current point along a set of directions. The step size is incrementally adapted as the search proceeds. Another method uses a simplex that adaptively expands and contracts as it traverses the design space in the apparent direction of improvement. Finally, we discuss a method motivated by Lipschitz continuity to increase resolution in areas deemed likely to contain the global minimum.

Chapter 8 introduces *stochastic* methods, where randomness is incorporated into the optimization process. We show how stochasticity can improve some of the algorithms discussed in earlier chapters, such as steepest descent and pattern search. Some of the methods involve incrementally traversing the search space, but others involve learning a probability distribution over the design space, assigning greater weight to regions that are more likely to contain an optimum.

Chapter 9 discusses *population* methods, where a collection of points is used to explore the design space. Having a large number of points distributed through the space can help reduce the risk of becoming stuck in a local minimum. Population methods generally rely upon stochasticity to encourage diversity in the population, and they can be combined with local descent methods.

Chapter 10 introduces the notion of *constraints* in optimization problems. We begin by discussing the mathematical conditions for optimality with constraints. We then introduce methods for incorporating constraints into the optimization algorithms discussed earlier through the use of penalty functions. We also discuss methods for ensuring that, if we start with a feasible point, the search will remain feasible.

Chapter 11 makes the assumption that both the objective function and constraints are *linear*. Although linearity may appear to be a strong assumption, many engineering problems can be framed as linear constrained optimization problems. Several methods have been developed for exploiting this linear structure. This chapter focuses on the simplex algorithm, which is guaranteed to result in a global minimum.

Chapter 12 shows how to address the problem of *multiobjective* optimization, where we have multiple objectives that we are trying to optimize simultaneously. Engineering often involves a tradeoff between multiple objectives, and it is often unclear how to prioritize different objectives. We discuss how to transform multi-objective problems into scalar-valued objective functions so that we can use the algorithms discussed in earlier chapters. We also discuss algorithms for finding the set of design points that represent the best tradeoff between objectives.

Chapter 13 discusses how to create *sampling plans* consisting of points that cover the design space. Random sampling of the design space often does not provide adequate coverage. We discuss methods for ensuring uniform coverage along each design dimension and methods for measuring and optimizing the coverage of the space. In addition, we discuss quasi-random sequences that can also be used to generate sampling plans.

Chapter 14 explains how to build *surrogate models* of the objective function. Surrogate models are often used for problems where evaluating the objective function is very expensive. An optimization algorithm can then use evaluations of the surrogate model instead of the actual objective function to improve the design. The evaluations can come from historical data, perhaps obtained through the use of a sampling plan introduced in the previous chapter. We discuss different types of surrogate models, how to fit them to data, and how to identify a suitable surrogate model.

Chapter 15 introduces *probabilistic surrogate models* that allow us to quantify our confidence in the predictions of the models. This chapter focuses on a particular type of surrogate model called a Gaussian process. We show how to use Gaussian processes for prediction, how to incorporate gradient measurements and noise, and how to estimate some of the parameters governing the Gaussian process from data.

Chapter 16 shows how to use the probabilistic models from the previous chapter to guide *surrogate optimization*. The chapter outlines several techniques for choosing which design point to evaluate next. We also discuss how surrogate models can be used to optimize an objective measure in a safe manner.

Chapter 17 explains how to perform *optimization under uncertainty*, relaxing the assumption made in previous chapters that the objective function is a deterministic function of the design variables. We discuss different approaches for representing uncertainty, including set-based and probabilistic approaches, and explain how to transform the problem to provide robustness to uncertainty.

Chapter 18 outlines approaches to *uncertainty propagation*, where known input distributions are used to estimate statistical quantities associated with the output distribution. Understanding the output distribution of an objective function is important to optimization under uncertainty. We discuss a variety of approaches, some based on mathematical concepts such as Monte Carlo, the Taylor series approximation, orthogonal polynomials, and Gaussian processes. They differ in the assumptions they make and the quality of their estimates.

Chapter 19 shows how to approach problems where the design variables are constrained to be *discrete*. A common approach is to relax the assumption that the variables are discrete, but this can result in infeasible designs. Another approach involves incrementally adding linear constraints until the optimal point is discrete. We also discuss branch and bound along with dynamic programming approaches, both of which guarantee optimality. The chapter also mentions a population-based method that often scales to large design spaces but does not provide guarantees.

Chapter 20 discusses how to search design spaces consisting of *expressions* defined by a grammar. For many problems, the number of variables is unknown, such as in the optimization of graphical structures or computer programs. We outline several algorithms that account for the grammatical structure of the design space to make the search more efficient.

Chapter 21 explains how to approach *multidisciplinary design optimization*. Many engineering problems involve complicated interactions between several disciplines, and optimizing disciplines individually may not lead to an optimal solution. This chapter discusses a variety of techniques for taking advantage of the structure of multidisciplinary problems to reduce the effort required for finding good designs.

The appendices contain supplementary material. Appendix A begins with a short introduction to the Julia programming language, focusing on the concepts used to specify the algorithms listed in this book. Appendix B specifies a variety of test functions used for evaluating the performance of different algorithms. Appendix C covers mathematical concepts used in the derivation and analysis of the optimization methods discussed in this text.

1.9 Summary

- Optimization in engineering is the process of finding the best system design subject to a set of constraints.

- Optimization is concerned with finding global minima of a function.

- Minima occur where the gradient is zero, but zero-gradient does not imply optimality.

1.10 Exercises

Exercise 1.1. Give an example of a function with a local minimum that is not a global minimum.

Exercise 1.2. What is the minimum of the function $f(x) = x^3 - x$?

Exercise 1.3. Does the first-order condition $f'(x) = 0$ hold when x is the optimal solution of a constrained problem?

Exercise 1.4. How many minima does $f(x, y) = x^2 + y$, subject to $x > y \geq 1$, have?

Exercise 1.5. How many inflection points does $f(x) = x^3 - 10$ have?

2 Derivatives and Gradients

Optimization is concerned with finding the design point that minimizes (or maximizes) an objective function. Knowing how the value of a function changes as its input is varied is useful because it tells us in which direction we can move to improve on previous points. The change in the value of the function is measured by the derivative in one dimension and the gradient in multiple dimensions. This chapter briefly reviews some essential elements from calculus.[1]

[1] For a more comprehensive review, see S. J. Colley, *Vector Calculus*, 4th ed. Pearson, 2011.

2.1 Derivatives

The *derivative* $f'(x)$ of a function f of a single variable x is the rate at which the value of f changes at x. It is often visualized, as shown in figure 2.1, using the tangent line to the graph of the function at x. The value of the derivative equals the slope of the tangent line.

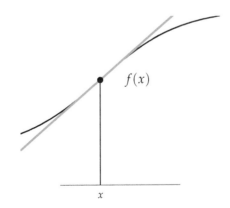

$f(x)$

x

Figure 2.1. The function f is drawn in black and the tangent line to $f(x)$ is drawn in blue. The derivative of f at x is the slope of the tangent line.

We can use the derivative to provide a linear approximation of the function near x:

$$f(x + \Delta x) \approx f(x) + f'(x)\Delta x \qquad (2.1)$$

The derivative is the ratio between the change in f and the change in x at the point x:

$$f'(x) = \frac{\Delta f(x)}{\Delta x} \qquad (2.2)$$

which is the change in $f(x)$ divided by the change in x as the step becomes infinitesimally small as illustrated by figure 2.2.

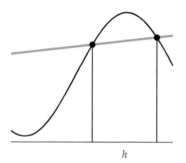
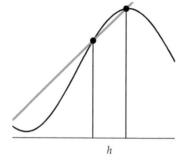
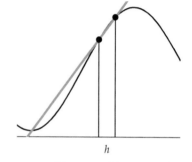

h h h

Figure 2.2. The tangent line is obtained by joining points with sufficiently small step differences.

The notation $f'(x)$ can be attributed to Lagrange. We also use the notation created by Leibniz,

$$f'(x) \equiv \frac{df(x)}{dx} \qquad (2.3)$$

which emphasizes the fact that the derivative is the ratio of the change in f to the change in x at the point x.

The limit equation defining the derivative can be presented in three different ways: the *forward difference*, the *central difference*, and the *backward difference*. Each method uses an infinitely small step size h:

$$f'(x) \equiv \underbrace{\lim_{h \to 0} \frac{f(x+h) - f(x)}{h}}_{\text{forward difference}} = \underbrace{\lim_{h \to 0} \frac{f(x+h/2) - f(x-h/2)}{h}}_{\text{central difference}} = \underbrace{\lim_{h \to 0} \frac{f(x) - f(x-h)}{h}}_{\text{backward difference}} \qquad (2.4)$$

If f can be represented symbolically, *symbolic differentiation* can often provide an exact analytic expression for f' by applying derivative rules from calculus. The analytic expression can then be evaluated at any point x. The process is illustrated in example 2.1.

The implementation details of symbolic differentiation is outside the scope of this text. Various software packages, such as `SymEngine.jl` in Julia and `SymPy` in Python, provide implementations. Here we use `SymEngine.jl` to compute the derivative of $x^2 + x/2 - \sin(x)/x$.

Example 2.1. Symbolic differentiation provides analytical derivatives.

```julia
julia> using SymEngine
julia> @vars x; # define x as a symbolic variable
julia> f = x^2 + x/2 - sin(x)/x;
julia> diff(f, x)
1/2 + 2*x + sin(x)/x^2 - cos(x)/x
```

2.2   Derivatives in Multiple Dimensions

The *gradient* is the generalization of the derivative to multivariate functions. It captures the local slope of the function, allowing us to predict the effect of taking a small step from a point in any direction. Recall that the derivative is the slope of the tangent line. The gradient points in the direction of steepest ascent of the tangent *hyperplane* as shown in figure 2.3. A hyperplane in an n-dimensional space is the set of points that satisfies

$$w_1 x_1 + \cdots + w_n x_n = b \qquad (2.5)$$

for some vector $\mathbf{w}$ and scalar b. A hyperplane has $n - 1$ dimensions.

The gradient of f at $\mathbf{x}$ is written $\nabla f(\mathbf{x})$ and is a vector. Each component of that vector is the *partial derivative*[2] of f with respect to that component:

$$\nabla f(\mathbf{x}) = \left[\frac{\partial f(\mathbf{x})}{\partial x_1}, \quad \frac{\partial f(\mathbf{x})}{\partial x_2}, \quad \ldots, \quad \frac{\partial f(\mathbf{x})}{\partial x_n} \right] \qquad (2.6)$$

We use the convention that vectors written with commas are column vectors. For example, we have $[a, b, c] = [a \ b \ c]^\top$. Example 2.2 shows how to compute the gradient of a function at a particular point.

The *Hessian* of a multivariate function is a matrix containing all of the second derivatives with respect to the input.[3] The second derivatives capture information about the local curvature of the function.

$$\nabla^2 f(\mathbf{x}) = \begin{bmatrix} \frac{\partial^2 f(\mathbf{x})}{\partial x_1 \partial x_1} & \frac{\partial^2 f(\mathbf{x})}{\partial x_1 \partial x_2} & \cdots & \frac{\partial^2 f(\mathbf{x})}{\partial x_1 \partial x_n} \\ & \vdots & & \\ \frac{\partial^2 f(\mathbf{x})}{\partial x_n \partial x_1} & \frac{\partial^2 f(\mathbf{x})}{\partial x_n \partial x_2} & \cdots & \frac{\partial^2 f(\mathbf{x})}{\partial x_n \partial x_n} \end{bmatrix} \qquad (2.7)$$

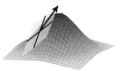

Figure 2.3. Each component of the gradient defines a local tangent line. These tangent lines define the local tangent hyperplane. The gradient vector points in the direction of greatest increase.

[2] The partial derivative of a function with respect to a variable is the derivative assuming all other input variables are held constant. It is denoted $\partial f / \partial x$.

[3] The Hessian is symmetric only if the second derivatives of f are all continuous in a neighborhood of the point at which it is being evaluated:

$$\frac{\partial^2 f}{\partial x_1 \partial x_2} = \frac{\partial^2 f}{\partial x_2 \partial x_1}$$

Compute the gradient of $f(\mathbf{x}) = x_1 \sin(x_2) + 1$ at $\mathbf{c} = [2,0]$.

$$f(\mathbf{x}) = x_1 \sin(x_2) + 1$$

$$\nabla f(\mathbf{x}) = \left[\frac{\partial f}{\partial x_1}, \frac{\partial f}{\partial x_2} \right] = [\sin(x_2), x_1 \cos(x_2)]$$

$$\nabla f(\mathbf{c}) = [0,2]$$

Example 2.2. Computing the gradient at a particular point.

The *directional derivative* $\nabla_{\mathbf{s}} f(\mathbf{x})$ of a multivariate function f is the instantaneous rate of change of $f(\mathbf{x})$ as $\mathbf{x}$ is moved with velocity $\mathbf{s}$. The definition is closely related to the definition of a derivative of a univariate function:[4]

[4] Some texts require that $\mathbf{s}$ be a unit vector. See, for example, G. B. Thomas, *Calculus and Analytic Geometry*, 9th ed. Addison-Wesley, 1968.

$$\nabla_{\mathbf{s}} f(\mathbf{x}) \equiv \underbrace{\lim_{h \to 0} \frac{f(\mathbf{x} + h\mathbf{s}) - f(\mathbf{x})}{h}}_{\text{forward difference}} = \underbrace{\lim_{h \to 0} \frac{f(\mathbf{x} + h\mathbf{s}/2) - f(\mathbf{x} - h\mathbf{s}/2)}{h}}_{\text{central difference}} = \underbrace{\lim_{h \to 0} \frac{f(\mathbf{x}) - f(\mathbf{x} - h\mathbf{s})}{h}}_{\text{backward difference}} \tag{2.8}$$

The directional derivative can be computed using the gradient of the function:

$$\nabla_{\mathbf{s}} f(\mathbf{x}) = \nabla f(\mathbf{x})^{\top} \mathbf{s} \tag{2.9}$$

Another way to compute the directional derivative $\nabla_{\mathbf{s}} f(\mathbf{x})$ is to define $g(\alpha) \equiv f(\mathbf{x} + \alpha \mathbf{s})$ and then compute $g'(0)$, as illustrated in example 2.3.

The directional derivative is highest in the gradient direction, and it is lowest in the direction opposite the gradient. This directional dependence arises from the dot product in the directional derivative's definition and from the fact that the gradient is a local tangent hyperplane.

Example 2.3. Computing a directional derivative.

We wish to compute the directional derivative of $f(\mathbf{x}) = x_1 x_2$ at $\mathbf{x} = [1,0]$ in the direction $\mathbf{s} = [-1, -1]$:

$$\nabla f(\mathbf{x}) = \left[\frac{\partial f}{\partial x_1}, \ \frac{\partial f}{\partial x_2} \right] = [x_2, x_1]$$

$$\nabla_{\mathbf{s}} f(\mathbf{x}) = \nabla f(\mathbf{x})^\top \mathbf{s} = \begin{bmatrix} 0 & 1 \end{bmatrix} \begin{bmatrix} -1 \\ -1 \end{bmatrix} = -1$$

We can also compute the directional derivative as follows:

$$g(\alpha) = f(\mathbf{x} + \alpha \mathbf{s}) = (1 - \alpha)(-\alpha) = \alpha^2 - \alpha$$
$$g'(\alpha) = 2\alpha - 1$$
$$g'(0) = -1$$

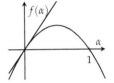

2.3 Numerical Differentiation

The process of estimating derivatives numerically is referred to as *numerical differentiation*. Estimates can be derived in different ways from function evaluations. This section discusses finite difference methods and the complex step method.[5]

2.3.1 Finite Difference Methods

As the name implies, *finite difference methods* compute the difference between two values that differ by a finite step size. They approximate the derivative definitions in equation (2.4) using small differences:

$$f'(x) \approx \underbrace{\frac{f(x+h) - f(x)}{h}}_{\text{forward difference}} \approx \underbrace{\frac{f(x+h/2) - f(x-h/2)}{h}}_{\text{central difference}} \approx \underbrace{\frac{f(x) - f(x-h)}{h}}_{\text{backward difference}} \qquad (2.10)$$

Mathematically, the smaller the step size h, the better the derivative estimate. Practically, values for h that are too small can result in numerical cancellation errors. This effect is shown later in figure 2.4. Algorithm 2.1 provides implementations for these methods.

[5] For a more comprehensive treatment of the topics discussed in the remainder of this chapter, see A. Griewank and A. Walther, *Evaluating Derivatives: Principles and Techniques of Algorithmic Differentiation*, 2nd ed. SIAM, 2008.

```
diff_forward(f, x; h=sqrt(eps(Float64))) = (f(x+h) - f(x))/h
diff_central(f, x; h=cbrt(eps(Float64))) = (f(x+h/2) - f(x-h/2))/h
diff_backward(f, x; h=sqrt(eps(Float64))) = (f(x) - f(x-h))/h
```

Algorithm 2.1. Finite difference methods for estimating the derivative of a function f at x with finite difference h. The default step sizes are the square root or cube root of the machine precision for floating point values. These step sizes balance machine round-off error with step size error.

The eps function provides the step size between 1.0 and the next larger representable floating-point value.

The finite difference methods can be derived using the Taylor expansion. We will derive the forward difference derivative estimate, beginning with the Taylor expansion of f about x:

$$f(x+h) = f(x) + \frac{f'(x)}{1!}h + \frac{f''(x)}{2!}h^2 + \frac{f'''(x)}{3!}h^3 + \cdots \qquad (2.11)$$

We can rearrange and solve for the first derivative:

$$f'(x)h = f(x+h) - f(x) - \frac{f''(x)}{2!}h^2 - \frac{f'''(x)}{3!}h^3 - \cdots \qquad (2.12)$$

$$f'(x) = \frac{f(x+h) - f(x)}{h} - \frac{f''(x)}{2!}h - \frac{f'''(x)}{3!}h^2 - \cdots \qquad (2.13)$$

$$f'(x) \approx \frac{f(x+h) - f(x)}{h} \qquad (2.14)$$

The forward difference approximates the true derivative for small h with error dependent on $\frac{f''(x)}{2!}h + \frac{f'''(x)}{3!}h^2 + \cdots$. This error term is $O(h)$, meaning the forward difference has linear error as h approaches zero.[6]

The central difference method has an error term of $O(h^2)$.[7] We can derive this error term using the Taylor expansion. The Taylor expansions about x for $f(x+h/2)$ and $f(x-h/2)$ are:

$$f(x+h/2) = f(x) + f'(x)\frac{h}{2} + \frac{f''(x)}{2!}\left(\frac{h}{2}\right)^2 + \frac{f'''(x)}{3!}\left(\frac{h}{2}\right)^3 + \cdots \qquad (2.15)$$

$$f(x-h/2) = f(x) - f'(x)\frac{h}{2} + \frac{f''(x)}{2!}\left(\frac{h}{2}\right)^2 - \frac{f'''(x)}{3!}\left(\frac{h}{2}\right)^3 + \cdots \qquad (2.16)$$

Subtracting these expansions produces:

$$f(x+h/2) - f(x-h/2) \approx 2f'(x)\frac{h}{2} + \frac{2}{3!}f'''(x)\left(\frac{h}{2}\right)^3 \qquad (2.17)$$

We rearrange to obtain:

$$f'(x) \approx \frac{f(x+h/2) - f(x-h/2)}{h} - \frac{f'''(x)h^2}{24} \qquad (2.18)$$

which shows that the approximation has quadratic error.

[6] Asymptotic notation is covered in appendix C.

[7] J. H. Mathews and K. D. Fink, *Numerical Methods Using MATLAB*, 4th ed. Pearson, 2004.

2.3.2   *Complex Step Method*

We often run into the problem of needing to choose a step size h small enough to provide a good approximation but not too small so as to lead to numerical subtractive cancellation issues. The *complex step method* bypasses the effect of subtractive cancellation by using a single function evaluation. We evaluate the function once after taking a step in the imaginary direction.[8]

The Taylor expansion for an imaginary step is:

$$f(x + ih) = f(x) + ihf'(x) - h^2\frac{f''(x)}{2!} - ih^3\frac{f'''(x)}{3!} + \cdots \qquad (2.19)$$

Taking only the imaginary component of each side produces a derivative approximation:

$$\text{Im}(f(x + ih)) = hf'(x) - h^3\frac{f'''(x)}{3!} + \cdots \qquad (2.20)$$

$$\Rightarrow f'(x) = \frac{\text{Im}(f(x + ih))}{h} + h^2\frac{f'''(x)}{3!} - \cdots \qquad (2.21)$$

$$= \frac{\text{Im}(f(x + ih))}{h} + O(h^2) \text{ as } h \to 0 \qquad (2.22)$$

An implementation is provided by algorithm 2.2. The real part approximates $f(x)$ to within $O(h^2)$ as $h \to 0$:

$$\text{Re}(f(x + ih)) = f(x) - h^2\frac{f''(x)}{2!} + \cdots \qquad (2.23)$$

$$\Rightarrow f(x) = \text{Re}(f(x + ih)) + h^2\frac{f''(x)}{2!} - \cdots \qquad (2.24)$$

Thus, we can evaluate both $f(x)$ and $f'(x)$ using a single evaluation of f with complex arguments. Example 2.4 shows the calculations involved for estimating the derivative of a function at a particular point. Algorithm 2.2 implements the complex step method. Figure 2.4 compares the numerical error of the complex step method to the forward and central difference methods as the step size is varied.

```
diff_complex(f, x; h=1e-20) = imag(f(x + h*im)) / h
```

[8] J. R. R. A. Martins, P. Sturdza, and J. J. Alonso, "The Complex-Step Derivative Approximation," *ACM Transactions on Mathematical Software*, vol. 29, no. 3, pp. 245–262, 2003. Special care must be taken to ensure that the implementation of f properly supports complex numbers as input.

Algorithm 2.2. The complex step method for estimating the derivative of a function f at x with finite difference h.

Consider $f(x) = \sin(x^2)$. The function value at $x = \pi/2$ is approximately 0.624266 and the derivative is $\pi \cos(\pi^2/4) \approx -2.45425$. We can arrive at this using the complex step method:

Example 2.4. The complex step method for estimating derivatives.

```julia
julia> f = x -> sin(x^2);
julia> v = f(π/2 + 0.001im);
julia> real(v) # f(x)
0.6242698144866649
julia> imag(v)/0.001 # f'(x)
-2.454251617038178
```

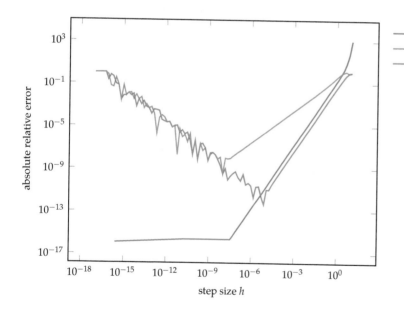

Figure 2.4. A comparison of the error in derivative estimate for the function $\sin(x)$ at $x = 1/2$ as the step size is varied. The linear error of the forward difference method and the quadratic error of the central difference and complex methods can be seen by the constant slopes on the right hand side. The complex step method avoids the subtractive cancellation error that occurs when differencing two function evaluations that are close together.

2.4   Automatic Differentiation

This section introduces algorithms for the numeric evaluation of derivatives of functions specified by a computer program. Key to these *automatic differentiation* techniques is the application of the chain rule:

$$\frac{d}{dx}f(g(x)) = \frac{d}{dx}(f \circ g)(x) = \frac{df}{dg}\frac{dg}{dx} \qquad (2.25)$$

A program is composed of elementary operations like addition, subtraction, multiplication, and division.

Consider the function $f(a,b) = \ln(ab + \max(a,2))$. If we want to compute the partial derivative with respect to a at a point, we need to apply the chain rule several times:[9]

$$\frac{\partial f}{\partial a} = \frac{\partial}{\partial a}\ln(ab + \max(a,2)) \qquad (2.26)$$

$$= \frac{1}{ab + \max(a,2)}\frac{\partial}{\partial a}(ab + \max(a,2)) \qquad (2.27)$$

$$= \frac{1}{ab + \max(a,2)}\left[\frac{\partial(ab)}{\partial a} + \frac{\partial \max(a,2)}{\partial a}\right] \qquad (2.28)$$

$$= \frac{1}{ab + \max(a,2)}\left[\left(b\frac{\partial a}{\partial a} + a\frac{\partial b}{\partial a}\right) + \left((2 > a)\frac{\partial 2}{\partial a} + (2 < a)\frac{\partial a}{\partial a}\right)\right] \qquad (2.29)$$

$$= \frac{1}{ab + \max(a,2)}[b + (2 < a)] \qquad (2.30)$$

This process can be automated through the use of a *computational graph*. A computational graph represents a function where the nodes are operations and the edges are input-output relations. The leaf nodes of a computational graph are input variables or constants, and terminal nodes are values output by the function. A computational graph is shown in figure 2.5.

There are two methods for automatically differentiating f using its computational graph. The *forward accumulation* method used by dual numbers traverses the tree from inputs to outputs, whereas *reverse accumulation* requires a backwards pass through the graph.

2.4.1   Forward Accumulation

Forward accumulation will automatically differentiate a function using a single forward pass through the function's computational graph. The method is equivalent

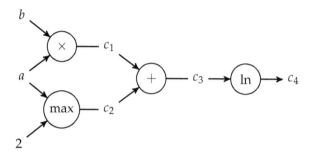

Figure 2.5. The computational graph for $\ln(ab + \max(a, 2))$.

to iteratively expanding the chain rule of the inner operation:

$$\frac{df}{dx} = \frac{df}{dc_4}\frac{dc_4}{dx} = \frac{df}{dc_4}\left(\frac{dc_4}{dc_3}\frac{dc_3}{dx}\right) = \frac{df}{dc_4}\left(\frac{dc_4}{dc_3}\left(\frac{dc_3}{dc_2}\frac{dc_2}{dx} + \frac{dc_3}{dc_1}\frac{dc_1}{dx}\right)\right) \qquad (2.31)$$

To illustrate forward accumulation, we apply it to the example function $f(a, b) = \ln(ab + \max(a, 2))$ to calculate the partial derivative at $a = 3, b = 2$ with respect to a.

1. The procedure starts at the graph's source nodes consisting of the function inputs and any constant values. For each of these nodes, we note both the value and the partial derivative with respect to our target variable, as shown in figure 2.6.[10]

2. Next we proceed down the tree, one node at a time, choosing as our next node one whose inputs have already been computed. We can compute the value by passing through the previous nodes' values, and we can compute the local partial derivative with respect to a using both the previous nodes' values and their partial derivatives. The calculations are shown in figure 2.7.

We end up with the correct result, $f(3, 2) = \ln 9$ and $\partial f / \partial a = 1/3$. This was done using one pass through the computational graph.

This process can be conveniently automated by a computer using a programming language which has overridden each operation to produce both the value and its derivative. Such pairs are called *dual numbers*.

[10] For compactness in this figure, we use *dot notation* or *Newton's notation* for derivatives. For example, if it is clear that we are taking the derivative with respect to a, we can write $\partial b / \partial a$ as $\dot{b}$.

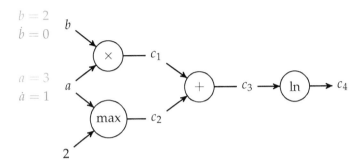

Figure 2.6.   The computational graph for $\ln(ab + \max(a, 2))$ being set up for forward accumulation to calculate $\partial f / \partial a$ with $a = 3$ and $b = 2$.

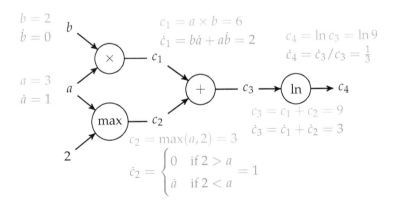

Figure 2.7.   The computational graph for $\ln(ab + \max(a, 2))$ after forward accumulation is applied to calculate $\partial f / \partial a$ with $a = 3$ and $b = 2$.

Dual numbers can be expressed mathematically by including the abstract quantity ϵ, where ϵ^2 is defined to be 0. Like a complex number, a dual number is written $a + b\epsilon$ where a and b are both real values. We have:

$$(a + b\epsilon) + (c + d\epsilon) = (a + c) + (b + d)\epsilon \tag{2.32}$$

$$(a + b\epsilon) \times (c + d\epsilon) = (ac) + (ad + bc)\epsilon \tag{2.33}$$

In fact, by passing a dual number into any smooth function f, we get the evaluation and its derivative. We can show this using the Taylor series:

$$f(x) = \sum_{k=0}^{\infty} \frac{f^{(k)}(a)}{k!}(x - a)^k \tag{2.34}$$

$$f(a + b\epsilon) = \sum_{k=0}^{\infty} \frac{f^{(k)}(a)}{k!}(a + b\epsilon - a)^k \tag{2.35}$$

$$= \sum_{k=0}^{\infty} \frac{f^{(k)}(a)b^k \epsilon^k}{k!} \tag{2.36}$$

$$= f(a) + bf'(a)\epsilon + \epsilon^2 \sum_{k=2}^{\infty} \frac{f^{(k)}(a)b^k}{k!}\epsilon^{(k-2)} \tag{2.37}$$

$$= f(a) + bf'(a)\epsilon \tag{2.38}$$

Example 2.5 shows an implementation.

2.4.2   Reverse Accumulation

Forward accumulation requires n passes in order to compute an n-dimensional gradient. *Reverse accumulation*[11] requires only a single run in order to compute a complete gradient but requires two passes through the graph: a *forward pass* during which necessary intermediate values are computed and a *backward pass* which computes the gradient. Reverse accumulation is often preferred over forward accumulation when gradients are needed, though care must be taken on memory-constrained systems when the computational graph is very large.[12]

Like forward accumulation, reverse accumulation will compute the partial derivative with respect to the chosen target variable but iteratively substitutes the outer function instead:

$$\frac{df}{dx} = \frac{df}{dc_4}\frac{dc_4}{dx} = \left(\frac{df}{dc_3}\frac{dc_3}{dc_4}\right)\frac{dc_4}{dx} = \left(\left(\frac{df}{dc_2}\frac{dc_2}{dc_3} + \frac{df}{dc_1}\frac{dc_1}{dc_3}\right)\frac{dc_3}{dc_4}\right)\frac{dc_4}{dx} \tag{2.39}$$

[11] S. Linnainmaa, "The Representation of the Cumulative Rounding Error of an Algorithm as a Taylor Expansion of the Local Rounding Errors," M.S. thesis, University of Helsinki, 1970.

[12] Reverse accumulation is central to the backpropagation algorithm used to train neural networks. D. E. Rumelhart, G. E. Hinton, and R. J. Williams, "Learning Representations by Back-Propagating Errors," *Nature*, vol. 323, pp. 533–536, 1986.

Dual numbers can be implemented by defining a struct `Dual` that contains two fields, the value `v` and the derivative ∂.

Example 2.5. An implementation of dual numbers allows for automatic forward accumulation.

```
struct Dual
    v
    ∂
end
```

We must then implement methods for each of the base operations required. These methods take in dual numbers and produce new dual numbers using that operation's chain rule logic.

```
Base.:+(a::Dual, b::Dual) = Dual(a.v + b.v, a.∂ + b.∂)
Base.:*(a::Dual, b::Dual) = Dual(a.v * b.v, a.v*b.∂ + b.v*a.∂)
Base.log(a::Dual) = Dual(log(a.v), a.∂/a.v)
function Base.max(a::Dual, b::Dual)
    v = max(a.v, b.v)
    ∂ = a.v > b.v ? a.∂ : a.v < b.v ? b.∂ : NaN
    return Dual(v, ∂)
end
function Base.max(a::Dual, b::Int)
    v = max(a.v, b)
    ∂ = a.v > b ? a.∂ : a.v < b ? 0 : NaN
    return Dual(v, ∂)
end
```

The `ForwardDiff.jl` package supports an extensive set of mathematical operations and additionally provides gradients and Hessians.

```
julia> using ForwardDiff
julia> a = ForwardDiff.Dual(3,1);
julia> b = ForwardDiff.Dual(2,0);
julia> log(a*b + max(a,2))
Dual{Nothing}(2.1972245773362196,0.3333333333333333)
```

This process is the reverse pass, the evaluation of which requires intermediate values that are obtained during a forward pass.

Reverse accumulation can be implemented through *operation overloading*[13] in a similar manner to the way dual numbers are used to implement forward accumulation. Two functions must be implemented for each fundamental operation: a forward operation that overloads the operation to store local gradient information during the forward pass and a backward operation that uses the information to propagate the gradient backwards. Packages like Tensorflow[14] or Zygote.jl can automatically construct the computational graph and the associated forward and backwards pass operations. Example 2.6 shows how Zygote.jl can be used.

[13] Operation overloading refers to providing implementations for common operations such as $+$, $-$, or $=$ for custom variable types. Overloading is discussed in appendix A.2.5.

[14] Tensorflow is an open source software library for numerical computation using data flow graphs and is often used for deep learning applications. It may be obtained from tensorflow.org.

The Zygote.jl package provides automatic differentiation in the form of reverse-accumulation. Here the gradient function is used to automatically generate the backwards pass through the source code of f to obtain the gradient.

```julia
julia> import Zygote: gradient
julia> f(a, b) = log(a*b + max(a,2));
julia> gradient(f, 3.0, 2.0)
(0.3333333333333333, 0.3333333333333333)
```

Example 2.6. Automatic differentiation using the Zygote.jl package. We find that the gradient at $[3, 2]$ is $[1/3, 1/3]$.

2.5 Summary

- Derivatives are useful in optimization because they provide information about how to change a given point in order to improve the objective function.

- For multivariate functions, various derivative-based concepts are useful for directing the search for an optimum, including the gradient, the Hessian, and the directional derivative.

- One approach to numerical differentiation includes finite difference approximations.

- The complex step method can eliminate the effect of subtractive cancellation error when taking small steps, resulting in high quality gradient estimates.

- Analytic differentiation methods include forward and reverse accumulation on computational graphs.

2.6   Exercises

Exercise 2.1. Adopt the forward difference method to approximate the Hessian of $f(\mathbf{x})$ using its gradient, $\nabla f(\mathbf{x})$.

Exercise 2.2. What is a drawback of the central difference method over other finite difference methods if we already know $f(\mathbf{x})$?

Exercise 2.3. Compute the gradient of $f(x) = \ln x + e^x + \frac{1}{x}$ for a point x close to zero. What term dominates in the expression?

Exercise 2.4. Suppose $f(x)$ is a real-valued function that is also defined for complex inputs. If $f(3 + ih) = 2 + 4ih$, what is $f'(3)$?

Exercise 2.5. Draw the computational graph for $f(x, y) = \sin(x + y^2)$. Use the computational graph with forward accumulation to compute $\partial f / \partial y$ at $(x, y) = (1, 1)$. Label the intermediate values and partial derivatives as they are propagated through the graph.

Exercise 2.6. Combine the forward and backward difference methods to obtain a difference method for estimating the second-order derivative of a function f at x using three function evaluations.

3 Bracketing

This chapter presents a variety of *bracketing* methods for univariate functions, or functions involving a single variable. Bracketing is the process of identifying an interval in which a local minimum lies and then successively shrinking the interval. For many functions, derivative information can be helpful in directing the search for an optimum, but, for some functions, this information may not be available or might not exist. This chapter outlines a wide variety of approaches that leverage different assumptions. Later chapters that consider multivariate optimization will build upon the concepts introduced here.

3.1 Unimodality

Several of the algorithms presented in this chapter assume *unimodality* of the objective function. A *unimodal function* f is one where there is a *unique* x^*, such that f is monotonically decreasing for $x \leq x^*$ and monotonically increasing for $x \geq x^*$. It follows from this definition that the unique global minimum is at x^*, and there are no other local minima.[1]

Given a unimodal function, we can *bracket* an interval $[a, c]$ containing the global minimum if we can find three points $a < b < c$, such that $f(a) > f(b) < f(c)$. Figure 3.1 shows an example.

3.2 Finding an Initial Bracket

When optimizing a function, we often start by first bracketing an interval containing a local minimum. We then successively reduce the size of the bracketed interval to converge on the local minimum. A simple procedure (algorithm 3.1) can be used to find an initial bracket. Starting at a given point, we take a step

[1] It is perhaps more conventional to define unimodal functions in the opposite sense, such that there is a unique global *maximum* rather than a minimum. However, in this text, we try to minimize functions, and so we use the definition in this paragraph.

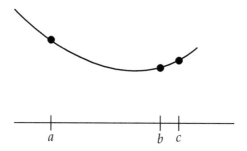

Figure 3.1. Three points shown bracketing a minimum.

in the positive direction. The distance we take is a *hyperparameter* to this algorithm,[2] but the algorithm provided defaults it to 1×10^{-2}. We then search in the downhill direction to find a new point that exceeds the lowest point. With each step, we expand the step size by some factor, which is another hyperparameter to this algorithm that is often set to 2. An example is shown in figure 3.2. Functions without local minima, such as $\exp(x)$, cannot be bracketed and will cause `bracket_minimum` to fail.

[2] A hyperparameter is a parameter that governs the function of an algorithm. It can be set by an expert or tuned using an optimization algorithm. Many of the algorithms in this text have hyperparameters. We often provide default values suggested in the literature. The success of an algorithm can be sensitive to the choice of hyperparameter.

```
function bracket_minimum(f, x=0; s=1e-2, k=2.0)
    a, ya = x, f(x)
    b, yb = a + s, f(a + s)
    if yb > ya
        a, b = b, a
        ya, yb = yb, ya
        s = -s
    end
    while true
        c, yc = b + s, f(b + s)
        if yc > yb
            return a < c ? (a, c) : (c, a)
        end
        a, ya, b, yb = b, yb, c, yc
        s *= k
    end
end
```

Algorithm 3.1. An algorithm for bracketing an interval in which a local minimum must exist. It takes as input a univariate function f and starting position x, which defaults to 0. The starting step size s and the expansion factor k can be specified. It returns a tuple containing the new interval $[a, b]$.

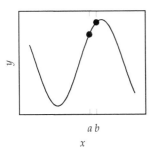 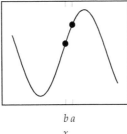 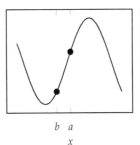 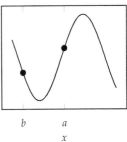

Figure 3.2. An example of running `bracket_minimum` on a function. The method reverses direction between the first and second iteration and then expands until a minimum is bracketed in the fourth iteration.

3.3 Fibonacci Search

Suppose we have a unimodal f bracketed by the interval $[a, b]$. Given a limit on the number of times we can query the objective function, *Fibonacci search* (algorithm 3.2) is guaranteed to maximally shrink the bracketed interval.

Suppose we can query f only twice. If we query f on the one-third and two-third points on the interval, then we are guaranteed to remove one-third of our interval, regardless of f, as shown in figure 3.3.

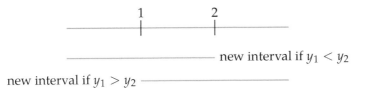

Figure 3.3. Our initial guess for two queries will remove one-third of the initial interval.

We can guarantee a tighter bracket by moving our guesses toward the center. In the limit as $\epsilon \to 0$, we are guaranteed to shrink our interval by a factor of two as shown in figure 3.4.

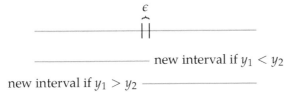

Figure 3.4. The most we can guarantee to shrink our interval is by just under a factor of two.

With three queries, we can shrink the interval by a factor of three. We first query f on the one-third and two-third points on the interval, discard one-third of the interval, and then sample just next to the better sample as shown in figure 3.5.

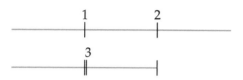

Figure 3.5. With three queries we can shrink the domain by a factor of three. The third query is made based on the result of the first two queries.

For n queries, the interval lengths are related to the Fibonacci sequence: 1, 1, 2, 3, 5, 8, and so forth. The first two terms are one, and the following terms are always the sum of the previous two:

$$F_n = \begin{cases} 1 & \text{if } n \le 2 \\ F_{n-1} + F_{n-2} & \text{otherwise} \end{cases} \tag{3.1}$$

Figure 3.6 shows the relationship between the intervals. Example 3.1 walks through an application to a univariate function.

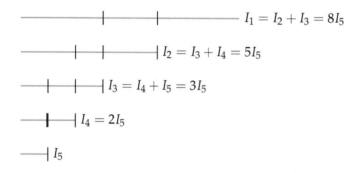

Figure 3.6. For n queries we are guaranteed to shrink our interval by a factor of F_{n+1}. The length of every interval constructed during Fibonacci search can be expressed in terms of the final interval times a Fibonacci number. If the final, smallest interval has length I_n, then the second smallest interval has length $I_{n-1} = F_3 I_n$, the third smallest interval has length $I_{n-2} = F_4 I_n$, and so forth.

The Fibonacci sequence can be determined analytically using *Binet's formula*:

$$F_n = \frac{\varphi^n - (1 - \varphi)^n}{\sqrt{5}}, \tag{3.2}$$

where $\varphi = (1 + \sqrt{5})/2 \approx 1.61803$ is the *golden ratio*.

The ratio between successive values in the Fibonacci sequence is:

$$\frac{F_n}{F_{n-1}} = \varphi \frac{1 - s^{n+1}}{1 - s^n} \tag{3.3}$$

where $s = (1 - \sqrt{5})/(1 + \sqrt{5}) \approx -0.382$.

```
function fibonacci_search(f, a, b, n; ϵ=0.01)
    s = (1-√5)/(1+√5)
    ρ = 1 / (φ*(1-s^(n+1))/(1-s^n))
    d = ρ*b + (1-ρ)*a
    yd = f(d)
    for i in 1 : n-1
        if i == n-1
            c = ϵ*a + (1-ϵ)*d
        else
            c = ρ*a + (1-ρ)*b
        end
        yc = f(c)
        if yc < yd
            b, d, yd = d, c, yc
        else
            a, b = b, c
        end
        ρ = 1 / (φ*(1-s^(n-i+1))/(1-s^(n-i)))
    end
    return a < b ? (a, b) : (b, a)
end
```

Algorithm 3.2. Fibonacci search to be run on univariate function f, with bracketing interval $[a, b]$, for n > 1 function evaluations. It returns the new interval (a, b). The optional parameter ϵ controls the lowest-level interval. The golden ratio φ is defined in Base.MathConstants.jl.

3.4 Golden Section Search

If we take the limit for large n, we see that the ratio between successive values of the Fibonacci sequence approaches the golden ratio:

$$\lim_{n\to\infty} \frac{F_n}{F_{n-1}} = \varphi. \tag{3.4}$$

Golden section search (algorithm 3.3) uses the golden ratio to approximate Fibonacci search. Figure 3.7 shows the relationship between the intervals. Figures 3.8 and 3.9 compare Fibonacci search with golden section search on unimodal and non-unimodal functions, respectively.

Consider using Fibonacci search with five function evaluations to minimize $f(x) = \exp(x-2) - x$ over the interval $[a,b] = [-2,6]$. The first two function evaluations are made at $\frac{F_5}{F_6}$ and $1 - \frac{F_5}{F_6}$, along the length of the initial bracketing interval:

Example 3.1.   Using Fibonacci search with five function evaluations to optimize a univariate function.

$$f(x^{(1)}) = f\left(a + (b-a)\left(1 - \frac{F_5}{F_6}\right)\right) \qquad = f(1) = -0.632$$

$$f(x^{(2)}) = f\left(a + (b-a)\frac{F_5}{F_6}\right) \qquad = f(3) = -0.282$$

The evaluation at $x^{(1)}$ is lower, yielding the new interval $[a,b] = [-2,3]$. Two evaluations are needed for the next interval split:

$$x_{\text{left}} = a + (b-a)\left(1 - \frac{F_4}{F_5}\right) = 0$$

$$x_{\text{right}} = a + (b-a)\frac{F_4}{F_5} = 1$$

A third function evaluation is thus made at x_{left}, as x_{right} has already been evaluated:

$$f(x^{(3)}) = f(0) = 0.135$$

The evaluation at $x^{(1)}$ is lower, yielding the new interval $[a,b] = [0,3]$. Two evaluations are needed for the next interval split:

$$x_{\text{left}} = a + (b-a)\left(1 - \frac{F_3}{F_4}\right) = 1$$

$$x_{\text{right}} = a + (b-a)\frac{F_3}{F_4} = 2$$

A fourth functional evaluation is thus made at x_{right}, as x_{left} has already been evaluated:

$$f(x^{(4)}) = f(2) = -1$$

The new interval is $[a,b] = [1,3]$. A final evaluation is made just next to the center of the interval at $2 + \epsilon$, and it is found to have a slightly higher value than $f(2)$. The final interval is $[1, 2 + \epsilon]$.

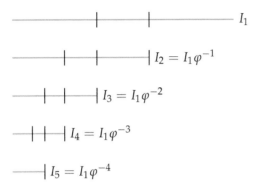

Figure 3.7. For n queries of a univariate function we are guaranteed to shrink a bracketing interval by a factor of φ^{n-1}.

```
function golden_section_search(f, a, b, n)
    ρ = φ-1
    d = ρ * b + (1 - ρ)*a
    yd = f(d)
    for i = 1 : n-1
        c = ρ*a + (1 - ρ)*b
        yc = f(c)
        if yc < yd
            b, d, yd = d, c, yc
        else
            a, b = b, c
        end
    end
    return a < b ? (a, b) : (b, a)
end
```

Algorithm 3.3. Golden section search to be run on a univariate function f, with bracketing interval $[a, b]$, for n > 1 function evaluations. It returns the new interval (a, b). Julia already has the golden ratio φ defined. Guaranteeing convergence to within ϵ requires $n = (b - a)/(\epsilon \ln \varphi)$ iterations.

Fibonacci Search

Golden Section Search

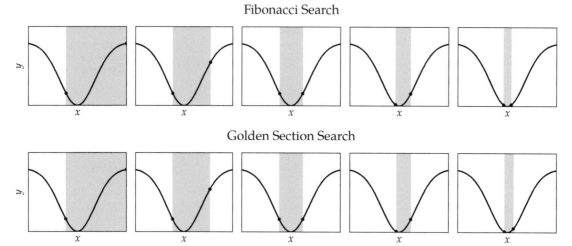

Figure 3.8. Fibonacci and golden section search on a unimodal function.

Fibonacci Search

Golden Section Search

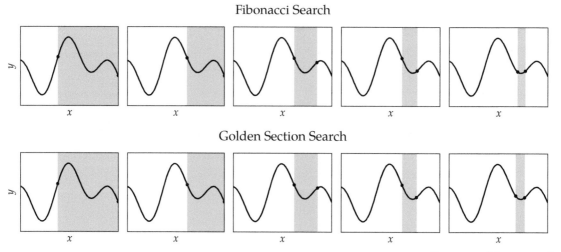

Figure 3.9. Fibonacci and golden section search on a nonunimodal function. Search is not guaranteed to find a global minimum.

3.5   Quadratic Fit Search

Quadratic fit search leverages our ability to analytically solve for the minimum of a quadratic function. Many local minima look quadratic when we zoom in close enough. Quadratic fit search iteratively fits a quadratic function to three bracketing points, solves for the minimum, chooses a new set of bracketing points, and repeats as shown in figure 3.10.

Given bracketing points $a < b < c$, we wish to find the coefficients p_1, p_2, and p_3 for the quadratic function q that goes through (a, y_a), (b, y_b), and (c, y_c):

$$q(x) = p_1 + p_2 x + p_3 x^2 \tag{3.5}$$
$$y_a = p_1 + p_2 a + p_3 a^2 \tag{3.6}$$
$$y_b = p_1 + p_2 b + p_3 b^2 \tag{3.7}$$
$$y_c = p_1 + p_2 c + p_3 c^2 \tag{3.8}$$

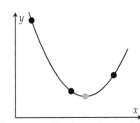

Figure 3.10. Quadratic fit search fits a quadratic function to three bracketing points (black dots) and uses the analytic minimum (blue dot) to determine the next set of bracketing points.

In matrix form, we have

$$\begin{bmatrix} y_a \\ y_b \\ y_c \end{bmatrix} = \begin{bmatrix} 1 & a & a^2 \\ 1 & b & b^2 \\ 1 & c & c^2 \end{bmatrix} \begin{bmatrix} p_1 \\ p_2 \\ p_3 \end{bmatrix} \tag{3.9}$$

We can solve for the coefficients through matrix inversion:

$$\begin{bmatrix} p_1 \\ p_2 \\ p_3 \end{bmatrix} = \begin{bmatrix} 1 & a & a^2 \\ 1 & b & b^2 \\ 1 & c & c^2 \end{bmatrix}^{-1} \begin{bmatrix} y_a \\ y_b \\ y_c \end{bmatrix} \tag{3.10}$$

Our quadratic function is then

$$q(x) = y_a \frac{(x-b)(x-c)}{(a-b)(a-c)} + y_b \frac{(x-a)(x-c)}{(b-a)(b-c)} + y_c \frac{(x-a)(x-b)}{(c-a)(c-b)} \tag{3.11}$$

We can solve for the unique minimum by finding where the derivative is zero:

$$x^* = \frac{1}{2} \frac{y_a(b^2 - c^2) + y_b(c^2 - a^2) + y_c(a^2 - b^2)}{y_a(b - c) + y_b(c - a) + y_c(a - b)} \tag{3.12}$$

Quadratic fit search is typically faster than golden section search. It may need safeguards for cases where the next point is very close to other points. A basic implementation is provided in algorithm 3.4. Figure 3.11 shows several iterations of the algorithm.

```
function quadratic_fit_search(f, a, b, c, n)
    ya, yb, yc = f(a), f(b), f(c)
    for i in 1:n-3
        x = 0.5*(ya*(b^2-c^2)+yb*(c^2-a^2)+yc*(a^2-b^2)) /
                (ya*(b-c)    +yb*(c-a)    +yc*(a-b))
        yx = f(x)
        if x > b
            if yx > yb
                c, yc = x, yx
            else
                a, ya, b, yb = b, yb, x, yx
            end
        elseif x < b
            if yx > yb
                a, ya = x, yx
            else
                c, yc, b, yb = b, yb, x, yx
            end
        end
    end
    return (a, b, c)
end
```

Algorithm 3.4. Quadratic fit search to be run on univariate function f, with bracketing interval $[a, c]$ with $a < b < c$. The method will run for n function evaluations. It returns the new bracketing values as a tuple, (a, b, c).

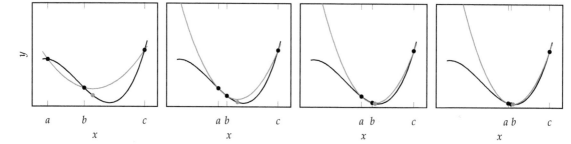

Figure 3.11. Four iterations of the quadratic fit method.

3.6   Shubert-Piyavskii Method

In contrast with previous methods in this chapter, the *Shubert-Piyavskii method*[3] is a *global optimization method* over a domain $[a, b]$, meaning it is guaranteed to converge on the global minimum of a function irrespective of any local minima or whether the function is unimodal. A basic implementation is provided by algorithm 3.5.

The Shubert-Piyavskii method requires that the function be *Lipschitz continuous*, meaning that it is continuous and there is an upper bound on the magnitude of its derivative. A function f is Lipschitz continuous on $[a, b]$ if there exists an $\ell > 0$ such that:[4]

$$|f(x) - f(y)| \le \ell |x - y| \text{ for all } x, y \in [a, b] \tag{3.13}$$

Intuitively, ℓ is as large as the largest unsigned instantaneous rate of change the function attains on $[a, b]$. Given a point $(x_0, f(x_0))$, we know that the lines $f(x_0) - \ell(x - x_0)$ for $x > x_0$ and $f(x_0) + \ell(x - x_0)$ for $x < x_0$ form a lower bound of f.

The Shubert-Piyavskii method iteratively builds a tighter and tighter lower bound on the function. Given a valid Lipschitz constant ℓ, the algorithm begins by sampling the midpoint, $x^{(1)} = (a + b)/2$. A sawtooth lower bound is constructed using lines of slope $\pm \ell$ from this point. These lines will always lie below f if ℓ is a valid Lipschitz constant as shown in figure 3.12.

[3] S. Piyavskii, "An Algorithm for Finding the Absolute Extremum of a Function," *USSR Computational Mathematics and Mathematical Physics*, vol. 12, no. 4, pp. 57–67, 1972. B. O. Shubert, "A Sequential Method Seeking the Global Maximum of a Function," *SIAM Journal on Numerical Analysis*, vol. 9, no. 3, pp. 379–388, 1972.

[4] We can extend the definition of Lipschitz continuity to multivariate functions, where **x** and **y** are vectors and the absolute value is replaced by any vector norm.

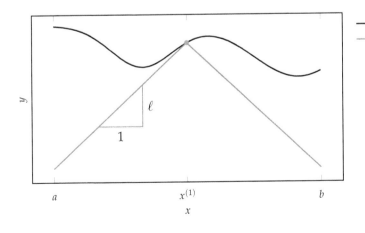

Figure 3.12. The first iteration of the Shubert-Piyavskii method.

Upper vertices in the sawtooth correspond to sampled points. Lower vertices correspond to intersections between the Lipschitz lines originating from each sampled point. Further iterations find the minimum point in the sawtooth, evaluate the function at that x value, and then use the result to update the sawtooth. Figure 3.13 illustrates this process.

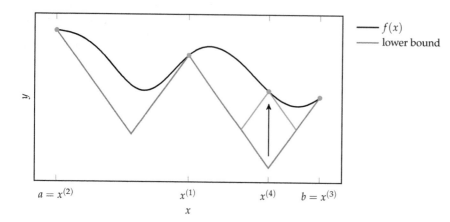

— $f(x)$
— lower bound

Figure 3.13. Updating the lower bound involves sampling a new point and intersecting the new lines with the existing sawtooth.

The algorithm is typically stopped when the difference in height between the minimum sawtooth value and the function evaluation at that point is less than a given tolerance ϵ. For the minimum peak $(x^{(n)}, y^{(n)})$ and function evaluation $f(x^{(n)})$, we thus terminate if $y^{(n)} - f(x^{(n)}) < \epsilon$.

The regions in which the minimum could lie can be computed using this update information. For every peak, an uncertainty region can be computed according to:

$$\left[x^{(i)} - \frac{1}{\ell}(y_{\min} - y^{(i)}), x^{(i)} + \frac{1}{\ell}(y_{\min} - y^{(i)}) \right] \tag{3.14}$$

for each sawtooth lower vertex $(x^{(i)}, y^{(i)})$ and the minimum sawtooth upper vertex $(x_{\min}, y_{\min})$. A point will contribute an uncertainty region only if $y^{(i)} < y_{\min}$. The minimum is located in one of these peak uncertainty regions.

The main drawback of the Shubert-Piyavskii method is that it requires knowing a valid Lipschitz constant. Large Lipschitz constants will result in poor lower bounds. Figure 3.14 shows several iterations of the Shubert-Piyavskii method.

```
struct Pt
    x
    y
end
function _get_sp_intersection(A, B, l)
    t = ((A.y - B.y) - l*(A.x - B.x)) / 2l
    return Pt(A.x + t, A.y - t*l)
end
function shubert_piyavskii(f, a, b, l, ϵ, δ=0.01)
    m = (a+b)/2
    A, M, B = Pt(a, f(a)), Pt(m, f(m)), Pt(b, f(b))
    pts = [A, _get_sp_intersection(A, M, l),
           M, _get_sp_intersection(M, B, l), B]
    Δ = Inf
    while Δ > ϵ
        i = argmin([P.y for P in pts])
        P = Pt(pts[i].x, f(pts[i].x))
        Δ = P.y - pts[i].y

        P_prev = _get_sp_intersection(pts[i-1], P, l)
        P_next = _get_sp_intersection(P, pts[i+1], l)

        deleteat!(pts, i)
        insert!(pts, i, P_next)
        insert!(pts, i, P)
        insert!(pts, i, P_prev)
    end

    intervals = []
    P_min = pts[2*(argmin([P.y for P in pts[1:2:end]])) - 1]
    y_min = P_min.y
    for i in 2:2:length(pts)
        if pts[i].y < y_min
            dy = y_min - pts[i].y
            x_lo = max(a, pts[i].x - dy/l)
            x_hi = min(b, pts[i].x + dy/l)
            if !isempty(intervals) && intervals[end][2] + δ ≥ x_lo
                intervals[end] = (intervals[end][1], x_hi)
            else
                push!(intervals, (x_lo, x_hi))
            end
        end
    end
    return (P_min, intervals)
end
```

Algorithm 3.5. The Shubert-Piyavskii method to be run on univariate function f, with bracketing interval a < b and Lipschitz constant l. The algorithm runs until the update is less than the tolerance ϵ. Both the best point and the set of uncertainty intervals are returned. The uncertainty intervals are returned as an array of (a,b) tuples. The parameter δ is a tolerance used to merge the uncertainty intervals.

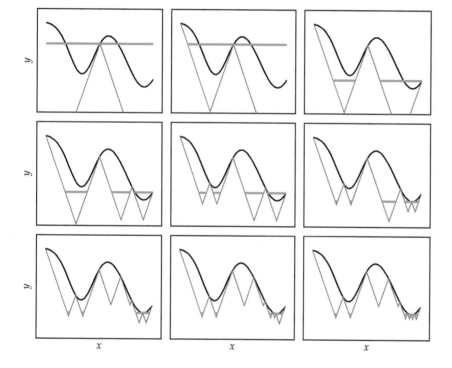

Figure 3.14. Nine iterations of the Shubert-Piyavskii method proceeding left to right and top to bottom. The blue lines are uncertainty regions in which the global minimum could lie.

3.7 Bisection Method

The *bisection method* (algorithm 3.6) can be used to find *roots* of a function, or points where the function is zero. Such *root-finding methods* can be used for optimization by applying them to the derivative of the objective, locating where $f'(x) = 0$. In general, we must ensure that the resulting points are indeed local minima.

The bisection method maintains a bracket $[a, b]$ in which at least one root is known to exist. If f is continuous on $[a, b]$, and there is some $y \in [f(a), f(b)]$, then the *intermediate value theorem* stipulates that there exists at least one $x \in [a, b]$, such that $f(x) = y$ as shown in figure 3.15. It follows that a bracket $[a, b]$ is guaranteed to contain a zero if $f(a)$ and $f(b)$ have opposite signs.

The bisection method cuts the bracketed region in half with every iteration. The midpoint $(a + b)/2$ is evaluated, and the new bracket is formed from the midpoint and whichever side that continues to bracket a zero. We can terminate immediately if the midpoint evaluates to zero. Otherwise we can terminate after a fixed number of iterations. Figure 3.16 shows four iterations of the bisection method. This method is guaranteed to converge within ϵ of x^* within $\lg\left(\frac{|b-a|}{\epsilon}\right)$ iterations, where lg denotes the base 2 logarithm.

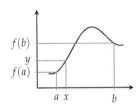

Figure 3.15. A horizontal line drawn from any $y \in [f(a), f(b)]$ must intersect the graph at least once.

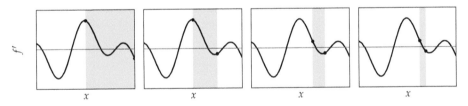

Figure 3.16. Four iterations of the bisection method. The horizontal line corresponds to $f'(x) = 0$. Note that multiple roots exist within the initial bracket.

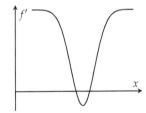

Figure 3.17. A bracketing method initialized such that it straddles the two roots in this figure will expand forever, never to find a sign change. Also, if the initial interval is between the two roots, doubling the interval can cause both ends of the interval to simultaneously pass the two roots.

Root-finding algorithms like the bisection method require starting intervals $[a, b]$ on opposite sides of a zero. That is, $\text{sign}(f'(a)) \neq \text{sign}(f'(b))$, or equivalently, $f'(a)f'(b) \leq 0$. Algorithm 3.7 provides a method for automatically determining such an interval. It starts with a guess interval $[a, b]$. So long as the interval is invalid, its width is increased by a constant factor. Doubling the interval size is a common choice. This method will not always succeed as shown in figure 3.17. Functions that have two nearby roots can be missed, causing the interval to infinitely increase without termination.

The *Brent-Dekker* method is an extension of the bisection method. It is a root-finding algorithm that combines elements of the secant method (section 6.2) and inverse quadratic interpolation. It has reliable and fast convergence properties, and

```
function bisection(f′, a, b, ε)
    if a > b; a,b = b,a; end # ensure a < b

    ya, yb = f′(a), f′(b)
    if ya == 0; b = a; end
    if yb == 0; a = b; end

    while b - a > ε
        x = (a+b)/2
        y = f′(x)
        if y == 0
            a, b = x, x
        elseif sign(y) == sign(ya)
            a = x
        else
            b = x
        end
    end

    return (a,b)
end
```

Algorithm 3.6. The bisection algorithm, where f′ is the derivative of the univariate function we seek to optimize. We have a < b that bracket a zero of f′. The interval width tolerance is ε. Calling bisection returns the new bracketed interval $[a, b]$ as a tuple.

The prime character ′ is not an apostrophe. Thus, f′ is a variable name rather than a transposed vector f. The symbol can be created by typing \prime and hitting tab.

```
function bracket_sign_change(f′, a, b; k=2)
    if a > b; a,b = b,a; end # ensure a < b

    center, half_width = (b+a)/2, (b-a)/2
    while f′(a)*f′(b) > 0
        half_width *= k
        a = center - half_width
        b = center + half_width
    end

    return (a,b)
end
```

Algorithm 3.7. An algorithm for finding an interval in which a sign change occurs. The inputs are the real-valued function f′ defined on the real numbers, and starting interval $[a, b]$. It returns the new interval as a tuple by expanding the interval width until there is a sign change between the function evaluated at the interval bounds. The expansion factor k defaults to 2.

it is the univariate optimization algorithm of choice in many popular numerical optimization packages.[5]

[5] The details of this algorithm can be found in R. P. Brent, *Algorithms for Minimization Without Derivatives*. Prentice Hall, 1973. The algorithm is an extension of the work by T. J. Dekker, "Finding a Zero by Means of Successive Linear Interpolation," in *Constructive Aspects of the Fundamental Theorem of Algebra*, B. Dejon and P. Henrici, eds., Interscience, 1969.

3.8 Summary

- Many optimization methods shrink a bracketing interval, including Fibonacci search, golden section search, and quadratic fit search.

- The Shubert-Piyavskii method outputs a set of bracketed intervals containing the global minima, given the Lipschitz constant.

- Root-finding methods like the bisection method can be used to find where the derivative of a function is zero.

3.9 Exercises

Exercise 3.1. Give an example of a problem when Fibonacci search is preferred over the bisection method.

Exercise 3.2. What is a drawback of the Shubert-Piyavskii method?

Exercise 3.3. Give an example of a nontrivial function where quadratic fit search would identify the minimum correctly once the function values at three distinct points are available.

Exercise 3.4. Suppose we have $f(x) = x^2/2 - x$. Apply the bisection method to find an interval containing the minimizer of f starting with the interval $[0, 1000]$. Execute three steps of the algorithm.

Exercise 3.5. Suppose we have a function $f(x) = (x + 2)^2$ on the interval $[0, 1]$. Is 2 a valid Lipschitz constant for f on that interval?

Exercise 3.6. Suppose we have a unimodal function defined on the interval $[1, 32]$. After three function evaluations of our choice, will we be able to narrow the optimum to an interval of at most length 10? Why or why not?

4 *Local Descent*

Up to this point, we have focused on optimization involving a single design variable. This chapter introduces a general approach to optimization involving *multivariate* functions, or functions with more than one variable. The focus of this chapter is on how to use *local models* to incrementally improve a design point until some convergence criterion is met. We begin by discussing methods that, at each iteration, choose a descent direction based on a local model and then choose a step size. We then discuss methods that restrict the step to be within a region where the local model is believed to be valid. This chapter concludes with a discussion of convergence conditions. The next two chapters will discuss how to use first- and second-order models built from gradient or Hessian information.

4.1 *Descent Direction Iteration*

A common approach to optimization is to incrementally improve a design point $\mathbf{x}$ by taking a step that minimizes the objective value based on a local model. The local model may be obtained, for example, from a first- or second-order Taylor approximation. Optimization algorithms that follow this general approach are referred to as *descent direction methods*. They start with a design point $\mathbf{x}^{(1)}$ and then generate a sequence of points, sometimes called *iterates*, to converge to a local minimum.[1]

The iterative descent direction procedure involves the following steps:

1. Check whether $\mathbf{x}^{(k)}$ satisfies the termination conditions. If it does, terminate; otherwise proceed to the next step.

2. Determine the *descent direction* $\mathbf{d}^{(k)}$ using local information such as the gradient or Hessian. Some algorithms assume $\|\mathbf{d}^{(k)}\| = 1$, but others do not.

[1] The choice of $\mathbf{x}^{(1)}$ can affect the success of the algorithm in finding a minimum. Domain knowledge is often used to choose a reasonable value. When that is not available, we can search over the design space using the techniques that will be covered in chapter 13.

3. Determine the step size or learning rate $\alpha^{(k)}$. Some algorithms attempt to optimize the step size so that the step maximally decreases f.[2]

4. Compute the next design point according to:

$$\mathbf{x}^{(k+1)} \leftarrow \mathbf{x}^{(k)} + \alpha^{(k)} \mathbf{d}^{(k)} \tag{4.1}$$

There are many different optimization methods, each with their own ways of determining α and $\mathbf{d}$.

[2] We use *step size* to refer to the magnitude of the overall step. Obtaining a new iterate using equation (4.1) with a step size $\alpha^{(k)}$ implies that the descent direction $\mathbf{d}^{(k)}$ has unit length. We use *learning rate* to refer to a scalar multiple used on a descent direction vector which does not necessarily have unit length.

4.2   Line Search

For the moment, assume we have chosen a descent direction $\mathbf{d}$, perhaps using one of the methods discussed in one of the subsequent chapters. We need to choose the step factor α to obtain our next design point. One approach is to use *line search*, which selects the step factor that minimizes the one-dimensional function:

$$\underset{\alpha}{\text{minimize}}\, f(\mathbf{x} + \alpha \mathbf{d}) \tag{4.2}$$

Line search is a univariate optimization problem, which was covered in chapter 3. We can apply the univariate optimization method of our choice.[3] To inform the search, we can use the derivative of the line search objective, which is simply the directional derivative along $\mathbf{d}$ at $\mathbf{x} + \alpha \mathbf{d}$. Line search is demonstrated in example 4.1 and implemented in algorithm 4.1.

[3] The Brent-Dekker method, mentioned in the previous chapter, is a commonly used univariate optimization method. It combines the robustness of the bisection method with the speed of the secant method.

```
function line_search(f, x, d)
    objective = α -> f(x + α*d)
    a, b = bracket_minimum(objective)
    α = minimize(objective, a, b)
    return x + α*d
end
```

Algorithm 4.1. A method for conducting a line search, which finds the optimal step factor along a descent direction d from design point x to minimize function f. The `minimize` function can be implemented using a univariate optimization algorithm such as the Brent-Dekker method.

One disadvantage of conducting a line search at each step is the computational cost of optimizing α to a high degree of precision. Instead, it is common to quickly find a reasonable value and then move on, selecting $\mathbf{x}^{(k+1)}$, and then picking a new direction $\mathbf{d}^{(k+1)}$.

Some algorithms use a fixed step factor. Large steps will tend to result in faster convergence but risk overshooting the minimum. Smaller steps tend to be more stable but can result in slower convergence. A fixed step factor α is sometimes referred to as a *learning rate*.

Another method is to use a *decaying step factor*:

$$\alpha^{(k)} = \alpha^{(1)}\gamma^{k-1} \quad \text{for } \gamma \in (0,1] \tag{4.3}$$

Decaying step factors are especially popular when minimizing noisy objective functions,[4] and are commonly used in machine learning applications.

[4] We will discuss optimization in the presence of noise and other forms of uncertainty in chapter 17.

Consider conducting a line search on $f(x_1, x_2, x_3) = \sin(x_1 x_2) + \exp(x_2 + x_3) - x_3$ from $\mathbf{x} = [1, 2, 3]$ in the direction $\mathbf{d} = [0, -1, -1]$. The corresponding optimization problem is:

$$\underset{\alpha}{\text{minimize}} \sin((1 + 0\alpha)(2 - \alpha)) + \exp((2 - \alpha) + (3 - \alpha)) - (3 - \alpha)$$

which simplifies to:

$$\underset{\alpha}{\text{minimize}} \sin(2 - \alpha) + \exp(5 - 2\alpha) + \alpha - 3$$

The minimum is at $\alpha \approx 3.127$ with $\mathbf{x} \approx [1, -1.126, -0.126]$.

Example 4.1. Line search used to minimize a function along a descent direction.

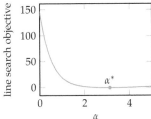

4.3 Approximate Line Search

It is often more computationally efficient to perform more iterations of a descent method than to do exact line search at each iteration, especially if the function and derivative calculations are expensive. Many of the methods discussed so far can benefit from using *approximate line search* to find a suitable step size with a small number of evaluations. Since descent methods must descend, a step size α may be suitable if it causes a decrease in the objective function value. However, a variety of other conditions may be enforced to encourage faster convergence.

The condition for *sufficient decrease*[5] requires that the step size cause a sufficient decrease in the objective function value:

[5] This condition is sometimes referred to as the *Armijo condition*.

$$f(\mathbf{x}^{(k+1)}) \leq f(\mathbf{x}^{(k)}) + \beta\alpha\nabla_{\mathbf{d}^{(k)}}f(\mathbf{x}^{(k)}) \tag{4.4}$$

with $\beta \in [0, 1]$ often set to $\beta = 1 \times 10^{-4}$. Figure 4.1 illustrates this condition. If $\beta = 0$, then any decrease is acceptable. If $\beta = 1$, then the decrease has to be at least as much as what would be predicted by a first-order approximation.

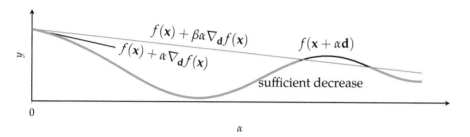

Figure 4.1. The sufficient decrease condition, the first Wolfe condition, can always be satisfied by a sufficiently small step size along a descent direction.

If $\mathbf{d}$ is a valid descent direction, then there must exist a sufficiently small step size that satisfies the sufficient decrease condition. We can thus start with a large step size and decrease it by a constant reduction factor until the sufficient decrease condition is satisfied. This algorithm is known as *backtracking line search*[6] because of how it backtracks along the descent direction. Backtracking line search is shown in figure 4.2 and implemented in algorithm 4.2. We walk through the procedure in example 4.2.

[6] Also known as *Armijo line search*, L. Armijo, "Minimization of Functions Having Lipschitz Continuous First Partial Derivatives," *Pacific Journal of Mathematics*, vol. 16, no. 1, pp. 1–3, 1966.

```
function backtracking_line_search(f, ∇f, x, d, α; p=0.5, β=1e-4)
    y, g = f(x), ∇f(x)
    while f(x + α*d) > y + β*α*(g·d)
        α *= p
    end
    α
end
```

Algorithm 4.2. The backtracking line search algorithm, which takes objective function f, its gradient ∇f, the current design point x, a descent direction d, and the maximum step size α. We can optionally specify the reduction factor p and the first Wolfe condition parameter β.

Note that the cdot character · aliases to the dot function such that a·b is equivalent to dot(a,b). The symbol can be created by typing \cdot and hitting tab.

The first condition is insufficient to guarantee convergence to a local minimum. Very small step sizes will satisfy the first condition but can prematurely converge. Backtracking line search avoids premature convergence by accepting the largest satisfactory step size obtained by sequential downscaling and is guaranteed to converge to a local minimum.

Another condition, called the *curvature condition*, requires the directional derivative at the next iterate to be shallower:

$$\nabla_{\mathbf{d}^{(k)}} f(\mathbf{x}^{(k+1)}) \geq \sigma \nabla_{\mathbf{d}^{(k)}} f(\mathbf{x}^{(k)}) \tag{4.5}$$

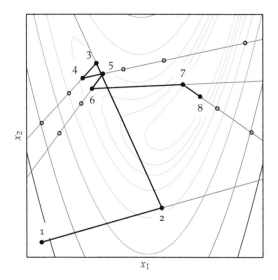

Figure 4.2.   Backtracking line search used on the Rosenbrock function (appendix B.6). The black lines show the seven iterations taken by the descent method and the red lines show the points considered during each line search.

where σ controls how shallow the next directional derivative must be. Figures 4.3 and 4.4 illustrate this condition. It is common to set $\beta < \sigma < 1$ with $\sigma = 0.1$ when approximate linear search is used with the conjugate gradient method and to 0.9 when used with Newton's method.[7]

An alternative to the curvature condition is the *strong curvature condition*, which is a more restrictive criterion in that the slope is also required not to be too positive:

$$|\nabla_{\mathbf{d}^{(k)}} f(\mathbf{x}^{(k+1)})| \leq -\sigma \nabla_{\mathbf{d}^{(k)}} f(\mathbf{x}^{(k)}) \qquad (4.6)$$

Figure 4.5 illustrates this condition.

Together, the sufficient decrease condition and the curvature condition form the *Wolfe conditions*. The sufficient decrease condition is often called the *first Wolfe condition* and the curvature condition is called the *second Wolfe condition*. The sufficient decrease condition with the strong curvature condition form the *strong Wolfe conditions*.

Satisfying the strong Wolfe conditions requires a more complicated algorithm, *strong backtracking line search* (algorithm 4.3).[8] The method operates in two phases. The first phase, the *bracketing phase*, tests successively larger step sizes to bracket an interval $[\alpha^{(k-1)}, \alpha^{(k)}]$ guaranteed to contain step lengths satisfying the Wolfe conditions.

[7] The conjugate gradient method is introduced in section 5.2, and Newton's method is introduced in section 6.1.

[8] J. Nocedal and S. J. Wright, *Numerical Optimization*, 2nd ed. Springer, 2006.

Consider approximate line search on $f(x_1, x_2) = x_1^2 + x_1x_2 + x_2^2$ from $\mathbf{x} = [1, 2]$ in the direction $\mathbf{d} = [-1, -1]$, using a maximum step size of 10, a reduction factor of 0.5, a first Wolfe condition parameter $\beta = 1 \times 10^{-4}$, and a second Wolfe condition parameter $\sigma = 0.9$.

We check whether the maximum step size satisfies the first Wolfe condition, where the gradient at $\mathbf{x}$ is $\mathbf{g} = [4, 5]$:

$$f(\mathbf{x} + \alpha\mathbf{d}) \leq f(\mathbf{x}) + \beta\alpha(\mathbf{g}^\top\mathbf{d})$$
$$f([1, 2] + 10 \cdot [-1, -1]) \leq 7 + 1 \times 10^{-4} \cdot 10 \cdot [4, 5]^\top [-1, -1]$$
$$217 \leq 6.991$$

It is not satisfied.

The step size is multiplied by 0.5 to obtain 5, and the first Wolfe condition is checked again:

$$f([1, 2] + 5 \cdot [-1, -1]) \leq 7 + 1 \times 10^{-4} \cdot 5 \cdot [4, 5]^\top [-1, -1]$$
$$37 \leq 6.996$$

It is not satisfied.

The step size is multiplied by 0.5 to obtain 2.5, and the first Wolfe condition is checked again:

$$f([1, 2] + 2.5 \cdot [-1, -1]) \leq 7 + 1 \times 10^{-4} \cdot 2.5 \cdot [4, 5]^\top [-1, -1]$$
$$3.25 \leq 6.998$$

The first Wolfe condition is satisfied.

The candidate design point $\mathbf{x}' = \mathbf{x} + \alpha\mathbf{d} = [-1.5, -0.5]$ is checked against the second Wolfe condition:

$$\nabla_{\mathbf{d}} f(\mathbf{x}') \geq \sigma\nabla_{\mathbf{d}} f(\mathbf{x})$$
$$[-3.5, -2.5]^\top [-1, -1] \geq \sigma [4, 5]^\top [-1, -1]$$
$$6 \geq -8.1$$

The second Wolfe condition is satisfied.

Approximate line search terminates with $\mathbf{x} = [-1.5, -0.5]$.

Example 4.2. An example of backtracking line search, an approximate line search method.

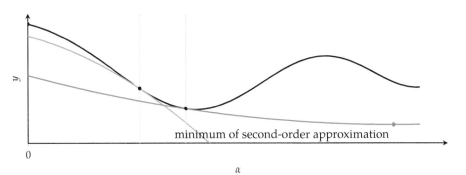

Figure 4.3. The curvature condition, the second Wolfe condition, is necessary to ensure that second-order function approximations have positive curvature, thereby having a unique global minimum.

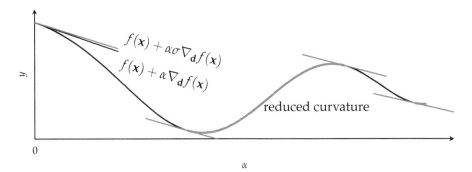

Figure 4.4. Regions where the curvature condition is satisfied.

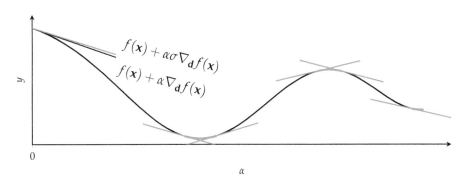

Figure 4.5. Regions where the strong curvature condition is satisfied.

An interval guaranteed to contain step lengths satisfying the Wolfe conditions is found when one of the following conditions hold:

$$f(\mathbf{x} + \alpha\mathbf{d}) \geq f(\mathbf{x}) \tag{4.7}$$

$$f(\mathbf{x} + \alpha\mathbf{d}) > f(\mathbf{x}) + \beta\alpha\nabla_{\mathbf{d}}f(\mathbf{x}) \tag{4.8}$$

$$\nabla f(\mathbf{x} + \alpha\mathbf{d}) \geq \mathbf{0} \tag{4.9}$$

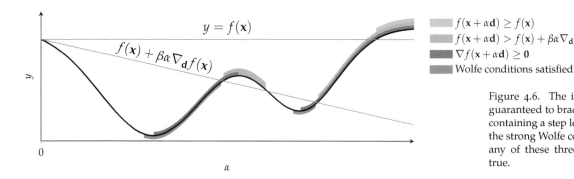

$f(\mathbf{x} + \alpha\mathbf{d}) \geq f(\mathbf{x})$
$f(\mathbf{x} + \alpha\mathbf{d}) > f(\mathbf{x}) + \beta\alpha\nabla_{\mathbf{d}}f(\mathbf{x})$
$\nabla f(\mathbf{x} + \alpha\mathbf{d}) \geq \mathbf{0}$
Wolfe conditions satisfied

Figure 4.6. The interval $[0, \alpha]$ is guaranteed to bracket an interval containing a step length satisfying the strong Wolfe conditions when any of these three conditions is true.

Satisfying equation (4.8) is equivalent to violating the first Wolfe condition, thereby ensuring that shrinking the step length will guarantee a satisfactory step length. Similarly, equation (4.7) and equation (4.9) guarantee that the descent step has overshot a local minimum, and the region between must therefore contain a satisfactory step length.

Figure 4.6 shows where each bracketing condition is true for an example line search. The figure shows bracket intervals $[0, \alpha]$, whereas advanced backtracking line search successively increases the step length to obtain a bracketing interval $[\alpha^{(k-1)}, \alpha^{(k)}]$.

In the *zoom phase*, we shrink the interval to find a step size satisfying the strong Wolfe conditions. The shrinking can be done using the bisection method (section 3.7), updating the interval boundaries according to the same interval conditions. This process is shown in figure 4.7.

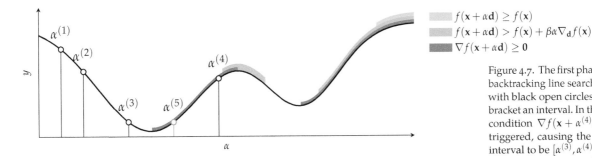

$\blacksquare\ f(\mathbf{x} + \alpha\mathbf{d}) \geq f(\mathbf{x})$
$\blacksquare\ f(\mathbf{x} + \alpha\mathbf{d}) > f(\mathbf{x}) + \beta\alpha\nabla_{\mathbf{d}}f(\mathbf{x})$
$\blacksquare\ \nabla f(\mathbf{x} + \alpha\mathbf{d}) \geq \mathbf{0}$

Figure 4.7. The first phase of strong backtracking line search, indicated with black open circles, is used to bracket an interval. In this case, the condition $\nabla f(\mathbf{x} + \alpha^{(4)}\mathbf{d}) \geq 0$ is triggered, causing the bracketing interval to be $[\alpha^{(3)}, \alpha^{(4)}]$. Then the zoom phase, indicated by the red open circle, shrinks the bracketed region until a suitable step length is found.

4.4   Trust Region Methods

Descent methods can place too much trust in their first- or second-order information, which can result in excessively large steps or premature convergence. A *trust region*[9] is the local area of the design space where the local model is believed to be reliable. A trust region method, or *restricted step method*, maintains a local model of the trust region that both limits the step taken by traditional line search and predicts the improvement associated with taking the step. If the improvement closely matches the predicted value, the trust region is expanded. If the improvement deviates from the predicted value, the trust region is contracted.[10] Figure 4.8 shows a design point centered within a circular trust region.

Trust region methods first choose the maximum step size and then the step direction, which is in contrast with line search methods that first choose a step direction and then optimize the step size. A trust region approach finds the next step by minimizing a model of the objective function $\hat{f}$ over a trust region centered on the current design point $\mathbf{x}$. An example of $\hat{f}$ is a second-order Taylor approximation (see appendix C.2). The radius of the trust region, δ, is expanded and contracted based on how well the model predicts function evaluations. The next design point $\mathbf{x}'$ is obtained by solving:

$$
\begin{aligned}
&\underset{\mathbf{x}'}{\text{minimize}} && \hat{f}(\mathbf{x}') \\
&\text{subject to} && \|\mathbf{x} - \mathbf{x}'\| \leq \delta
\end{aligned}
\tag{4.10}
$$

where the trust region is defined by the positive radius δ and a vector norm.[11] The equation above is a constrained optimization problem, which is covered in chapter 10.

[9] K. Levenberg, "A Method for the Solution of Certain Non-Linear Problems in Least Squares," *Quarterly of Applied Mathematics*, vol. 2, no. 2, pp. 164–168, 1944.

[10] A recent review of trust region methods is provided by Y. X. Yuan, "Recent Advances in Trust Region Algorithms," *Mathematical Programming*, vol. 151, no. 1, pp. 249–281, 2015.

[11] There are a variety of efficient methods for solving equation (4.10) efficiently. For an overview of the trust region method applied to quadratic models, see D. C. Sorensen, "Newton's Method with a Model Trust Region Modification," *SIAM Journal on Numerical Analysis*, vol. 19, no. 2, pp. 409–426, 1982.

```
function strong_backtracking(f, ∇, x, d; α=1, β=1e-4, σ=0.1)
    y0, g0, y_prev, α_prev = f(x), ∇(x)·d, NaN, 0
    αlo, αhi = NaN, NaN

    # bracket phase
    while true
        y = f(x + α*d)
        if y > y0 + β*α*g0 || (!isnan(y_prev) && y ≥ y_prev)
            αlo, αhi = α_prev, α
            break
        end
        g = ∇(x + α*d)·d
        if abs(g) ≤ -σ*g0
            return α
        elseif g ≥ 0
            αlo, αhi = α, α_prev
            break
        end
        y_prev, α_prev, α = y, α, 2α
    end

    # zoom phase
    ylo = f(x + αlo*d)
    while true
        α = (αlo + αhi)/2
        y = f(x + α*d)
        if y > y0 + β*α*g0 || y ≥ ylo
            αhi = α
        else
            g = ∇(x + α*d)·d
            if abs(g) ≤ -σ*g0
                return α
            elseif g*(αhi - αlo) ≥ 0
                αhi = αlo
            end
            αlo = α
        end
    end
end
```

Algorithm 4.3. Strong backtracking approximate line search for satisfying the strong Wolfe conditions. It takes as input the objective function f, the gradient function ∇, the design point x and direction d from which line search is conducted, an initial step size α, and the Wolfe condition parameters β and σ. The algorithm's bracket phase first brackets an interval containing a step size that satisfies the strong Wolfe conditions. It then reduces this bracketed interval in the zoom phase until a suitable step size is found. We interpolate with bisection, but other schemes can be used.

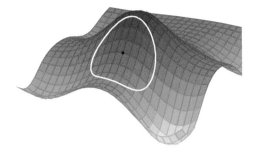

Figure 4.8. Trust region methods constrain the next step to lie within a local region. The trusted region is expanded and contracted based on the predictive performance of models of the objective function.

The trust region radius δ is expanded or contracted based on the local model's predictive performance. Trust region methods compare the predicted improvement $\Delta y_{\text{pred}} = f(\mathbf{x}) - \hat{f}(\mathbf{x}')$ to the actual improvement $\Delta y_{\text{act}} = f(\mathbf{x}) - f(\mathbf{x}')$:

$$\eta = \frac{\text{actual improvement}}{\text{predicted improvement}} = \frac{f(\mathbf{x}) - f(\mathbf{x}')}{f(\mathbf{x}) - \hat{f}(\mathbf{x}')} \tag{4.11}$$

The ratio η is close to 1 when the predicted step size matches the actual step size. If the ratio is too small, such as below a threshold η_1, then the improvement is considered sufficiently less than expected, and the trust region radius is scaled down by a factor $\gamma_1 < 1$. If the ratio is sufficiently large, such as above a threshold η_2, then our prediction is considered accurate, and the trust region radius is scaled up by a factor $\gamma_2 > 1$. Algorithm 4.4 provides an implementation and figure 4.9 visualizes the optimization procedure. Example 4.3 shows how to construct noncircular trust regions.

4.5   Termination Conditions

There are four common termination conditions for descent direction methods:

- *Maximum iterations.* We may want to terminate when the number of iterations k exceeds some threshold k_{max}. Alternatively, we might want to terminate once a maximum amount of elapsed time is exceeded.

$$k > k_{\text{max}} \tag{4.12}$$

```
function trust_region_descent(f, ∇f, H, x, k_max;
    η1=0.25, η2=0.5, γ1=0.5, γ2=2.0, δ=1.0)
    y = f(x)
    for k in 1 : k_max
        x′, y′ = solve_trust_region_subproblem(∇f, H, x, δ)
        r = (y - f(x′)) / (y - y′)
        if r < η1
            δ *= γ1
        else
            x, y = x′, y′
            if r > η2
                δ *= γ2
            end
        end
    end
    return x
end
```

```
using Convex, SCS
function solve_trust_region_subproblem(∇f, H, x0, δ)
    x = Variable(length(x0))
    p = minimize(∇f(x0)·(x-x0) + quadform(x-x0, H(x0))/2)
    p.constraints += norm(x-x0) <= δ
    solve!(p, SCS.Optimizer)
    return (x.value, p.optval)
end
```

Algorithm 4.4. The trust region descent method, where f is the objective function, ∇f produces the derivative, H produces the Hessian, x is an initial design point, and k_max is the number of iterations. The optional parameters η1 and η2 determine when the trust region radius δ is increased or decreased, and γ1 and γ2 control the magnitude of the change. An implementation for solve_trust_region_subproblem must be provided that solves equation (4.10). We have provided an example implementation that uses a second-order Taylor approximation about x0 with a circular trust region.

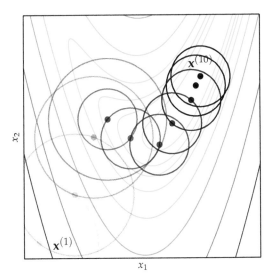

Figure 4.9. Trust region optimization used on the Rosenbrock function (appendix B.6).

Trust regions need not be circular. In some cases, certain directions may have higher trust than others.

A norm can be constructed to produce elliptical regions:

$$\|\mathbf{x} - \mathbf{x}_0\|_{\mathbf{E}} = (\mathbf{x} - \mathbf{x}_0)^{\top} \mathbf{E}(\mathbf{x} - \mathbf{x}_0)$$

with $\|\mathbf{x} - \mathbf{x}_0\|_{\mathbf{E}} \leq 1$ where $\mathbf{E}$ is a symmetric matrix that defines the ellipse.

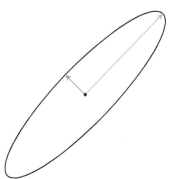

The ellipse matrix $\mathbf{E}$ can be updated with each descent iteration, which can involve more complicated adjustments than scaling the trusted region.

Example 4.3.   Trust region optimization need not use circular trust regions. Additional detail is provided by J. Nocedal and S. J. Wright, "Trust-Region Methods," in *Numerical Optimization*. Springer, 2006, pp. 66–100.

- *Absolute improvement.* This termination condition looks at the change in the function value over subsequent steps. If the change is smaller than a given threshold, it will terminate:

$$f(\mathbf{x}^{(k)}) - f(\mathbf{x}^{(k+1)}) < \epsilon_a \qquad (4.13)$$

- *Relative improvement.* This termination condition also looks at the change in function value but uses the step factor relative to the current function value:

$$f(\mathbf{x}^{(k)}) - f(\mathbf{x}^{(k+1)}) < \epsilon_r |f(\mathbf{x}^{(k)})| \qquad (4.14)$$

- *Gradient magnitude.* We can also terminate based on the magnitude of the gradient:

$$\|\nabla f(\mathbf{x}^{(k+1)})\| < \epsilon_g \qquad (4.15)$$

In cases where multiple local minima are likely to exist, it can be beneficial to incorporate *random restarts* after our termination conditions are met where we restart our local descent method from randomly selected initial points.

4.6   Summary

- Descent direction methods incrementally descend toward a local optimum.

- Univariate optimization can be applied during line search.

- Approximate line search can be used to identify appropriate descent step sizes.

- Trust region methods constrain the step to lie within a local region that expands or contracts based on predictive accuracy.

- Termination conditions for descent methods can be based on criteria such as the change in the objective function value or magnitude of the gradient.

4.7   Exercises

Exercise 4.1. Why is it important to have more than one termination condition?

Exercise 4.2. The first Wolfe condition requires

$$f(\mathbf{x}^{(k)} + \alpha \mathbf{d}^{(k)}) \leq f(\mathbf{x}^{(k)}) + \beta \alpha \nabla_{\mathbf{d}^{(k)}} f(\mathbf{x}^{(k)}) \tag{4.16}$$

What is the maximum step length α that satisfies this condition, given that $f(\mathbf{x}) = 5 + x_1^2 + x_2^2$, $\mathbf{x}^{(k)} = [-1, -1]$, $\mathbf{d} = [1, 0]$, and $\beta = 10^{-4}$?

5 *First-Order Methods*

The previous chapter introduced the general concept of descent direction methods. This chapter discusses a variety of algorithms that use *first-order* methods to select the appropriate descent direction. First-order methods rely on gradient information to help direct the search for a minimum, which can be obtained using methods outlined in chapter 2.

5.1 *Gradient Descent*

An intuitive choice for descent direction $\mathbf{d}$ is the direction of steepest descent. Following the direction of steepest descent is guaranteed to lead to improvement, provided that the objective function is smooth, the step size is sufficiently small, and we are not already at a point where the gradient is zero.[1] The direction of steepest descent is the direction opposite the gradient ∇f, hence the name *gradient descent*. For convenience in this chapter, we define

$$\mathbf{g}^{(k)} = \nabla f(\mathbf{x}^{(k)}) \tag{5.1}$$

[1] A point where the gradient is zero is called a *stationary point*.

where $\mathbf{x}^{(k)}$ is our design point at descent iteration k.

In gradient descent, we typically normalize the direction of steepest descent (see example 5.1):

$$\mathbf{d}^{(k)} = -\frac{\mathbf{g}^{(k)}}{\|\mathbf{g}^{(k)}\|} \tag{5.2}$$

Jagged search paths result if we choose a step size that leads to the maximal decrease in f. In fact, the next direction will always be orthogonal to the current direction. We can show this as follows:

Suppose we have $f(\mathbf{x}) = x_1 x_2^2$. The gradient is $\nabla f = [x_2^2, 2x_1 x_2]$. For $\mathbf{x}^{(k)} = [1, 2]$ we get an unnormalized direction of steepest descent $\mathbf{d} = [-4, -4]$, which is normalized to $\mathbf{d} = [-\frac{1}{\sqrt{2}}, -\frac{1}{\sqrt{2}}]$.

Example 5.1. Computing the gradient descent direction.

If we optimize the step size at each step, we have

$$\alpha^{(k)} = \arg\min_\alpha f(\mathbf{x}^{(k)} + \alpha \mathbf{d}^{(k)}) \tag{5.3}$$

The optimization above implies that the directional derivative equals zero. Using equation (2.9), we have

$$\nabla f(\mathbf{x}^{(k)} + \alpha \mathbf{d}^{(k)})^\top \mathbf{d}^{(k)} = 0 \tag{5.4}$$

We know

$$\mathbf{d}^{(k+1)} = -\frac{\nabla f(\mathbf{x}^{(k)} + \alpha \mathbf{d}^{(k)})}{\|\nabla f(\mathbf{x}^{(k)} + \alpha \mathbf{d}^{(k)})\|} \tag{5.5}$$

Hence,

$$\mathbf{d}^{(k+1)\top} \mathbf{d}^{(k)} = 0 \tag{5.6}$$

which means that $\mathbf{d}^{(k+1)}$ and $\mathbf{d}^{(k)}$ are orthogonal.

Narrow valleys aligned with a descent direction are not an issue. When the descent directions cross over the valley, many steps must be taken in order to make progress along the valley's floor as shown in figure 5.1. An implementation of gradient descent is provided by algorithm 5.1.

5.2   Conjugate Gradient

Gradient descent can perform poorly in narrow valleys. The *conjugate gradient* method overcomes this issue by borrowing inspiration from methods for optimizing quadratic functions:

$$\underset{\mathbf{x}}{\text{minimize}} f(\mathbf{x}) = \frac{1}{2}\mathbf{x}^\top \mathbf{A}\mathbf{x} + \mathbf{b}^\top \mathbf{x} + c \tag{5.7}$$

where $\mathbf{A}$ is symmetric and positive definite, and thus f has a unique local minimum (section 1.6.2).

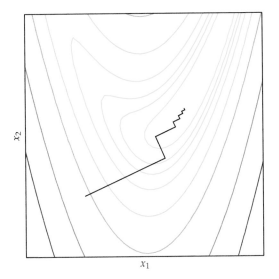

Figure 5.1. Gradient descent can result in zig-zagging in narrow canyons. Here we see the effect on the Rosenbrock function (appendix B.6).

```
abstract type DescentMethod end
struct GradientDescent <: DescentMethod
    α
end
init!(M::GradientDescent, f, ∇f, x) = M
function step!(M::GradientDescent, f, ∇f, x)
    α, g = M.α, ∇f(x)
    return x - α*g
end
```

Algorithm 5.1. The gradient descent method, which follows the direction of gradient descent with a fixed learning rate. The step! function produces the next iterate whereas the init function does nothing.

The conjugate gradient method can optimize n-dimensional quadratic functions in n steps as shown in figure 5.2. Its directions are *mutually conjugate* with respect to $\mathbf{A}$:

$$\mathbf{d}^{(i)\top}\mathbf{A}\,\mathbf{d}^{(j)} = 0 \text{ for all } i \neq j \tag{5.8}$$

The mutually conjugate vectors are the basis vectors of $\mathbf{A}$. They are generally not orthogonal to one another.

The successive conjugate directions are computed using gradient information and the previous descent direction. The algorithm starts with the direction of steepest descent:

$$\mathbf{d}^{(1)} = -\mathbf{g}^{(1)} \tag{5.9}$$

We then use line search to find the next design point. For quadratic functions, the step factor α can be computed exactly (example 5.2). The update is then:

$$\mathbf{x}^{(2)} = \mathbf{x}^{(1)} + \alpha^{(1)}\mathbf{d}^{(1)} \tag{5.10}$$

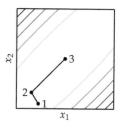

Figure 5.2. Conjugate gradient descent converges in n steps when applied to an n-dimensional quadratic function.

Suppose we want to derive the optimal step factor for a line search on a quadratic function:

$$\underset{\alpha}{\text{minimize}}\, f(\mathbf{x} + \alpha\mathbf{d})$$

We can compute the derivative with respect to α:

$$\frac{\partial f(\mathbf{x} + \alpha\mathbf{d})}{\partial \alpha} = \frac{\partial}{\partial \alpha}\left[\frac{1}{2}(\mathbf{x} + \alpha\mathbf{d})^\top\mathbf{A}(\mathbf{x} + \alpha\mathbf{d}) + \mathbf{b}^\top(\mathbf{x} + \alpha\mathbf{d}) + c\right]$$

$$= \mathbf{d}^\top\mathbf{A}(\mathbf{x} + \alpha\mathbf{d}) + \mathbf{d}^\top\mathbf{b}$$

$$= \mathbf{d}^\top(\mathbf{A}\mathbf{x} + \mathbf{b}) + \alpha\mathbf{d}^\top\mathbf{A}\mathbf{d}$$

Setting $\frac{\partial f(\mathbf{x} + \alpha\mathbf{d})}{\partial \alpha} = 0$ results in:

$$\alpha = -\frac{\mathbf{d}^\top(\mathbf{A}\mathbf{x} + \mathbf{b})}{\mathbf{d}^\top\mathbf{A}\mathbf{d}}$$

Example 5.2. The optimal step factor for a line search on a quadratic function.

Subsequent iterations choose $\mathbf{d}^{(k+1)}$ based on the current gradient and a contribution from the previous descent direction:

$$\mathbf{d}^{(k)} = -\mathbf{g}^{(k)} + \beta^{(k)}\mathbf{d}^{(k-1)} \tag{5.11}$$

for scalar parameter β. Larger values of β indicate that the previous descent direction contributes more strongly.

We can derive the best value for β for a known $\mathbf{A}$, using the fact that $\mathbf{d}^{(k)}$ is conjugate to $\mathbf{d}^{(k-1)}$:

$$\mathbf{d}^{(k)\top}\mathbf{A}\mathbf{d}^{(k-1)} = 0 \tag{5.12}$$

$$\Rightarrow (-\mathbf{g}^{(k)} + \beta^{(k)}\mathbf{d}^{(k-1)})^\top \mathbf{A}\mathbf{d}^{(k-1)} = 0 \tag{5.13}$$

$$\Rightarrow -\mathbf{g}^{(k)\top}\mathbf{A}\mathbf{d}^{(k-1)} + \beta^{(k)}\mathbf{d}^{(k-1)\top}\mathbf{A}\mathbf{d}^{(k-1)} = 0 \tag{5.14}$$

$$\Rightarrow \beta^{(k)} = \frac{\mathbf{g}^{(k)\top}\mathbf{A}\mathbf{d}^{(k-1)}}{\mathbf{d}^{(k-1)\top}\mathbf{A}\mathbf{d}^{(k-1)}} \tag{5.15}$$

The conjugate gradient method can be applied to nonquadratic functions as well. Smooth, continuous functions behave like quadratic functions close to a local minimum, and the conjugate gradient method will converge very quickly in such regions.

Unfortunately, we do not know the value of $\mathbf{A}$ that best approximates f around $\mathbf{x}^{(k)}$. Instead, several choices for $\beta^{(k)}$ tend to work well:

Fletcher-Reeves:[2]

$$\beta^{(k)} = \frac{\mathbf{g}^{(k)\top}\mathbf{g}^{(k)}}{\mathbf{g}^{(k-1)\top}\mathbf{g}^{(k-1)}} \tag{5.16}$$

Polak-Ribière:[3]

$$\beta^{(k)} = \frac{\mathbf{g}^{(k)\top}\left(\mathbf{g}^{(k)} - \mathbf{g}^{(k-1)}\right)}{\mathbf{g}^{(k-1)\top}\mathbf{g}^{(k-1)}} \tag{5.17}$$

Convergence for the Polak-Ribière method (algorithm 5.2) can be guaranteed if we modify it to allow for automatic resets:

$$\beta \leftarrow \max(\beta, 0) \tag{5.18}$$

Figure 5.3 shows an example search using this method.

[2] R. Fletcher and C. M. Reeves, "Function Minimization by Conjugate Gradients," *The Computer Journal*, vol. 7, no. 2, pp. 149–154, 1964.

[3] E. Polak and G. Ribière, "Note sur la Convergence de Méthodes de Directions Conjuguées," *Revue Française d'informatique et de Recherche Opérationnelle, Série Rouge*, vol. 3, no. 1, pp. 35–43, 1969.

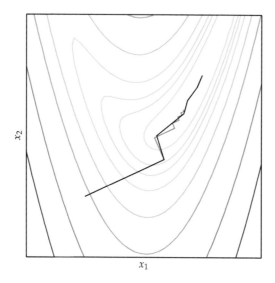

x_2

x_1

Figure 5.3. The conjugate gradient method with the Polak-Ribière update. Gradient descent is shown in gray.

```
mutable struct ConjugateGradientDescent <: DescentMethod
    d
    g
end
function init!(M::ConjugateGradientDescent, f, ∇f, x)
    M.g = ∇f(x)
    M.d = -M.g
    return M
end
function step!(M::ConjugateGradientDescent, f, ∇f, x)
    d, g = M.d, M.g
    g′ = ∇f(x)
    β = max(0, dot(g′, g′-g)/(g⋅g))
    d′ = -g′ + β*d
    x′ = line_search(f, x, d′)
    M.d, M.g = d′, g′
    return x′
end
```

Algorithm 5.2. The conjugate gradient method with the Polak-Ribière update, where d is the previous search direction and g is the previous gradient.

5.3   Momentum

Gradient descent will take a long time to traverse a nearly flat surface as shown in figure 5.4. Allowing momentum to accumulate is one way to speed progress. We can modify gradient descent to incorporate momentum.

The *momentum* update equations are:

$$\mathbf{v}^{(k+1)} = \beta \mathbf{v}^{(k)} - \alpha \mathbf{g}^{(k)} \tag{5.19}$$

$$\mathbf{x}^{(k+1)} = \mathbf{x}^{(k)} + \mathbf{v}^{(k+1)} \tag{5.20}$$

For $\beta = 0$, we recover gradient descent. Momentum can be interpreted as a ball rolling down a nearly horizontal incline. The ball naturally gathers momentum as gravity causes it to accelerate, just as the gradient causes momentum to accumulate in this descent method. An implementation is provided in algorithm 5.3. Momentum descent is compared to gradient descent in figure 5.5.

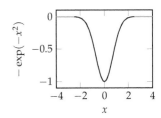

Figure 5.4. Regions that are nearly flat have gradients with small magnitudes and can thus require many iterations of gradient descent to traverse.

```
mutable struct Momentum <: DescentMethod
    α # learning rate
    β # momentum decay
    v # momentum
end
function init!(M::Momentum, f, ∇f, x)
    M.v = zeros(length(x))
    return M
end
function step!(M::Momentum, f, ∇f, x)
    α, β, v, g = M.α, M.β, M.v, ∇f(x)
    v[:] = β*v - α*g
    return x + v
end
```

Algorithm 5.3. The momentum method for accelerated descent.

The first line in step! makes copies of the scalars α and β, but creates a reference to the vector v. Thus, the following line v[:] = β*v - α*g modifies the original momentum vector in the struct M.

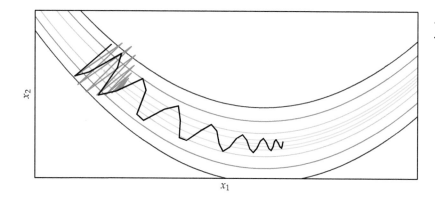

- gradient descent
- momentum

Figure 5.5. Gradient descent and the momentum method compared on the Rosenbrock function with $b = 100$; see appendix B.6.

5.4   Nesterov Momentum

One issue of momentum is that the steps do not slow down enough at the bottom of a valley and tend to overshoot the valley floor. *Nesterov momentum*[4] modifies the momentum algorithm to use the gradient at the projected future position:

[4] Y. Nesterov, "A Method of Solving a Convex Programming Problem with Convergence Rate $O(1/k^2)$," *Soviet Mathematics Doklady*, vol. 27, no. 2, pp. 543–547, 1983.

$$\mathbf{v}^{(k+1)} = \beta\mathbf{v}^{(k)} - \alpha\nabla f(\mathbf{x}^{(k)} + \beta\mathbf{v}^{(k)}) \tag{5.21}$$

$$\mathbf{x}^{(k+1)} = \mathbf{x}^{(k)} + \mathbf{v}^{(k+1)} \tag{5.22}$$

An implementation is provided by algorithm 5.4. The Nesterov momentum and momentum descent methods are compared in figure 5.6.

```julia
mutable struct NesterovMomentum <: DescentMethod
    α # learning rate
    β # momentum decay
    v # momentum
end
function init!(M::NesterovMomentum, f, ∇f, x)
    M.v = zeros(length(x))
    return M
end
function step!(M::NesterovMomentum, f, ∇f, x)
    α, β, v = M.α, M.β, M.v
    v[:] = β*v - α*∇f(x + β*v)
    return x + v
end
```

Algorithm 5.4. Nesterov's momentum method of accelerated descent.

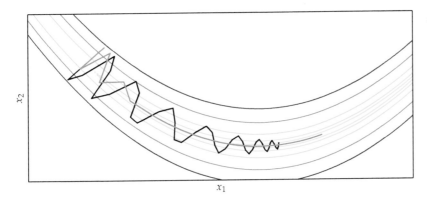

— momentum
— Nesterov momentum

Figure 5.6. The momentum and Nesterov momentum methods compared on the Rosenbrock function with $b = 100$; see appendix B.6.

5.5 Adagrad

Momentum and Nesterov momentum update all components of **x** with the same learning rate. The *adaptive subgradient* method, or *Adagrad*,[5] adapts a learning rate for each component of **x**. Adagrad dulls the influence of parameters with consistently high gradients, thereby increasing the influence of parameters with infrequent updates.[6]

The Adagrad update step is:

$$x_i^{(k+1)} = x_i^{(k)} - \frac{\alpha}{\epsilon + \sqrt{s_i^{(k)}}} g_i^{(k)} \tag{5.23}$$

where $\mathbf{s}^{(k)}$ is a vector whose ith entry is the sum of the squares of the partials, with respect to x_i, up to time step k,

$$s_i^{(k)} = \sum_{j=1}^{k} \left(g_i^{(j)} \right)^2 \tag{5.24}$$

and ϵ is a small value, on the order of 1×10^{-8}, to prevent division by zero.

Adagrad is far less sensitive to the learning rate parameter α. The learning rate parameter is typically set to a default value of 0.01. Adagrad's primary weakness is that the components of **s** are each strictly nondecreasing. The accumulated sum causes the effective learning rate to decrease during training, often becoming infinitesimally small before convergence. An implementation is provided by algorithm 5.5.

[5] J. Duchi, E. Hazan, and Y. Singer, "Adaptive Subgradient Methods for Online Learning and Stochastic Optimization," *Journal of Machine Learning Research*, vol. 12, pp. 2121–2159, 2011.

[6] Adagrad excels when the gradient is sparse. The original paper was motivated by stochastic gradient descent, which picks a random batch of training data for each iteration from which to compute a gradient. Many deep learning datasets for real-world problems produce sparse gradients where some features occur far less frequently than others.

```
mutable struct Adagrad <: DescentMethod
    α # learning rate
    ϵ # small value
    s # sum of squared gradient
end
function init!(M::Adagrad, f, ∇f, x)
    M.s = zeros(length(x))
    return M
end
function step!(M::Adagrad, f, ∇f, x)
    α, ϵ, s, g = M.α, M.ϵ, M.s, ∇f(x)
    s[:] += g.*g
    return x - α*g ./ (sqrt.(s) .+ ϵ)
end
```

Algorithm 5.5. The Adagrad accelerated descent method.

5.6   RMSProp

RMSProp[7] extends Adagrad to avoid the effect of a monotonically decreasing learning rate. RMSProp maintains a decaying average of squared gradients. This average is updated according to:[8]

$$\mathbf{s}^{(k+1)} = \gamma \mathbf{s}^{(k)} + (1 - \gamma)\left(\mathbf{g}^{(k)} \odot \mathbf{g}^{(k)}\right) \tag{5.25}$$

where the decay $\gamma \in [0, 1]$ is typically close to 0.9.

The decaying average of past squared gradients can be substituted into RMSProp's update equation:[9]

$$x_i^{(k+1)} = x_i^{(k)} - \frac{\alpha}{\epsilon + \sqrt{\hat{s}_i^{(k+1)}}} g_i^{(k)} \tag{5.26}$$

$$= x_i^{(k)} - \frac{\alpha}{\epsilon + \text{RMS}(g_i)} g_i^{(k)} \tag{5.27}$$

An implementation is provided by algorithm 5.6.

5.7   Adadelta

Adadelta[10] is another method for overcoming Adagrad's monotonically decreasing learning rate. After independently deriving the RMSProp update, the authors noticed that the units in the update equations for gradient descent, momentum,

[7] RMSProp is unpublished and comes from Lecture 6e of Geoff Hinton's Coursera class.

[8] The operation $\mathbf{a} \odot \mathbf{b}$ is the element-wise product between vectors $\mathbf{a}$ and $\mathbf{b}$.

[9] The denominator is similar to the root mean square (RMS) of the gradient component. In this chapter we use RMS(x) to refer to the decaying root mean square of the time series of x.

[10] M. D. Zeiler, "ADADELTA: An Adaptive Learning Rate Method," *ArXiv*, no. 1212.5701, 2012.

```
mutable struct RMSProp <: DescentMethod
    α # learning rate
    γ # decay
    ϵ # small value
    s # sum of squared gradient
end
function init!(M::RMSProp, f, ∇f, x)
    M.s = zeros(length(x))
    return M
end
function step!(M::RMSProp, f, ∇f, x)
    α, γ, ϵ, s, g = M.α, M.γ, M.ϵ, M.s, ∇f(x)
    s[:] = γ*s + (1-γ)*(g.*g)
    return x - α*g ./ (sqrt.(s) .+ ϵ)
end
```

Algorithm 5.6. The RMSProp accelerated descent method.

and Adagrad do not match. To fix this, they use an exponentially decaying average of the square updates:

$$x_i^{(k+1)} = x_i^{(k)} - \frac{\text{RMS}(\Delta x_i)}{\epsilon + \text{RMS}(g_i)} g_i^{(k)} \tag{5.28}$$

which eliminates the learning rate parameter entirely. An implementation is provided by algorithm 5.7.

5.8 Adam

The *adaptive moment estimation* method, or *Adam*,[11] also adapts learning rates to each parameter (algorithm 5.8). It stores both an exponentially decaying squared gradient like RMSProp and Adadelta, but also an exponentially decaying gradient like momentum.

Initializing the gradient and squared gradient to zero introduces a bias. A bias correction step helps alleviate the issue.[12] The equations applied during each

[11] D. Kingma and J. Ba, "Adam: A Method for Stochastic Optimization," in *International Conference on Learning Representations (ICLR)*, 2015.

[12] According to the original paper, good default settings are $\alpha = 0.001$, $\gamma_v = 0.9$, $\gamma_s = 0.999$, and $\epsilon = 1 \times 10^{-8}$.

```julia
mutable struct Adadelta <: DescentMethod
    γs # gradient decay
    γx # update decay
    ϵ # small value
    s # sum of squared gradients
    u # sum of squared updates
end
function init!(M::Adadelta, f, ∇f, x)
    M.s = zeros(length(x))
    M.u = zeros(length(x))
    return M
end
function step!(M::Adadelta, f, ∇f, x)
    γs, γx, ϵ, s, u, g = M.γs, M.γx, M.ϵ, M.s, M.u, ∇f(x)
    s[:] = γs*s + (1-γs)*g.*g
    Δx = - (sqrt.(u) .+ ϵ) ./ (sqrt.(s) .+ ϵ) .* g
    u[:] = γx*u + (1-γx)*Δx.*Δx
    return x + Δx
end
```

Algorithm 5.7. The Adadelta accelerated descent method. The small constant ϵ is added to the numerator as well to prevent progress from entirely decaying to zero and to start off the first iteration where $\Delta x = 0$.

iteration for Adam are:

$$\text{biased decaying momentum: } \mathbf{v}^{(k+1)} = \gamma_v \mathbf{v}^{(k)} + (1 - \gamma_v)\mathbf{g}^{(k)} \tag{5.29}$$

$$\text{biased decaying sq. gradient: } \mathbf{s}^{(k+1)} = \gamma_s \mathbf{s}^{(k)} + (1 - \gamma_s)\left(\mathbf{g}^{(k)} \odot \mathbf{g}^{(k)}\right) \tag{5.30}$$

$$\text{corrected decaying momentum: } \hat{\mathbf{v}}^{(k+1)} = \mathbf{v}^{(k+1)}/(1 - \gamma_v^k) \tag{5.31}$$

$$\text{corrected decaying sq. gradient: } \hat{\mathbf{s}}^{(k+1)} = \mathbf{s}^{(k+1)}/(1 - \gamma_s^k) \tag{5.32}$$

$$\text{next iterate: } \mathbf{x}^{(k+1)} = \mathbf{x}^{(k)} - \alpha\hat{\mathbf{v}}^{(k+1)}/\left(\epsilon + \sqrt{\hat{\mathbf{s}}^{(k+1)}}\right) \tag{5.33}$$

5.9   Hypergradient Descent

The accelerated descent methods are either extremely sensitive to the learning rate or go to great lengths to adapt the learning rate during execution. The learning rate dictates how sensitive the method is to the gradient signal. A rate that is too high or too low often drastically affects performance.

```
mutable struct Adam <: DescentMethod
    α # learning rate
    γv # decay
    γs # decay
    ε # small value
    k # step counter
    v # 1st moment estimate
    s # 2nd moment estimate
end
function init!(M::Adam, f, ∇f, x)
    M.k = 0
    M.v = zeros(length(x))
    M.s = zeros(length(x))
    return M
end
function step!(M::Adam, f, ∇f, x)
    α, γv, γs, ε, k = M.α, M.γv, M.γs, M.ε, M.k
    s, v, g = M.s, M.v, ∇f(x)
    v[:] = γv*v + (1-γv)*g
    s[:] = γs*s + (1-γs)*g.*g
    M.k = k += 1
    v_hat = v ./ (1 - γv^k)
    s_hat = s ./ (1 - γs^k)
    return x - α*v_hat ./ (sqrt.(s_hat) .+ ε)
end
```

Algorithm 5.8. The Adam accelerated descent method.

Hypergradient descent[13] was developed with the understanding that the derivative of the learning rate should be useful for improving optimizer performance. A *hypergradient* is a derivative taken with respect to a hyperparameter. Hypergradient algorithms reduce the sensitivity to the hyperparameter, allowing it to adapt more quickly.

[13] A. G. Baydin, R. Cornish, D. M. Rubio, M. Schmidt, and F. Wood, "Online Learning Rate Adaptation with Hypergradient Descent," in *International Conference on Learning Representations (ICLR)*, 2018.

Hypergradient descent applies gradient descent to the learning rate of an underlying descent method. The method requires the partial derivative of the objective function with respect to the learning rate. For gradient descent, this partial derivative is:

$$\frac{\partial f(\mathbf{x}^{(k)})}{\partial \alpha} = (\mathbf{g}^{(k)})^\top \frac{\partial}{\partial \alpha}\left(\mathbf{x}^{(k-1)} - \alpha \mathbf{g}^{(k-1)}\right) \tag{5.34}$$

$$= (\mathbf{g}^{(k)})^\top \left(-\mathbf{g}^{(k-1)}\right) \tag{5.35}$$

Computing the hypergradient thus requires keeping track of the last gradient. The resulting update rule is:

$$\alpha^{(k+1)} = \alpha^{(k)} - \mu \frac{\partial f(\mathbf{x}^{(k)})}{\partial \alpha} \tag{5.36}$$

$$= \alpha^{(k)} + \mu (\mathbf{g}^{(k)})^\top \mathbf{g}^{(k-1)} \tag{5.37}$$

where μ is the hypergradient learning rate.

This derivation can be applied to any gradient-based descent method that follows equation (4.1). Implementations are provided for the hypergradient versions of gradient descent (algorithm 5.9) and Nesterov momentum (algorithm 5.10). These methods are visualized in figure 5.7.

```julia
mutable struct HyperGradientDescent <: DescentMethod
    α0 # initial learning rate
    μ # learning rate of the learning rate
    α # current learning rate
    g_prev # previous gradient
end
function init!(M::HyperGradientDescent, f, ∇f, x)
    M.α = M.α0
    M.g_prev = zeros(length(x))
    return M
end
function step!(M::HyperGradientDescent, f, ∇f, x)
    α, μ, g, g_prev = M.α, M.μ, ∇f(x), M.g_prev
    α = α + μ*(g·g_prev)
    M.g_prev, M.α = g, α
    return x - α*g
end
```

Algorithm 5.9. The hypergradient form of gradient descent.

```julia
mutable struct HyperNesterovMomentum <: DescentMethod
    α0 # initial learning rate
    μ # learning rate of the learning rate
    β # momentum decay
    v # momentum
    α # current learning rate
    g_prev # previous gradient
end
function init!(M::HyperNesterovMomentum, f, ∇f, x)
    M.α = M.α0
    M.v = zeros(length(x))
    M.g_prev = zeros(length(x))
    return M
end
function step!(M::HyperNesterovMomentum, f, ∇f, x)
    α, β, μ = M.α, M.β, M.μ
    v, g, g_prev = M.v, ∇f(x), M.g_prev
    α = α - μ*(g·(-g_prev - β*v))
    v[:] = β*v + g
    M.g_prev, M.α = g, α
    return x - α*(g + β*v)
end
```

Algorithm 5.10. The hypergradient form of the Nesterov momentum descent method.

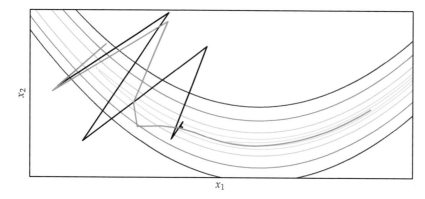

— hypergradient
— hyper-Nesterov

Figure 5.7. Hypergradient versions of gradient descent and Nesterov momentum compared on the Rosenbrock function with $b = 100$; see appendix B.6.

5.10   Summary

- Gradient descent follows the direction of steepest descent.

- The conjugate gradient method can automatically adjust to local valleys.

- Descent methods with momentum build up progress in favorable directions.

- A wide variety of accelerated descent methods use special techniques to speed up descent.

- Hypergradient descent applies gradient descent to the learning rate of an underlying descent method.

5.11   Exercises

Exercise 5.1. Compute the gradient of $\mathbf{x}^\top \mathbf{A} \mathbf{x} + \mathbf{b}^\top \mathbf{x}$ when $\mathbf{A}$ is symmetric.

Exercise 5.2. Apply gradient descent with a unit step size to $f(x) = x^4$ from a starting point of your choice. Compute two iterations.

Exercise 5.3. Apply one step of gradient descent to $f(x) = e^x + e^{-x}$ from $x^{(1)} = 10$ with both a unit step factor and with exact line search.

Exercise 5.4. The conjugate gradient method can also be used to find a search direction $\mathbf{d}$ when a local quadratic model of a function is available at the current point. With $\mathbf{d}$ as search direction, let the model be

$$q(\mathbf{d}) = \mathbf{d}^\top \mathbf{H} \mathbf{d} + \mathbf{b}^\top \mathbf{d} + \mathbf{c}$$

for a symmetric matrix $\mathbf{H}$. What is the Hessian in this case? What is the gradient of q when $\mathbf{d} = \mathbf{0}$? What can go wrong if the conjugate gradient method is applied to the quadratic model to get the search direction $\mathbf{d}$?

Exercise 5.5. How is Nesterov momentum an improvement over momentum?

Exercise 5.6. In what way is the conjugate gradient method an improvement over steepest descent?

Exercise 5.7. In conjugate gradient descent, what is the normalized descent direction at the first iteration for the function $f(x, y) = x^2 + xy + y^2 + 5$ when initialized at $(x, y) = (1, 1)$? What is the resulting point after two steps of the conjugate gradient method?

Exercise 5.8. We have a polynomial function f such that $f(\mathbf{x}) > 2$ for all $\mathbf{x}$ in three-dimensional Euclidean space. Suppose we are using steepest descent with step lengths optimized at each step, and we want to find a local minimum of f. If our unnormalized descent direction is $[1, 2, 3]$ at step k, is it possible for our unnormalized descent direction at step $k + 1$ to be $[0, 0, -3]$? Why or why not?

6 Second-Order Methods

The previous chapter focused on optimization methods that involve first-order approximations of the objective function using the gradient. This chapter focuses on leveraging *second-order* approximations that use the second derivative in univariate optimization or the Hessian in multivariate optimization to direct the search. This additional information can help improve the local model used for informing the selection of directions and step lengths in descent algorithms.

6.1 Newton's Method

Knowing the function value and gradient for a design point can help determine the direction to travel, but this first-order information does not directly help determine how far to step to reach a local minimum. Second-order information, on the other hand, allows us to make a quadratic approximation of the objective function and approximate the right step size to reach a local minimum as shown in figure 6.1. As we have seen with quadratic fit search in chapter 3, we can analytically obtain the location where a quadratic approximation has a zero gradient. We can then use that location as the next iteration to approach a local minimum.

In univariate optimization, the quadratic approximation about a point $x^{(k)}$ comes from the second-order Taylor expansion:

$$q(x) = f(x^{(k)}) + (x - x^{(k)})f'(x^{(k)}) + \frac{(x - x^{(k)})^2}{2}f''(x^{(k)}) \qquad (6.1)$$

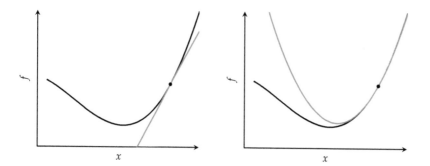

Figure 6.1. A comparison of first-order and second-order approximations. Bowl-shaped quadratic approximations have unique locations where the derivative is zero.

Setting the derivative to zero and solving for the root yields the update equation for *Newton's method*:

$$\frac{\partial}{\partial x}q(x) = f'(x^{(k)}) + (x - x^{(k)})f''(x^{(k)}) = 0 \qquad (6.2)$$

$$x^{(k+1)} = x^{(k)} - \frac{f'(x^{(k)})}{f''(x^{(k)})} \qquad (6.3)$$

This update is shown in figure 6.2.

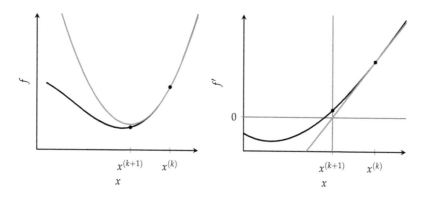

Figure 6.2. Newton's method can be interpreted as a root-finding method applied to f' that iteratively improves a univariate design point by taking the tangent line at $(x, f'(x))$, finding the intersection with the x-axis, and using that x value as the next design point.

The update rule in Newton's method involves dividing by the second derivative. The update is undefined if the second derivative is zero, which occurs when the quadratic approximation is a horizontal line. Instability also occurs when the second derivative is very close to zero, in which case the next iterate will lie very far from the current design point, far from where the local quadratic approximation is valid. Poor local approximations can lead to poor performance with Newton's method. Figure 6.3 shows three kinds of failure cases.

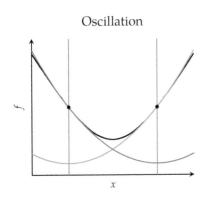

Oscillation

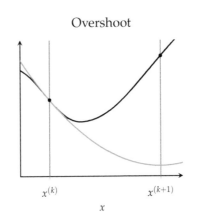

Overshoot

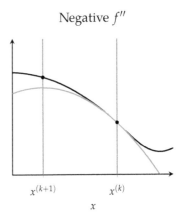

Negative f''

Figure 6.3. Examples of failure cases with Newton's method.

Newton's method does tend to converge quickly when in a bowl-like region that is sufficiently close to a local minimum. It has *quadratic convergence*, meaning the difference between the minimum and the iterate is approximately squared with every iteration. This rate of convergence holds for Newton's method starting from $x^{(1)}$ within a distance δ of a root x^* if[1]

- $f''(x) \neq 0$ for all points in I,

- $f'''(x)$ is continuous on I, and

- $\frac{1}{2}\left|\frac{f'''(x^{(1)})}{f''(x^{(1)})}\right| < c\left|\frac{f'''(x^*)}{f''(x^*)}\right|$ for some $c < \infty$

for an interval $I = [x^* - \delta, x^* + \delta]$. The final condition guards against overshoot.

Newton's method can be extended to multivariate optimization (algorithm 6.1). The multivariate second-order Taylor expansion at $\mathbf{x}^{(k)}$ is:

$$f(\mathbf{x}) \approx q(\mathbf{x}) = f(\mathbf{x}^{(k)}) + (\mathbf{g}^{(k)})^\top (\mathbf{x} - \mathbf{x}^{(k)}) + \frac{1}{2}(\mathbf{x} - \mathbf{x}^{(k)})^\top \mathbf{H}^{(k)} (\mathbf{x} - \mathbf{x}^{(k)}) \tag{6.4}$$

where $\mathbf{g}^{(k)}$ and $\mathbf{H}^{(k)}$ are the gradient and Hessian at $\mathbf{x}^{(k)}$, respectively.

We evaluate the gradient and set it to zero:

$$\nabla q(\mathbf{x}^{(k)}) = \mathbf{g}^{(k)} + \mathbf{H}^{(k)}(\mathbf{x} - \mathbf{x}^{(k)}) = \mathbf{0} \tag{6.5}$$

We then solve for the next iterate, thereby obtaining Newton's method in multivariate form:

$$\mathbf{x}^{(k+1)} = \mathbf{x}^{(k)} - (\mathbf{H}^{(k)})^{-1}\mathbf{g}^{(k)} \tag{6.6}$$

[1] The final condition enforces *sufficient closeness*, ensuring that the function is sufficiently approximated by the Taylor expansion. J. Stoer and R. Bulirsch, *Introduction to Numerical Analysis*, 3rd ed. Springer, 2002.

If f is quadratic and its Hessian is positive definite, then the update converges to the global minimum in one step. For general functions, Newton's method is often terminated once x ceases to change by more than a given tolerance.[2] Example 6.1 shows how Newton's method can be used to minimize a function.

[2] Termination conditions for descent methods are given in chapter 5.

With $\mathbf{x}^{(1)} = [9, 8]$, we will use Newton's method to minimize Booth's function:

$$f(\mathbf{x}) = (x_1 + 2x_2 - 7)^2 + (2x_1 + x_2 - 5)^2$$

The gradient of Booth's function is:

$$\nabla f(\mathbf{x}) = [10x_1 + 8x_2 - 34, 8x_1 + 10x_2 - 38]$$

The Hessian of Booth's function is:

$$\mathbf{H}(\mathbf{x}) = \begin{bmatrix} 10 & 8 \\ 8 & 10 \end{bmatrix}$$

The first iteration of Newton's method yields:

$$\mathbf{x}^{(2)} = \mathbf{x}^{(1)} - \left(\mathbf{H}^{(1)}\right)^{-1}\mathbf{g}^{(1)} = \begin{bmatrix} 9 \\ 8 \end{bmatrix} - \begin{bmatrix} 10 & 8 \\ 8 & 10 \end{bmatrix}^{-1} \begin{bmatrix} 10 \cdot 9 + 8 \cdot 8 - 34 \\ 8 \cdot 9 + 10 \cdot 8 - 38 \end{bmatrix}$$

$$= \begin{bmatrix} 9 \\ 8 \end{bmatrix} - \begin{bmatrix} 10 & 8 \\ 8 & 10 \end{bmatrix}^{-1} \begin{bmatrix} 120 \\ 114 \end{bmatrix} = \begin{bmatrix} 1 \\ 3 \end{bmatrix}$$

The gradient at $\mathbf{x}^{(2)}$ is zero, so we have converged after a single iteration. The Hessian is positive definite everywhere, so $\mathbf{x}^{(2)}$ is the global minimum.

Example 6.1. Newton's method used to minimize Booth's function; see appendix B.2.

Newton's method can also be used to supply a descent direction to line search or can be modified to use a step factor.[3] Smaller steps toward the minimum or line searches along the descent direction can increase the method's robustness. The descent direction is:[4]

$$\mathbf{d}^{(k)} = -(\mathbf{H}^{(k)})^{-1}\mathbf{g}^{(k)} \tag{6.7}$$

[3] See chapter 5.

[4] The descent direction given by Newton's method is similar to the *natural gradient* or *covariant gradient*. S. Amari, "Natural Gradient Works Efficiently in Learning," *Neural Computation*, vol. 10, no. 2, pp. 251–276, 1998.

```
function newtons_method(∇f, H, x, ϵ, k_max)
    k, Δ = 1, fill(Inf, length(x))
    while norm(Δ) > ϵ && k ≤ k_max
        Δ = H(x) \ ∇f(x)
        x -= Δ
        k += 1
    end
    return x
end
```

Algorithm 6.1. Newton's method, which takes the gradient of the function ∇f, the Hessian of the objective function H, an initial point x, a step size tolerance ϵ, and a maximum number of iterations k_max.

6.2   Secant Method

Newton's method for univariate function minimization requires the first and second derivatives f' and f''. In many cases, f' is known but the second derivative is not. The *secant method* (algorithm 6.2) applies Newton's method using estimates of the second derivative and thus only requires f'. This property makes the secant method more convenient to use in practice.

The secant method uses the last two iterates to approximate the second derivative:

$$f''(x^{(k)}) \approx \frac{f'(x^{(k)}) - f'(x^{(k-1)})}{x^{(k)} - x^{(k-1)}} \tag{6.8}$$

This estimate is substituted into Newton's method:

$$x^{(k+1)} \leftarrow x^{(k)} - \frac{x^{(k)} - x^{(k-1)}}{f'(x^{(k)}) - f'(x^{(k-1)})} f'(x^{(k)}) \tag{6.9}$$

The secant method requires an additional initial design point. It suffers from the same problems as Newton's method and may take more iterations to converge due to approximating the second derivative.

6.3   Quasi-Newton Methods

Just as the secant method approximates f'' in the univariate case, *quasi-Newton* methods approximate the inverse Hessian. Quasi-Newton method updates have the form:

$$\mathbf{x}^{(k+1)} \leftarrow \mathbf{x}^{(k)} - \alpha^{(k)} \mathbf{Q}^{(k)} \mathbf{g}^{(k)} \tag{6.10}$$

where $\alpha^{(k)}$ is a scalar step factor and $\mathbf{Q}^{(k)}$ approximates the inverse of the Hessian at $\mathbf{x}^{(k)}$.

```
function secant_method(f′, x0, x1, ϵ)
    g0 = f′(x0)
    Δ = Inf
    while abs(Δ) > ϵ
        g1 = f′(x1)
        Δ = (x1 - x0)/(g1 - g0)*g1
        x0, x1, g0 = x1, x1 - Δ, g1
    end
    return x1
end
```

Algorithm 6.2. The secant method for univariate function minimization. The inputs are the first derivative f′ of the target function, two initial points x0 and x1, and the desired tolerance ϵ. The final x-coordinate is returned.

These methods typically set $\mathbf{Q}^{(1)}$ to the identity matrix, and they then apply updates to reflect information learned with each iteration. To simplify the equations for the various quasi-Newton methods, we define the following:

$$\gamma^{(k+1)} \equiv \mathbf{g}^{(k+1)} - \mathbf{g}^{(k)} \tag{6.11}$$

$$\delta^{(k+1)} \equiv \mathbf{x}^{(k+1)} - \mathbf{x}^{(k)} \tag{6.12}$$

The *Davidon-Fletcher-Powell* (DFP) method (algorithm 6.3) uses:[5]

$$\mathbf{Q} \leftarrow \mathbf{Q} - \frac{\mathbf{Q}\gamma\gamma^\top\mathbf{Q}}{\gamma^\top\mathbf{Q}\gamma} + \frac{\delta\delta^\top}{\delta^\top\gamma} \tag{6.13}$$

where all terms on the right hand side are evaluated at iteration k.

The update for $\mathbf{Q}$ in the DFP method has three properties:

1. $\mathbf{Q}$ remains symmetric and positive definite.

2. If $f(\mathbf{x}) = \frac{1}{2}\mathbf{x}^\top\mathbf{A}\mathbf{x} + \mathbf{b}^\top\mathbf{x} + c$, then $\mathbf{Q} = \mathbf{A}^{-1}$. Thus the DFP has the same convergence properties as the conjugate gradient method.

3. For high-dimensional problems, storing and updating $\mathbf{Q}$ can be significant compared to other methods like the conjugate gradient method.

An alternative to DFP, the *Broyden-Fletcher-Goldfarb-Shanno* (BFGS) method (algorithm 6.4), uses:[6]

$$\mathbf{Q} \leftarrow \mathbf{Q} - \left(\frac{\delta\gamma^\top\mathbf{Q} + \mathbf{Q}\gamma\delta^\top}{\delta^\top\gamma}\right) + \left(1 + \frac{\gamma^\top\mathbf{Q}\gamma}{\delta^\top\gamma}\right)\frac{\delta\delta^\top}{\delta^\top\gamma} \tag{6.14}$$

[5] The original concept was presented in a technical report, W. C. Davidon, "Variable Metric Method for Minimization," Argonne National Laboratory, Tech. Rep. ANL-5990, 1959. It was later published: W. C. Davidon, "Variable Metric Method for Minimization," *SIAM Journal on Optimization*, vol. 1, no. 1, pp. 1–17, 1991. The method was modified by R. Fletcher and M. J. D. Powell, "A Rapidly Convergent Descent Method for Minimization," *The Computer Journal*, vol. 6, no. 2, pp. 163–168, 1963.

[6] R. Fletcher, *Practical Methods of Optimization*, 2nd ed. Wiley, 1987.

```
mutable struct DFP <: DescentMethod
    Q
end
function init!(M::DFP, f, ∇f, x)
    m = length(x)
    M.Q = Matrix(1.0I, m, m)
    return M
end
function step!(M::DFP, f, ∇f, x)
    Q, g = M.Q, ∇f(x)
    x′ = line_search(f, x, -Q*g)
    g′ = ∇f(x′)
    δ = x′ - x
    γ = g′ - g
    Q[:] = Q - Q*γ*γ'*Q/(γ'*Q*γ) + δ*δ'/(δ'*γ)
    return x′
end
```

Algorithm 6.3. The Davidon-Fletcher-Powell descent method.

```
mutable struct BFGS <: DescentMethod
    Q
end
function init!(M::BFGS, f, ∇f, x)
    m = length(x)
    M.Q = Matrix(1.0I, m, m)
    return M
end
function step!(M::BFGS, f, ∇f, x)
    Q, g = M.Q, ∇f(x)
    x′ = line_search(f, x, -Q*g)
    g′ = ∇f(x′)
    δ = x′ - x
    γ = g′ - g
    Q[:] = Q - (δ*γ'*Q + Q*γ*δ')/(δ'*γ) +
               (1 + (γ'*Q*γ)/(δ'*γ))[1]*(δ*δ')/(δ'*γ)
    return x′
end
```

Algorithm 6.4. The Broyden-Fletcher-Goldfarb-Shanno descent method.

BFGS does better than DFP with approximate line search but still uses an $n \times n$ dense matrix. For very large problems where space is a concern, the *Limited-memory BFGS* method (algorithm 6.5), or *L-BFGS*, can be used to approximate BFGS.[7] L-BFGS stores the last m values for δ and γ rather than the full inverse Hessian, where $i = 1$ indexes the oldest value and $i = m$ indexes the most recent.

The process for computing the descent direction $\mathbf{d}$ at $\mathbf{x}$ begins by computing $\mathbf{q}^{(m)} = \nabla f(\mathbf{x})$. The remaining vectors $\mathbf{q}^{(i)}$ for i from $m - 1$ down to 1 are computed using

[7] J. Nocedal, "Updating Quasi-Newton Matrices with Limited Storage," *Mathematics of Computation*, vol. 35, no. 151, pp. 773–782, 1980.

$$\mathbf{q}^{(i)} = \mathbf{q}^{(i+1)} - \frac{\left(\boldsymbol{\delta}^{(i+1)}\right)^\top \mathbf{q}^{(i+1)}}{\left(\boldsymbol{\gamma}^{(i+1)}\right)^\top \boldsymbol{\delta}^{(i+1)}} \boldsymbol{\gamma}^{(i+1)} \tag{6.15}$$

These vectors are used to compute another $m + 1$ vectors, starting with

$$\mathbf{z}^{(0)} = \frac{\boldsymbol{\gamma}^{(m)} \odot \boldsymbol{\delta}^{(m)} \odot \mathbf{q}^{(m)}}{\left(\boldsymbol{\gamma}^{(m)}\right)^\top \boldsymbol{\gamma}^{(m)}} \tag{6.16}$$

and proceeding with $\mathbf{z}^{(i)}$ for i from 1 to m according to

$$\mathbf{z}^{(i)} = \mathbf{z}^{(i-1)} + \boldsymbol{\delta}^{(i-1)} \left(\frac{\left(\boldsymbol{\delta}^{(i-1)}\right)^\top \mathbf{q}^{(i-1)}}{\left(\boldsymbol{\gamma}^{(i-1)}\right)^\top \boldsymbol{\delta}^{(i-1)}} - \frac{\left(\boldsymbol{\gamma}^{(i-1)}\right)^\top \mathbf{z}^{(i-1)}}{\left(\boldsymbol{\gamma}^{(i-1)}\right)^\top \boldsymbol{\delta}^{(i-1)}} \right) \tag{6.17}$$

The descent direction is $\mathbf{d} = -\mathbf{z}^{(m)}$.

For minimization, the inverse Hessian $\mathbf{Q}$ must remain positive definite. The initial Hessian is often set to the diagonal of

$$\mathbf{Q}^{(1)} = \frac{\boldsymbol{\gamma}^{(1)} \left(\boldsymbol{\delta}^{(1)}\right)^\top}{\left(\boldsymbol{\gamma}^{(1)}\right)^\top \boldsymbol{\gamma}^{(1)}} \tag{6.18}$$

Computing the diagonal for the above expression and substituting the result into $\mathbf{z}^{(1)} = \mathbf{Q}^{(1)} \mathbf{q}^{(1)}$ results in the equation for $\mathbf{z}^{(1)}$.

The quasi-Newton methods discussed in this section are compared in figure 6.4. They often perform quite similarly.

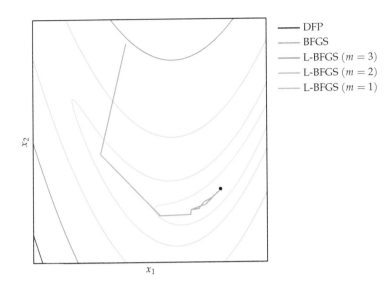

Figure 6.4. Several quasi-Newton methods compared on the Rosenbrock function; see appendix B.6. All methods have nearly identical updates, with L-BFGS noticeably deviating only when its history, m, is 1.

Legend:
- DFP
- BFGS
- L-BFGS ($m = 3$)
- L-BFGS ($m = 2$)
- L-BFGS ($m = 1$)

6.4 Summary

- Incorporating second-order information in descent methods often speeds convergence.

- Newton's method is a root-finding method that leverages second-order information to quickly descend to a local minimum.

- The secant method and quasi-Newton methods approximate Newton's method when the second-order information is not directly available.

6.5 Exercises

Exercise 6.1. What advantage does second-order information provide about convergence that first-order information lacks?

Exercise 6.2. When finding roots in one dimension, when would we use Newton's method instead of the bisection method?

Exercise 6.3. Apply Newton's method to $f(x) = x^2$ from a starting point of your choice. How many steps do we need to converge?

```
mutable struct LimitedMemoryBFGS <: DescentMethod
    m
    δs
    γs
    qs
end
function init!(M::LimitedMemoryBFGS, f, ∇f, x)
    M.δs = []
    M.γs = []
    M.qs = []
    return M
end
function step!(M::LimitedMemoryBFGS, f, ∇f, x)
    δs, γs, qs, g = M.δs, M.γs, M.qs, ∇f(x)
    m = length(δs)
    if m > 0
        q = g
        for i in m : -1 : 1
            qs[i] = copy(q)
            q -= (δs[i]·q)/(γs[i]·δs[i])*γs[i]
        end
        z = (γs[m] .* δs[m] .* q) / (γs[m]·γs[m])
        for i in 1 : m
            z += δs[i]*(δs[i]·qs[i] - γs[i]·z)/(γs[i]·δs[i])
        end
        x′ = line_search(f, x, -z)
    else
        x′ = line_search(f, x, -g)
    end
    g′ = ∇f(x′)
    push!(δs, x′ - x); push!(γs, g′ - g)
    push!(qs, zeros(length(x)))
    while length(δs) > M.m
        popfirst!(δs); popfirst!(γs); popfirst!(qs)
    end
    return x′
end
```

Algorithm 6.5. The Limited-memory BFGS descent method, which avoids storing the approximate inverse Hessian. The parameter m determines the history size. The LimitedMemoryBFGS type also stores the step differences δs, the gradient changes γs, and storage vectors qs.

Exercise 6.4. Apply Newton's method to $f(x) = \frac{1}{2}x^\top Hx$ starting from $x^{(1)} = [1,1]$. What have you observed? Use H as follows:

$$H = \begin{bmatrix} 1 & 0 \\ 0 & 1000 \end{bmatrix} \tag{6.19}$$

Next, apply gradient descent to the same optimization problem by stepping with the unnormalized gradient. Do two steps of the algorithm. What have you observed? Finally, apply the conjugate gradient method. How many steps do you need to converge?

Exercise 6.5. Compare Newton's method and the secant method on $f(x) = x^2 + x^4$, with $x^{(1)} = -3$ and $x^{(0)} = -4$. Run each method for 10 iterations. Make two plots:

1. Plot f vs. the iteration for each method.

2. Plot f' vs. x. Overlay the progression of each method, drawing lines from $(x^{(i)}, f'(x^{(i)}))$ to $(x^{(i+1)}, 0)$ to $(x^{(i+1)}, f'(x^{(i+1)}))$ for each transition.

What can we conclude about this comparison?

Exercise 6.6. Give an example of a sequence of points $x^{(1)}, x^{(2)}, \ldots$ and a function f such that $f(x^{(1)}) > f(x^{(2)}) > \cdots$ and yet the sequence does not converge to a local minimum. Assume f is bounded from below.

Exercise 6.7. What is the advantage of a Quasi-Newton method over Newton's method?

Exercise 6.8. Give an example where the BFGS update does not exist. What would you do in this case?

Exercise 6.9. Suppose we have a function $f(\mathbf{x}) = (x_1 + 1)^2 + (x_2 + 3)^2 + 4$. If we start at the origin, what is the resulting point after one step of Newton's method?

Exercise 6.10. In this problem we will derive the optimization problem from which the Davidon-Fletcher-Powell update is obtained. Start with a quadratic approximation at $\mathbf{x}^{(k)}$:

$$f^{(k)}(\mathbf{x}) = y^{(k)} + \left(\mathbf{g}^{(k)}\right)^\top \left(\mathbf{x} - \mathbf{x}^{(k)}\right) + \frac{1}{2}\left(\mathbf{x} - \mathbf{x}^{(k)}\right)^\top H^{(k)}\left(\mathbf{x} - \mathbf{x}^{(k)}\right)$$

where $y^{(k)}$, $\mathbf{g}^{(k)}$, and $\mathbf{H}^{(k)}$ are the objective function value, the true gradient, and a positive definite Hessian approximation at $\mathbf{x}^{(k)}$.

The next iterate is chosen using line search to obtain:

$$\mathbf{x}^{(k+1)} \leftarrow \mathbf{x}^{(k)} - \alpha^{(k)} \left(\mathbf{H}^{(k)} \right)^{-1} \mathbf{g}^{(k)}$$

We can construct a new quadratic approximation $f^{(k+1)}$ at $\mathbf{x}^{(k+1)}$. The approximation should enforce that the local function evaluation is correct:

$$f^{(k+1)}(\mathbf{x}^{(k+1)}) = y^{(k+1)}$$

and that the local gradient is correct:

$$\nabla f^{(k+1)}(\mathbf{x}^{(k+1)}) = \mathbf{g}^{(k+1)}$$

and that the previous gradient is correct:

$$\nabla f^{(k+1)}(\mathbf{x}^{(k)}) = \mathbf{g}^{(k)}$$

Show that updating the Hessian approximation to obtain $\mathbf{H}^{(k+1)}$ requires:[8]

$$\mathbf{H}^{(k+1)} \delta^{(k+1)} = \gamma^{(k+1)}$$

Then, show that in order for $\mathbf{H}^{(k+1)}$ to be positive definite, we require:[9]

$$\left(\delta^{(k+1)} \right)^{\top} \gamma^{(k+1)} > 0$$

Finally, assuming that the curvature condition is enforced, explain why one then solves the following optimization problem to obtain $\mathbf{H}^{(k+1)}$:[10]

$$
\begin{aligned}
& \underset{\mathbf{H}}{\text{minimize}} && \left\| \mathbf{H} - \mathbf{H}^{(k)} \right\| \\
& \text{subject to} && \mathbf{H} = \mathbf{H}^{\top} \\
& && \mathbf{H}\delta^{(k+1)} = \gamma^{(k+1)}
\end{aligned}
$$

where $\left\| \mathbf{H} - \mathbf{H}^{(k)} \right\|$ is a *matrix norm* that defines a distance between $\mathbf{H}$ and $\mathbf{H}^{(k)}$.

[8] This condition is called the *secant equation*. The vectors δ and γ are defined in equation (6.11).

[9] This condition is called the *curvature condition*. It can be enforced using the Wolfe conditions during line search.

[10] The Davidon-Fletcher-Powell update is obtained by solving such an optimization problem to obtain an analytical solution and then finding the corresponding update equation for the inverse Hessian approximation.

7 Direct Methods

Direct methods rely solely on the objective function f. These methods are also called *zero-order*, *black box*, *pattern search*, or *derivative-free* methods. Direct methods do not rely on derivative information to guide them toward a local minimum or identify when they have reached a local minimum. They use other criteria to choose the next search direction and to judge when they have converged.

7.1 Cyclic Coordinate Search

Cyclic coordinate search, also known as *coordinate descent* or *taxicab search*, simply alternates between coordinate directions for its line search. The search starts from an initial $\mathbf{x}^{(1)}$ and optimizes the first input:

$$\mathbf{x}^{(2)} = \arg\min_{x_1} f(x_1, x_2^{(1)}, x_3^{(1)}, \ldots, x_n^{(1)}) \tag{7.1}$$

Having solved this, it optimizes the next coordinate:

$$\mathbf{x}^{(3)} = \arg\min_{x_2} f(x_1^{(2)}, x_2, x_3^{(2)}, \ldots, x_n^{(2)}) \tag{7.2}$$

This process is equivalent to doing a sequence of line searches along the set of n basis vectors, where the ith basis vector is all zero except for the ith component, which has value 1 (algorithm 7.1). For example, the third basis function, denoted $\mathbf{e}^{(3)}$, in a four-dimensional space is:

$$\mathbf{e}^{(3)} = [0, 0, 1, 0] \tag{7.3}$$

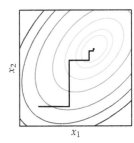

Figure 7.1. Cyclic coordinate descent alternates between coordinate directions.

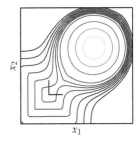

Figure 7.2. Above is an example of how cyclic coordinate search can get stuck. Moving in either of the coordinate directions will result only in increasing f, but moving diagonally, which is not allowed in cyclic coordinate search, can result in lowering f.

```
basis(i, n) = [k == i ? 1.0 : 0.0 for k in 1 : n]
```

Algorithm 7.1. A function for constructing the ith basis vector of length n.

Figure 7.1 shows an example of a search through a two-dimensional space.

Like steepest descent, cyclic coordinate search is guaranteed either to improve or to remain the same with each iteration. No significant improvement after a full cycle over all coordinates indicates that the method has converged. Algorithm 7.2 provides an implementation. As figure 7.2 shows, cyclic coordinate search can fail to find even a local minimum.

Algorithm 7.2. The cyclic coordinate descent method takes as input the objective function f and a starting point x, and it runs until the step size over a full cycle is less than a given tolerance ϵ.

```
function cyclic_coordinate_descent(f, x, ϵ)
    Δ, n = Inf, length(x)
    while abs(Δ) > ϵ
        x′ = copy(x)
        for i in 1 : n
            d = basis(i, n)
            x = line_search(f, x, d)
        end
        Δ = norm(x - x′)
    end
    return x
end
```

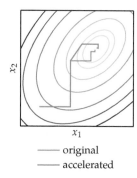

The method can be augmented with an acceleration step to help traverse diagonal valleys. For every full cycle starting with optimizing $\mathbf{x}^{(1)}$ along $\mathbf{e}^{(1)}$ and ending with $\mathbf{x}^{(n+1)}$ after optimizing along $\mathbf{e}^{(n)}$, an additional line search is conducted along the direction $\mathbf{x}^{(n+1)} - \mathbf{x}^{(1)}$. An implementation is provided in algorithm 7.3 and an example search trajectory is shown in figure 7.3.

—— original
—— accelerated

Figure 7.3. Adding the acceleration step to cyclic coordinate descent helps traverse valleys. Six steps are shown for both the original and accelerated versions.

7.2 Powell's Method

Powell's method[1] can search in directions that are not orthogonal to each other. The method can automatically adjust for long, narrow valleys that might otherwise require a large number of iterations for cyclic coordinate descent or other methods that search in axis-aligned directions.

[1] Powell's method was first introduced by M. J. D. Powell, "An Efficient Method for Finding the Minimum of a Function of Several Variables Without Calculating Derivatives," *Computer Journal*, vol. 7, no. 2, pp. 155–162, 1964. An overview is presented by W. H. Press, S. A. Teukolsky, W. T. Vetterling, and B. P. Flannery, *Numerical Recipes in C: The Art of Scientific Computing*. Cambridge University Press, 1982, vol. 2.

```
function cyclic_coordinate_descent_with_acceleration_step(f, x, ε)
    Δ, n = Inf, length(x)
    while abs(Δ) > ε
        x' = copy(x)
        for i in 1 : n
            d = basis(i, n)
            x = line_search(f, x, d)
        end
        x = line_search(f, x, x - x') # acceleration step
        Δ = norm(x - x')
    end
    return x
end
```

Algorithm 7.3. The cyclic coordinate descent method with an acceleration step takes as input the objective function f and a starting point x, and it runs until the step size over a full cycle is less than a given tolerance ε.

The algorithm maintains a list of search directions $\mathbf{u}^{(1)}, \ldots, \mathbf{u}^{(n)}$, which are initially the coordinate basis vectors, $\mathbf{u}^{(i)} = \mathbf{e}^{(i)}$ for all i. Starting at $\mathbf{x}^{(1)}$, Powell's method conducts a line search for each search direction in succession, updating the design point each time:

$$\mathbf{x}^{(i+1)} \leftarrow \text{line\_search}(f, \mathbf{x}^{(i)}, \mathbf{u}^{(i)}) \text{ for all } i \text{ in } \{1, \ldots, n\} \quad (7.4)$$

Next, all search directions are shifted down by one index, dropping the oldest search direction, $\mathbf{u}^{(1)}$:

$$\mathbf{u}^{(i)} \leftarrow \mathbf{u}^{(i+1)} \text{ for all } i \text{ in } \{1, \ldots, n-1\} \quad (7.5)$$

The last search direction is replaced with the direction from $\mathbf{x}^{(1)}$ to $\mathbf{x}^{(n+1)}$, which is the overall direction of progress over the last cycle:

$$\mathbf{u}^{(n)} \leftarrow \mathbf{x}^{(n+1)} - \mathbf{x}^{(1)} \quad (7.6)$$

and another line search is conducted along the new direction to obtain a new $\mathbf{x}^{(1)}$. This process is repeated until convergence. Algorithm 7.4 provides an implementation. Figure 7.4 shows an example search trajectory.

Powell showed that for quadratic functions, after k full iterations the last k directions will be mutually conjugate. Recall that n line searches along mutually conjugate directions will optimize a quadratic function. Thus, n full iterations of Powell's method, totaling $n(n+1)$ line searches, will minimize a quadratic function.

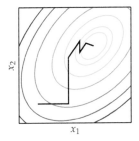

Figure 7.4. Powell's method starts the same as cyclic coordinate descent but iteratively learns conjugate directions.

```
function powell(f, x, ϵ)
    n = length(x)
    U = [basis(i,n) for i in 1 : n]
    Δ = Inf
    while Δ > ϵ
        x′ = x
        for i in 1 : n
            d = U[i]
            x′ = line_search(f, x′, d)
        end
        for i in 1 : n-1
            U[i] = U[i+1]
        end
        U[n] = d = x′ - x
        x′ = line_search(f, x′, d)
        Δ = norm(x′ - x)
        x = x′
    end
    return x
end
```

Algorithm 7.4. Powell's method, which takes the objective function f, a starting point x, and a tolerance ϵ.

The procedure of dropping the oldest search direction in favor of the overall direction of progress can lead the search directions to become linearly dependent. Without search vectors that are linearly independent, the search directions can no longer cover the full design space, and the method may not be able to find the minimum. This weakness can be mitigated by periodically resetting the search directions to the basis vectors. One recommendation is to reset every n or $n + 1$ iterations.

7.3   Hooke-Jeeves

The *Hooke-Jeeves method* (algorithm 7.5) traverses the search space based on evaluations at small steps in each coordinate direction.[2] At every iteration, the Hooke-Jeeves method evaluates $f(\mathbf{x})$ and $f(\mathbf{x} \pm \alpha \mathbf{e}^{(i)})$ for a given step size α in every coordinate direction from an *anchoring point* **x**. It accepts any improvement it may find. If no improvements are found, it will decrease the step size. The process repeats until the step size is sufficiently small. Figure 7.5 shows a few iterations of the algorithm.

[2] R. Hooke and T. A. Jeeves, "Direct Search Solution of Numerical and Statistical Problems," *Journal of the ACM (JACM)*, vol. 8, no. 2, pp. 212–229, 1961.

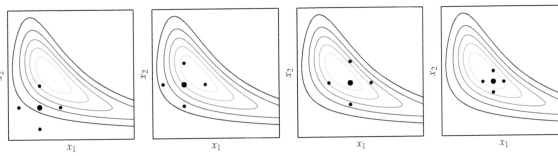

Figure 7.5. The Hooke-Jeeves method, proceeding left to right. It begins with a large step size but then reduces it once it cannot improve by taking a step in any coordinate direction.

One step of the Hooke-Jeeves method requires $2n$ function evaluations for an n-dimensional problem, which can be expensive for problems with many dimensions. The Hooke-Jeeves method is susceptible to local minima. The method has been proven to converge on certain classes of functions.[3]

[3] E. D. Dolan, R. M. Lewis, and V. Torczon, "On the Local Convergence of Pattern Search," *SIAM Journal on Optimization*, vol. 14, no. 2, pp. 567–583, 2003.

7.4   Generalized Pattern Search

In contrast with the Hooke-Jeeves method, which searches in the coordinate directions, *generalized pattern search* can search in arbitrary directions. A *pattern* $\mathcal{P}$ can be constructed from a set of directions $\mathcal{D}$ about an anchoring point $\mathbf{x}$ with a step size α according to:

$$\mathcal{P} = \{\mathbf{x} + \alpha\mathbf{d} \text{ for each } \mathbf{d} \text{ in } \mathcal{D}\} \qquad (7.7)$$

The Hooke-Jeeves method uses $2n$ directions for problems in n dimensions, but generalized pattern search can use as few as $n + 1$.

For generalized pattern search to converge to a local minimum, certain conditions must be met. The set of directions must be a *positive spanning set*, which means that we can construct any point in $\mathbb{R}^n$ using a nonnegative linear combination of the directions in $\mathcal{D}$. A positive spanning set ensures that at least one of the directions is a descent direction from a location with a nonzero gradient.[4]

We can determine whether a given set of directions $\mathcal{D} = \{\mathbf{d}^{(1)}, \mathbf{d}^{(2)}, \dots, \mathbf{d}^{(m)}\}$ in $\mathbb{R}^n$ is a positive spanning set. First, we construct the matrix $\mathbf{D}$ whose columns are the directions in $\mathcal{D}$ (see figure 7.6). The set of directions $\mathcal{D}$ is a positive spanning set if $\mathbf{D}$ has full row rank and if $\mathbf{Dx} = -\mathbf{D1}$ with $\mathbf{x} \geq \mathbf{0}$ has a solution.[5] This optimization problem is identical to the initialization phase of a linear program, which is covered in chapter 11.

[4] Convergence guarantees for generalized pattern search require that all sampled points fall on a scaled lattice. Each direction must thus be a product $\mathbf{d}^{(j)} = \mathbf{Gz}^{(j)}$ for a fixed nonsingular $n \times n$ matrix $\mathbf{G}$ and integer vector $\mathbf{z}$. V. Torczon, "On the Convergence of Pattern Search Algorithms," *SIAM Journal of Optimization*, vol. 7, no. 1, pp. 1–25, 1997.

[5] R. G. Regis, "On the Properties of Positive Spanning Sets and Positive Bases," *Optimization and Engineering*, vol. 17, no. 1, pp. 229–262, 2016.

```
function hooke_jeeves(f, x, α, ϵ, γ=0.5)
    y, n = f(x), length(x)
    while α > ϵ
        improved = false
        x_best, y_best = x, y
        for i in 1 : n
            for sgn in (-1,1)
                x′ = x + sgn*α*basis(i, n)
                y′ = f(x′)
                if y′ < y_best
                    x_best, y_best, improved = x′, y′, true
                end
            end
        end
        x, y = x_best, y_best

        if !improved
            α *= γ
        end
    end
    return x
end
```

Algorithm 7.5. The Hooke-Jeeves method, which takes the target function f, a starting point x, a starting step size α, a tolerance ϵ, and a step decay γ. The method runs until the step size is less than ϵ and the points sampled along the coordinate directions do not provide an improvement. Based on the implementation from A. F. Kaupe Jr, "Algorithm 178: Direct Search," *Communications of the ACM*, vol. 6, no. 6, pp. 313–314, 1963.

only positively spans the cone

only positively spans 1d space

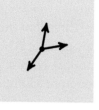

positively spans $\mathbb{R}^2$

Figure 7.6. A valid pattern for generalized pattern search requires a positive spanning set. These directions are stored in the set $\mathcal{D}$.

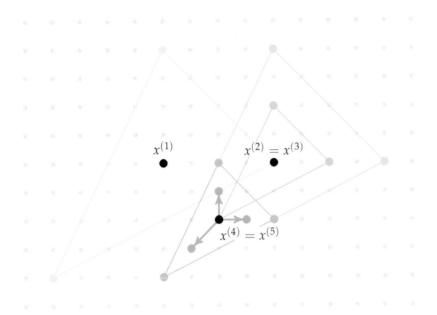

Figure 7.7. All previous points in generalized pattern search lie on a scaled lattice, or mesh. The lattice is not explicitly constructed and need not be axis-aligned.

The implementation of generalized pattern search in algorithm 7.6 contains additional enhancements over the original Hooke-Jeeves method.[6] First, the implementation is *opportunistic*—as soon as an evaluation improves the current best design, it is accepted as the anchoring design point for the next iteration. Second, the implementation uses *dynamic ordering* to accelerate convergence—a direction that leads to an improvement is promoted to the beginning of the list of directions. Figure 7.7 shows a few iterations of the algorithm.

7.5 Nelder-Mead Simplex Method

The *Nelder-Mead simplex method*[7] uses a simplex to traverse the space in search of a minimum. A *simplex* is a generalization of a tetrahedron to n-dimensional space. A simplex in one dimension is a line, and in two dimensions it is a triangle (see figure 7.8). The simplex derives its name from the fact that it is the simplest possible polytope in any given space.

[6] These enhancements are presented in C. Audet and J. E. Dennis Jr., "Mesh Adaptive Direct Search Algorithms for Constrained Optimization," *SIAM Journal on Optimization*, vol. 17, no. 1, pp. 188–217, 2006.

[7] The original simplex method is covered in J. A. Nelder and R. Mead, "A Simplex Method for Function Minimization," *The Computer Journal*, vol. 7, no. 4, pp. 308–313, 1965. We incorporate the improvement in J. C. Lagarias, J. A. Reeds, M. H. Wright, and P. E. Wright, "Convergence Properties of the Nelder–Mead Simplex Method in Low Dimensions," *SIAM Journal on Optimization*, vol. 9, no. 1, pp. 112–147, 1998.

```
function generalized_pattern_search(f, x, α, D, ϵ, γ=0.5)
    y, n = f(x), length(x)
    while α > ϵ
        improved = false
        for (i,d) in enumerate(D)
            x′ = x + α*d
            y′ = f(x′)
            if y′ < y
                x, y, improved = x′, y′, true
                D = pushfirst!(deleteat!(D, i), d)
                break
            end
        end
        if !improved
            α *= γ
        end
    end
    return x
end
```

Algorithm 7.6. Generalized pattern search, which takes the target function f, a starting point x, a starting step size α, a set of search directions D, a tolerance ϵ, and a step decay γ. The method runs until the step size is less than ϵ and the points sampled along the coordinate directions do not provide an improvement.

The Nelder-Mead method uses a series of rules that dictate how the simplex is updated based on evaluations of the objective function at its vertices. A flowchart outlines the procedure in figure 7.9, and algorithm 7.7 provides an implementation. Like the Hooke-Jeeves method, the simplex can move around while roughly maintaining its size, and it can shrink down as it approaches an optimum.

The simplex consist of the points $x^{(1)}, \ldots, x^{(n+1)}$. Let x_h be the vertex with the highest function value, let x_s be the vertex with the second highest function value, and let x_ℓ be the vertex with the lowest function value. Let $\bar{x}$ be the mean of all vertices except the highest point x_h. Finally, for any design point x_θ, let $y_\theta = f(x_\theta)$ be its objective function value. A single iteration then evaluates four simplex operations:

Figure 7.8. A simplex in two dimensions is a triangle. In order for the simplex to be valid, it must have a nonzero area.

Reflection. $x_r = \bar{x} + \alpha(\bar{x} - x_h)$, reflects the highest-valued point over the centroid. This typically moves the simplex from high regions toward lower regions. Here, $\alpha > 0$ and is typically set to 1.

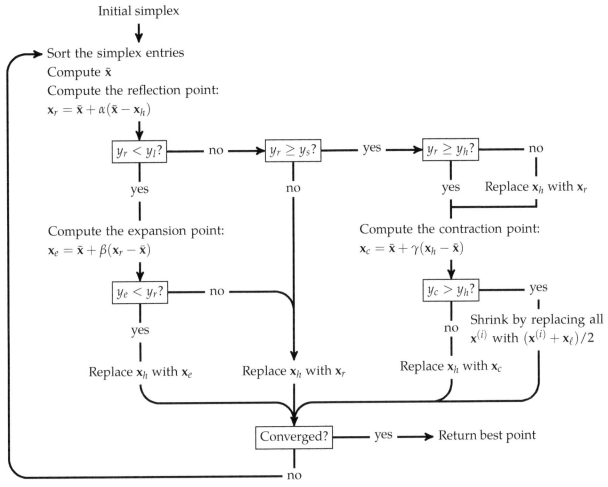

Figure 7.9. Flowchart for the Nelder-Mead algorithm.

Expansion. $x_e = \bar{x} + \beta(x_r - \bar{x})$, like reflection, but the reflected point is sent even further. This is done when the reflected point has an objective function value less than all points in the simplex. Here, $\beta > \max(1, \alpha)$ and is typically set to 2.

Contraction. $x_c = \bar{x} + \gamma(x_h - \bar{x})$, the simplex is shrunk down by moving away from the worst point. It is parameterized by $\gamma \in (0, 1)$ which is typically set to 0.5.

Shrinkage.   All points are moved toward the best point, typically halving the separation distance.

Figure 7.10 shows the four simplex operations. Figure 7.11 shows several iterations of the algorithm.

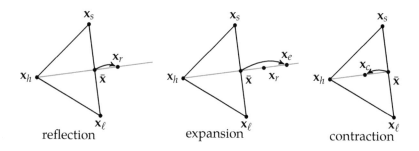

reflection   expansion   contraction   shrinkage

Figure 7.10. The Nelder-Mead simplex operations visualized in two-dimensions.

The convergence criterion for the Nelder-Mead simplex method is unlike Powell's method in that it considers the variation in the function values rather than the changes to the points in the design space. It compares the standard deviation of the sample[8] $y^{(1)}, \ldots, y^{(n+1)}$ to a tolerance ϵ. This value is high for a simplex over a highly curved region, and it is low for a simplex over a flat region. A highly curved region indicates that there is still further optimization possible.

[8] The *standard deviation of the sample* is also called the *uncorrected sample standard deviation*. In our case, it is $\sqrt{\frac{1}{n+1}\sum_{i=1}^{n+1}\left(y^{(i)} - \bar{y}\right)^2}$, where $\bar{y}$ is the mean of $y^{(1)}, \ldots, y^{(n+1)}$.

7.6   Divided Rectangles

The *divided rectangles* algorithm, or *DIRECT* for DIvided RECTangles, is a Lipschitzian optimization approach, similar in some ways to the Shubert-Piyavskii method described in section 3.6.[9] However, it eliminates the need for specifying a Lipschitz constant, and it can be more efficiently extended to multiple dimensions.

[9] D. R. Jones, C. D. Perttunen, and B. E. Stuckman, "Lipschitzian Optimization Without the Lipschitz Constant," *Journal of Optimization Theory and Application*, vol. 79, no. 1, pp. 157–181, 1993.

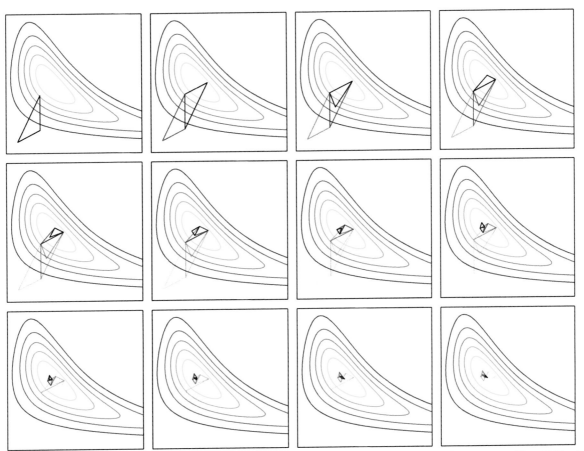

Figure 7.11. The Nelder-Mead method, proceeding left to right and top to bottom.

```
function nelder_mead(f, S, ε; α=1.0, β=2.0, γ=0.5)
    Δ, y_arr = Inf, f.(S)
    while Δ > ε
        p = sortperm(y_arr) # sort lowest to highest
        S, y_arr = S[p], y_arr[p]
        xl, yl = S[1], y_arr[1] # lowest
        xh, yh = S[end], y_arr[end] # highest
        xs, ys = S[end-1], y_arr[end-1] # second-highest
        xm = mean(S[1:end-1]) # centroid
        xr = xm + α*(xm - xh) # reflection point
        yr = f(xr)

        if yr < yl
            xe = xm + β*(xr-xm) # expansion point
            ye = f(xe)
            S[end],y_arr[end] = ye < yr ? (xe, ye) : (xr, yr)
        elseif yr ≥ ys
            if yr < yh
                xh, yh, S[end], y_arr[end] = xr, yr, xr, yr
            end
            xc = xm + γ*(xh - xm) # contraction point
            yc = f(xc)
            if yc > yh
                for i in 2 : length(y_arr)
                    S[i] = (S[i] + xl)/2
                    y_arr[i] = f(S[i])
                end
            else
                S[end], y_arr[end] = xc, yc
            end
        else
            S[end], y_arr[end] = xr, yr
        end

        Δ = std(y_arr, corrected=false)
    end
    return S[argmin(y_arr)]
end
```

Algorithm 7.7. The Nelder-Mead simplex method, which takes the objective function f, a starting simplex S consisting of a list of vectors, and a tolerance ε. The Nelder-Mead parameters can be specified as well and default to recommended values.

The notion of Lipschitz continuity can be extended to multiple dimensions. If f is Lipschitz continuous over a domain $\mathcal{X}$ with Lipschitz constant $\ell > 0$, then for a given design $\mathbf{x}^{(1)}$ and $y = f(\mathbf{x}^{(1)})$, the circular cone

$$f(\mathbf{x}^{(1)}) - \ell\|\mathbf{x} - \mathbf{x}^{(1)}\|_2 \tag{7.8}$$

forms a lower bound of f. Given m function evaluations with design points $\{\mathbf{x}^{(1)}, \ldots, \mathbf{x}^{(m)}\}$, we can construct a superposition of these lower bounds by taking their maximum:

$$\underset{i}{\text{maximize}}\ f(\mathbf{x}^{(i)}) - \ell\|\mathbf{x} - \mathbf{x}^{(i)}\|_2 \tag{7.9}$$

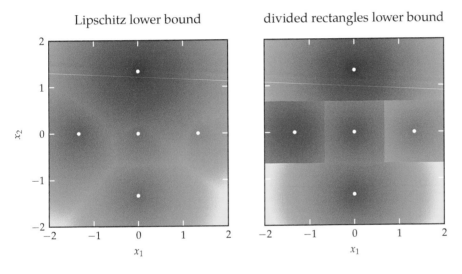

Figure 7.12. The Lipschitz lower bound is an intersection of cones, which creates complicated surfaces in multidimensional space. The divided rectangle lower bound isolates each lower-bound cone to its own hyper-rectangular region, making it trivial to compute the minimum value in each region given a Lipschitz constant.

The Shubert-Piyavskii method samples at a point where the bound derived from a known Lipschitz constant is lowest. Unfortunately, a Lipschitz lower bound has intricate geometry whose complexity increases with the dimensionality of the design space. The left contour plot in figure 7.12 shows such a lower bound using five function evaluations. The right contour plot shows the approximation made by DIRECT, which divides the region into hyper-rectangles—one centered about each design point. Making this assumption allows for the rapid calculation of the minimum of the lower bound.

The DIRECT method does not assume a Lipschitz constant is known. Figures 7.13 and 7.14 show lower bounds constructed using Lipschitz continuity and the DIRECT approximation, respectively, for several different Lipschitz constants. Notice that the location of the minimum changes as the Lipschitz constant changes, with small values of ℓ leading to designs near the lowest function evaluations and larger values of ℓ leading to designs the farthest from previous function evaluations.

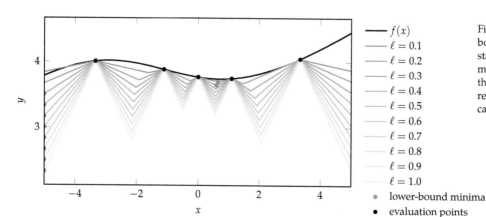

Figure 7.13. The Lipschitz lower bound for different Lipschitz constants ℓ. Not only does the estimated minimum change locally as the Lipschitz constant is varied, the region in which the minimum lies can vary as well.

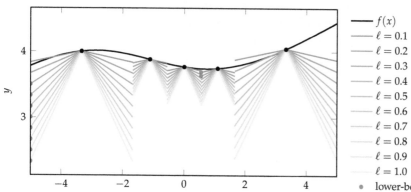

Figure 7.14. The DIRECT lower bound for different Lipschitz constants ℓ. The lower bound is not continuous. The minimum does not change locally but can change regionally as the Lipschitz constant changes.

7.6.1 Univariate DIRECT

In one dimension, DIRECT recursively divides intervals into thirds and then samples the objective function at the center of the intervals as shown in figure 7.15. This scheme is in contrast with the Shubert-Piyavskii method, where sampling occurs at a point where the bound derived from a known Lipschitz constant is lowest.

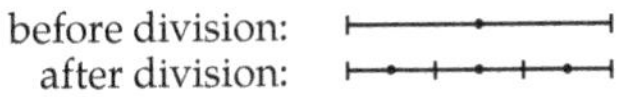

Figure 7.15. Center-point sampling, using the DIRECT scheme, divides intervals into thirds.

For an interval $[a, b]$ with center $c = (a + b)/2$, the lower bound based on $f(c)$ is:

$$f(x) \geq f(c) - \ell|x - c| \tag{7.10}$$

where ℓ is a Lipschitz constant that is unknown to us. The lowest value that the bound obtains on that interval is $f(c) - \ell(b - a)/2$, and it occurs at the edges of the interval.

Even though we do not know ℓ, we can deduce that the lower bound of some intervals are lower than others. If we have two intervals of the same length, for example, and the first one has a lower evaluation at the center point than the second one, then the lower bound of the first interval is lower than that of the second. Although this does not entail that the first interval contains a minimizer, it is an indication that we may want to focus our search in that interval.

During our search, we will have many intervals $[a_1, b_1], \ldots, [a_n, b_n]$ of different widths. We can plot our intervals according to their center value and interval width, as we do in figure 7.16. The lower bound for each interval is the vertical intercept of a line of slope ℓ through its center point. The center of the interval with the lowest lower bound will be the first point intersected as we shift a line with slope ℓ upwards from below.

DIRECT splits all intervals for which a Lipschitz constant exists such that they have the lowest lower bound, as shown in figure 7.17. We refer to these selected intervals as potentially optimal. Any interval may technically contain the optimum, though the selected points heuristically have the best chance of containing the optimum.

One iteration of the one-dimensional DIRECT algorithm consists of identifying the set of potentially optimal intervals and then dividing each interval into thirds. Example 7.1 demonstrates the univariate DIRECT method.

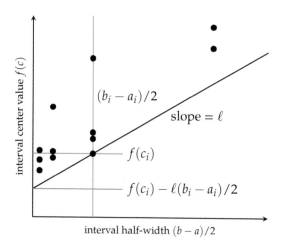

Figure 7.16. Interval selection for a particular Lipschitz constant ℓ. Black dots represent DIRECT intervals and the corresponding function evaluations at their centers. A black line of slope ℓ is drawn through the dot belonging to the selected interval. All other dots must lie on or above this line.

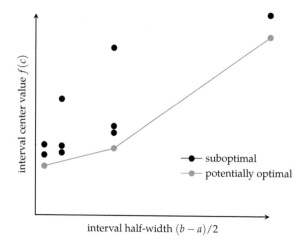

Figure 7.17. The potentially optimal intervals for the DIRECT method form a piecewise boundary that encloses all intervals along the lower-right. Each dot corresponds to an interval.

Consider the function $f(x) = \sin(x) + \sin(2x) + \sin(4x) + \sin(8x)$ on the interval $[-2, 2]$ with a global minimizer near -0.272. Optimization is difficult because of multiple local minima.

The figure below shows the progression of the univariate DIRECT method, with intervals chosen for splitting rendered in blue. The left side shows the intervals overlaid on the objective function. The right side shows the intervals as scatter points in the interval half-width versus center value space.

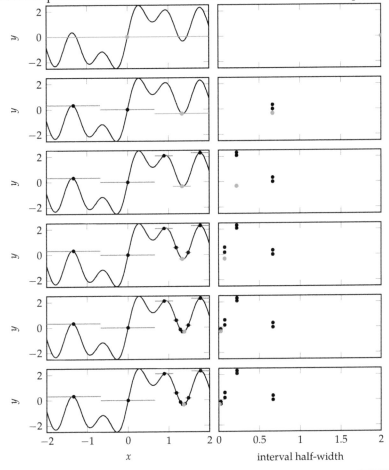

Example 7.1. The DIRECT method applied to a univariate function.

7.6.2   Multivariate DIRECT

In multiple dimensions, we divide rectangles (or hyper-rectangles in more than two dimensions) instead of intervals. Similar to the univariate case, we divide the rectangles into thirds along the axis directions. Before commencing the division of the rectangles, DIRECT normalizes the search space to be the unit hypercube.

As illustrated in figure 7.18, the choice of ordering of the directions when splitting the unit hypercube matters. DIRECT prioritizes the assignment of larger rectangles for the points with lower function evaluations. Larger rectangles are prioritized for additional splitting.

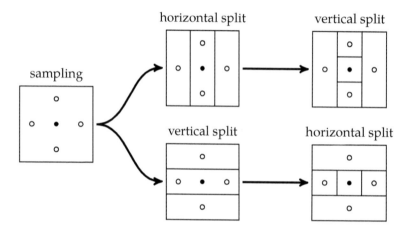

Figure 7.18. Interval splitting in multiple dimensions (using DI-RECT) requires choosing an ordering for the split dimensions.

When splitting a region without equal side lengths, only the longest dimensions are split (figure 7.19). Splitting then proceeds on these dimensions in the same manner as with a hypercube.

The set of potentially optimal intervals is obtained as it is with one dimension. The lower bound for each hyper-rectangle can be computed based on the longest center-to-vertex distance and center value. We can construct a diagram similar to figure 7.16 to identify the potentially optimal rectangles.[10]

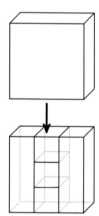

Figure 7.19. DIRECT will split only the longest dimensions of hyper-rectangles.

[10] As an additional requirement for selection, DIRECT also requires that the lower bound for the interval improve the current best value by a nontrivial amount.

7.6.3   Implementation

DIRECT (algorithm 7.8) is best understood when it is broken down into subroutines. We present these subroutines below.

Normalization to the unit hypercube is done by algorithm 7.9.

```
function direct(f, a, b, ϵ, k_max)
    g = reparameterize_to_unit_hypercube(f, a, b)
    intervals = Intervals()
    n = length(a)
    c = fill(0.5, n)
    interval = Interval(c, g(c), fill(0, n))
    add_interval!(intervals, interval)
    c_best, y_best = copy(interval.c), interval.y

    for k in 1 : k_max
        S = get_opt_intervals(intervals, ϵ, y_best)
        to_add = Interval[]
        for interval in S
            append!(to_add, divide(g, interval))
            dequeue!(intervals[vertex_dist(interval)])
        end
        for interval in to_add
            add_interval!(intervals, interval)
            if interval.y < y_best
                c_best, y_best = copy(interval.c), interval.y
            end
        end
    end

    return rev_unit_hypercube_parameterization(c_best, a, b)
end
```

Algorithm 7.8. DIRECT, which takes the multidimensional objective function f, vector of lower bounds a, vector of upper bounds b, tolerance parameter ϵ, and number of iterations k_max. It returns the best coordinate.

```
rev_unit_hypercube_parameterization(x, a, b) = x.*(b-a) + a
function reparameterize_to_unit_hypercube(f, a, b)
    Δ = b-a
    return x->f(x.*Δ + a)
end
```

Algorithm 7.9. A function that creates a function defined over the unit hypercube that is a reparameterized version of the function f defined over the hypercube with lower and upper bounds a and b.

Computing the set of potentially optimal rectangles can be done efficiently (algorithm 7.10). We use the fact that the interval width can only take on powers of one-third, and thus many of the points will share the same x coordinate. For any given x coordinate, only the one with the lowest y value can be a potentially optimal interval. We store the rectangular intervals according to their depth and then in a priority queue according to their centerpoint value.

```
using DataStructures
struct Interval
    c
    y
    depths
end
min_depth(interval) = minimum(interval.depths)
vertex_dist(interval) = norm(0.5*3.0.^(-interval.depths), 2)
const Intervals =
        OrderedDict{Float64,PriorityQueue{Interval, Float64}}
function add_interval!(intervals, interval)
    d = vertex_dist(interval)
    if !haskey(intervals, d)
        intervals[d] = PriorityQueue{Interval, Float64}()
    end
    return enqueue!(intervals[d], interval, interval.y)
end
```

Algorithm 7.10. The data structure used in DIRECT. Here, Interval has three fields: the interval center c, the center point value y = f(c), and the number of divisions in each dimension depths. The function add_interval! inserts a new Interval into the data structure.

We can use this data structure to obtain all potentially optimal intervals (algorithm 7.11). The algorithm proceeds from lowest interval width to highest interval width. For each point, we first determine whether it is above or below the line joining the previous two points. If it is below, we skip it. The same determination is then made for the next point.

Finally, we need a method for dividing the intervals. This is implemented by the divide method (algorithm 7.12).

The intervals obtained from running DIRECT in two dimensions are visualized in figure 7.20. Two iterations of DIRECT in two dimensions are worked out in example 7.2.

```
function get_opt_intervals(intervals, ε, y_best)
    stack = Interval[]
    for (x, pq) in intervals
        if !isempty(pq)
            interval = DataStructures.peek(pq)[1]
            y = interval.y

            while length(stack) > 1
                interval1 = stack[end]
                interval2 = stack[end-1]
                x1, y1 = vertex_dist(interval1), interval1.y
                x2, y2 = vertex_dist(interval2), interval2.y
                ℓ = (y2 - y) / (x2 - x)
                if y1 <= ℓ*(x1-x) + y + ε
                    break
                end
                # remove previous interval
                pop!(stack)
            end

            if !isempty(stack) && interval.y > stack[end].y + ε
                # skip new interval
                continue
            end

            push!(stack, interval) # add new interval
        end
    end
    return stack
end
```

Algorithm 7.11. A routine for obtaining the potentially optimal intervals, where intervals is of type Intervals, ε is a tolerance parameter, and y_best is the best function evaluation.

```
function divide(f, interval)
    c, d, n = interval.c, min_depth(interval), length(interval.c)
    dirs = findall(interval.depths .== d)
    cs = [(c + 3.0^(-d-1)*basis(i,n),
           c - 3.0^(-d-1)*basis(i,n)) for i in dirs]
    vs = [(f(C[1]), f(C[2])) for C in cs]
    minvals = [min(V[1], V[2]) for V in vs]

    intervals = Interval[]
    depths = copy(interval.depths)
    for j in sortperm(minvals)
        depths[dirs[j]] += 1
        C, V = cs[j], vs[j]
        push!(intervals, Interval(C[1], V[1], copy(depths)))
        push!(intervals, Interval(C[2], V[2], copy(depths)))
    end
    push!(intervals, Interval(c, interval.y, copy(depths)))
    return intervals
end
```

Algorithm 7.12. The divide routine for dividing an interval, where f is the objective function and interval is the interval to be divided. It returns a list of the resulting smaller intervals.

7.7   Summary

- Direct methods rely solely on the objective function and do not use derivative information.

- Cyclic coordinate search optimizes one coordinate direction at a time.

- Powell's method adapts the set of search directions based on the direction of progress.

- Hooke-Jeeves searches in each coordinate direction from the current point using a step size that is adapted over time.

- Generalized pattern search is similar to Hooke-Jeeves, but it uses fewer search directions that positively span the design space.

- The Nelder-Mead simplex method uses a simplex to search the design space, adaptively expanding and contracting the size of the simplex in response to evaluations of the objective function.

- The divided rectangles algorithm extends the Shubert-Piyavskii approach to multiple dimensions and does not require specifying a valid Lipschitz constant.

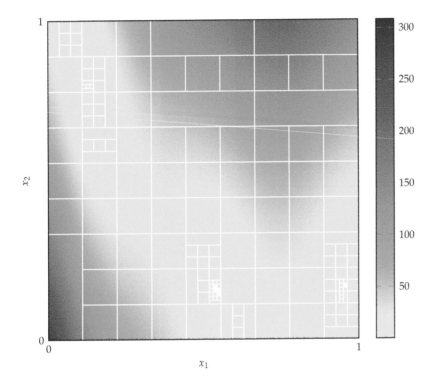

Figure 7.20. The DIRECT method after 16 iterations on the Branin function, appendix B.3. Each cell is bordered by white lines. The cells are much denser around the minima of the Branin function, as the DIRECT method procedurally increases its resolution in those regions.

Consider using DIRECT to optimize the flower function (appendix B.4) over $x_1 \in [-1, 3]$, $x_2 \in [-2, 1]$. The function is first normalized to the unit hypercube such that we optimize $x_1', x_2' \in [0, 1]$:

$$f(x_1', x_2') = \text{flower}(4x_1' - 1, 3x_2' - 2)$$

Example 7.2. The first two iterations of DIRECT worked out in detail.

The objective function is sampled at $[0.5, 0.5]$ to obtain 0.158. We have a single interval with center $[0.5, 0.5]$ and side lengths $[1, 1]$. The interval is divided twice, first into thirds in x_1' and then the center interval is divided into thirds in x_2'.

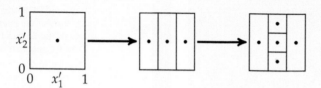

We now have five intervals:

interval	center	side lengths	vertex distance	center value
1	$[1/6, 3/6]$	$[1/3, 1]$	0.527	0.500
2	$[5/6, 3/6]$	$[1/3, 1]$	0.527	1.231
3	$[3/6, 3/6]$	$[1/3, 1/3]$	0.236	0.158
4	$[3/6, 1/6]$	$[1/3, 1/3]$	0.236	2.029
5	$[3/6, 5/6]$	$[1/3, 1/3]$	0.236	1.861

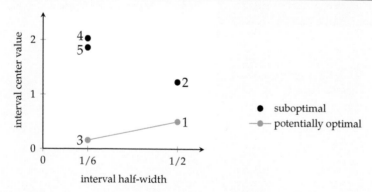

We next split on the two intervals centered at $[1/6, 3/6]$ and $[3/6, 3/6]$.

7.8   Exercises

Exercise 7.1. Previous chapters covered methods that leverage the derivative to descend toward a minimum. Direct methods are able to use only zero-order information—evaluations of f. How many evaluations are needed to approximate the derivative and the Hessian of an n-dimensional objective function using finite difference methods? Why do you think it is important to have zero-order methods?

Exercise 7.2. Design an objective function and a starting point x_0 such that Hooke-Jeeves will fail to attain a reduction in the objective function. You need to choose x_0 such that it is not a local minimizer.

Exercise 7.3. Is the design point obtained using the Hooke-Jeeves method guaranteed to be within ϵ of a local minimum?

Exercise 7.4. Give an example of a concrete engineering problem where you may not be able to compute analytical derivatives.

Exercise 7.5. State a difference between the divided rectangles algorithm in one dimension and the Shubert-Piyavskii method.

Exercise 7.6. Suppose our search algorithm has us transition from $\mathbf{x}^{(k)} = [1,2,3,4]$ to $\mathbf{x}^{(k+1)} = [2,2,2,4]$. Could our search algorithm be (a) cyclic coordinate search, (b) Powell's method, (c) both a and b, or (d) neither a nor b? Why?

8 Stochastic Methods

This chapter presents a variety of *stochastic methods* that use randomization strategically to help explore the design space for an optimum. Randomness can help escape local optima and increase the chances of finding a global optimum. Stochastic methods typically use *pseudo-random* number generators to ensure repeatability.[1] A large amount of randomness is generally ineffective because it prevents us from effectively using previous evaluation points to help guide the search. This chapter discusses a variety of ways to control the degree of randomness in our search.

[1] Although pseudo-random number generators produce numbers that appear random, they are actually a result of a deterministic process. Pseudo-random numbers can be produced through calls to the rand function. The process can be reset to an initial state using the seed! function from the Random.jl package.

8.1 Noisy Descent

Adding stochasticity to gradient descent can be beneficial in large nonlinear optimization problems. Saddle points, where the gradient is very close to zero, can cause descent methods to select step sizes that are too small to be useful. One approach is to add Gaussian noise at each descent step:[2]

$$\mathbf{x}^{(k+1)} \leftarrow \mathbf{x}^{(k)} + \alpha \mathbf{g}^{(k)} + \boldsymbol{\epsilon}^{(k)} \tag{8.1}$$

[2] G. Hinton and S. Roweis, "Stochastic Neighbor Embedding," in *Advances in Neural Information Processing Systems* (NIPS), 2003.

where $\boldsymbol{\epsilon}^{(k)}$ is zero-mean Gaussian noise with standard deviation σ. The amount of noise is typically reduced over time. The standard deviation of the noise is typically a decreasing sequence $\sigma^{(k)}$ such as $1/k$.[3] Algorithm 8.1 provides an implementation of this method. Figure 8.1 compares descent with and without noise on a saddle function.

[3] The Hinton and Roweis paper used a fixed standard deviation for the first 3,500 iterations and set the standard deviation to zero thereafter.

A common approach for training neural networks is *stochastic gradient descent*, which uses a noisy gradient approximation. In addition to helping traverse past saddle points, evaluating noisy gradients using randomly chosen subsets of the training data[4] is significantly less expensive computationally than calculating the true gradient at every iteration.

[4] These subsets are called *batches*.

```
mutable struct NoisyDescent <: DescentMethod
    submethod
    σ
    k
end
function init!(M::NoisyDescent, f, ∇f, x)
    init!(M.submethod, f, ∇f, x)
    M.k = 1
    return M
end
function step!(M::NoisyDescent, f, ∇f, x)
    x = step!(M.submethod, f, ∇f, x)
    σ = M.σ(M.k)
    x += σ.*randn(length(x))
    M.k += 1
    return x
end
```

Algorithm 8.1. A noisy descent method, which augments another descent method with additive Gaussian noise. The method takes another DescentMethod submethod, a noise sequence σ, and stores the iteration count k.

Convergence guarantees for stochastic gradient descent require that the positive step sizes be chosen such that:

$$\sum_{k=1}^{\infty} \alpha^{(k)} = \infty \qquad \sum_{k=1}^{\infty} \left(\alpha^{(k)}\right)^2 < \infty \tag{8.2}$$

These conditions ensure that the step sizes decrease and allow the method to converge, but not too quickly so as to become stuck away from a local minimum.

8.2 Mesh Adaptive Direct Search

The generalized pattern search methods covered in section 7.4 restricted local exploration to a fixed set of directions. In contrast, *mesh adaptive direct search* uses random positive spanning directions.[5]

The procedure used to sample positive spanning sets (see example 8.1) begins by constructing an initial linearly spanning set in the form of a lower triangular matrix $\mathbf{L}$. The diagonal terms in $\mathbf{L}$ are sampled from $\pm 1/\sqrt{\alpha^{(k)}}$, where $\alpha^{(k)}$ is the step size at iteration k. The lower components of $\mathbf{L}$ are sampled from

$$\left\{ -1/\sqrt{\alpha^{(k)}} + 1, -1/\sqrt{\alpha^{(k)}} + 2, \ldots, 1/\sqrt{\alpha^{(k)}} - 1 \right\} \tag{8.3}$$

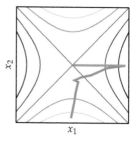

—— stochastic gradient descent
—— steepest descent

Figure 8.1. Adding stochasticity to a descent method helps with traversing saddle points such as $f(\mathbf{x}) = x_1^2 - x_2^2$ shown here. Due to the initialization, the steepest descent method converges to the saddle point where the gradient is zero.

[5] This section follows the lower triangular mesh adaptive direct search given by C. Audet and J. E. Dennis Jr., "Mesh Adaptive Direct Search Algorithms for Constrained Optimization," *SIAM Journal on Optimization*, vol. 17, no. 1, pp. 188–217, 2006.

Consider positive spanning sets constructed from the nonzero directions $d_1, d_2 \in \{-1, 0, 1\}$. There are 8 positive spanning sets with 3 elements that can be constructed from these directions:

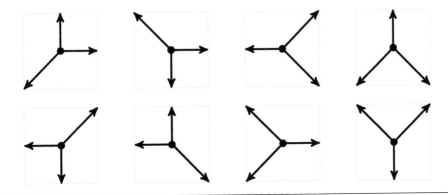

Example 8.1. Positive spanning sets for $\mathbb{R}^2$. Note that the lower triangular generation strategy can only generate the first two columns of spanning sets.

The rows and columns of **L** are then randomly permuted to obtain a matrix **D** whose columns correspond to n directions that linearly span $\mathbb{R}^n$. The maximum magnitude among these directions is $1/\sqrt{\alpha^{(k)}}$.

Two common methods for obtaining a positive spanning set from the linearly spanning set are to add one additional direction $\mathbf{d}^{(n+1)} = -\sum_{i=1}^{n} \mathbf{d}^{(i)}$ and to add n additional directions $\mathbf{d}^{(n+j)} = -\mathbf{d}^{(j)}$ for j in $\{1, \ldots, n\}$. We use the first method in algorithm 8.2.

The step size α starts at 1, is always a power of 4, and never exceeds 1. Using a power of 4 causes the maximum possible step size taken in each iteration to be scaled by a factor of 2, as the maximum step size $\alpha/\sqrt{\alpha}$ has length $4^m/\sqrt{4^m} = 2^m$ for integer $m < 1$. The step size is updated according to:

$$\alpha^{(k+1)} \leftarrow \begin{cases} \alpha^{(k)}/4 & \text{if no improvement was found in this iteration} \\ \min(1, 4\alpha^{(k)}) & \text{otherwise} \end{cases}$$

(8.4)

Mesh adaptive direct search is opportunistic but cannot support dynamic ordering[6] since, after a successful iteration, the step size is increased, and another step in the successful direction would lie outside of the mesh. The algorithm queries a new design point along the accepted descent direction. If $f(\mathbf{x}^{(k)} = \mathbf{x}^{(k-1)} + \alpha\mathbf{d}) < f(\mathbf{x}^{(k-1)})$, then the queried point is $\mathbf{x}^{(k-1)} + 4\alpha\mathbf{d} = \mathbf{x}^{(k)} + 3\alpha\mathbf{d}$. The

[6] See section 7.4.

```
function rand_positive_spanning_set(α, n)
    δ = round(Int, 1/sqrt(α))
    L = Matrix(Diagonal(δ*rand([1,-1], n)))
    for i in 1 : n-1
        for j in 1:i-1
            L[i,j] = rand(-δ+1:δ-1)
        end
    end
    D = L[randperm(n),:]
    D = D[:,randperm(n)]
    D = hcat(D, -sum(D,dims=2))
    return [D[:,i] for i in 1 : n+1]
end
```

Algorithm 8.2. Randomly sampling a positive spanning set of $n+1$ directions according to mesh adaptive direct search with step size α and number of dimensions n.

procedure is outlined in algorithm 8.3. Figure 8.2 illustrates how this algorithm explores the search space.

8.3 Simulated Annealing

Simulated annealing[7] borrows inspiration from metallurgy.[8] *Temperature* is used to control the degree of stochasticity during the randomized search. The temperature starts high, allowing the process to freely move about the search space, with the hope that in this phase the process will find a good region with the best local minimum. The temperature is then slowly brought down, reducing the stochasticity and forcing the search to converge to a minimum. Simulated annealing is often used on functions with many local minima due to its ability to escape local minima.

At every iteration, a candidate transition from $\mathbf{x}$ to $\mathbf{x}'$ is sampled from a transition distribution T and is accepted with probability

$$\begin{cases} 1 & \text{if } \Delta y \leq 0 \\ e^{-\Delta y/t} & \text{if } \Delta y > 0 \end{cases} \quad (8.5)$$

where $\Delta y = f(\mathbf{x}') - f(\mathbf{x})$ is the difference in the objective and t is the temperature. It is this acceptance probability, known as the *Metropolis criterion*, that allows the algorithm to escape from local minima when the temperature is high.

[7] S. Kirkpatrick, C. D. Gelatt Jr., and M. P. Vecchi, "Optimization by Simulated Annealing," *Science*, vol. 220, no. 4598, pp. 671–680, 1983.

[8] Annealing is a process in which a material is heated and then cooled, making it more workable. When hot, the atoms in the material are more free to move around, and, through random motion, tend to settle into better positions. A slow cooling brings the material to an ordered, crystalline state. A fast, abrupt quenching causes defects because the material is forced to settle in its current condition.

```
function mesh_adaptive_direct_search(f, x, ε)
    α, y, n = 1, f(x), length(x)
    while α > ε
        improved = false
        for (i,d) in enumerate(rand_positive_spanning_set(α, n))
            x´ = x + α*d
            y´ = f(x´)
            if y´ < y
                x, y, improved = x´, y´, true
                x´ = x + 3α*d
                y´ = f(x´)
                if y´ < y
                    x, y = x´, y´
                end
                break
            end
        end
        α = improved ? min(4α, 1) : α/4
    end
    return x
end
```

Algorithm 8.3. Mesh adaptive direct search for an objective function f, an initial design x, and a tolerance ϵ.

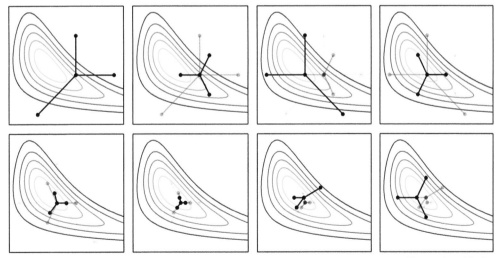

Figure 8.2. Mesh adaptive direct search proceeding left to right and top to bottom.

The temperature parameter t controls the acceptance probability. An annealing schedule is used to slowly bring down the temperature as the algorithm progresses, as illustrated by figure 8.3. The temperature must be brought down to ensure convergence. If it is brought down too quickly, the search method may not cover the portion of the search space containing the global minimum.

It can be shown that a *logarithmic annealing schedule* of $t^{(k)} = t^{(1)} \ln(2) / \ln(k+1)$ for the kth iteration is guaranteed to asymptotically reach the global optimum under certain conditions,[9] but it can be very slow in practice. The *exponential annealing schedule*, which is more common, uses a simple decay factor:

$$t^{(k+1)} = \gamma t^{(k)} \tag{8.6}$$

for some $\gamma \in (0, 1)$. Another common annealing schedule, *fast annealing*,[10] uses a temperature of

$$t^{(k)} = \frac{t^{(1)}}{k} \tag{8.7}$$

A basic implementation of simulated annealing is provided by algorithm 8.4. Example 8.2 shows the effect different transition distributions and annealing schedules have on the optimization process.

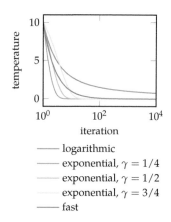

Figure 8.3. Several annealing schedules commonly used in simulated annealing. The schedules have an initial temperature of 10.

[9] B. Hajek, "Cooling Schedules for Optimal Annealing," *Mathematics of Operations Research*, vol. 13, no. 2, pp. 311–329, 1988.

```
function simulated_annealing(f, x, T, t, k_max)
    y = f(x)
    x_best, y_best = x, y
    for k in 1 : k_max
        x′ = x + rand(T)
        y′ = f(x′)
        Δy = y′ - y
        if Δy ≤ 0 || rand() < exp(-Δy/t(k))
            x, y = x′, y′
        end
        if y′ < y_best
            x_best, y_best = x′, y′
        end
    end
    return x_best
end
```

Algorithm 8.4. Simulated annealing, which takes as input an objective function f, an initial point x, a transition distribution T, an annealing schedule t, and the number of iterations k_max.

[10] H. Szu and R. Hartley, "Fast Simulated Annealing," *Physics Letters A*, vol. 122, no. 3-4, pp. 157–162, 1987.

A more sophisticated algorithm was introduced by Corana et al. in 1987 that allows for the step size to change during the search.[11] Rather than using a fixed

[11] A. Corana, M. Marchesi, C. Martini, and S. Ridella, "Minimizing Multimodal Functions of Continuous Variables with the 'Simulated Annealing' Algorithm," *ACM Transactions on Mathematical Software*, vol. 13, no. 3, pp. 262–280, 1987.

We can use simulated annealing to optimize Ackley's function, appendix B.1. Ackley's function has many local minima, making it easy for gradient-based methods to get stuck.

Suppose we start at $\mathbf{x}^{(1)} = [15, 15]$ and run 100 iterations. Below we show the distribution over iterations for multiple runs with different combinations of three zero-mean, diagonal covariance ($\sigma \mathbf{I}$) Gaussian transition distributions, and three different temperature schedules $t^{(k)} = t^{(1)}/k$.

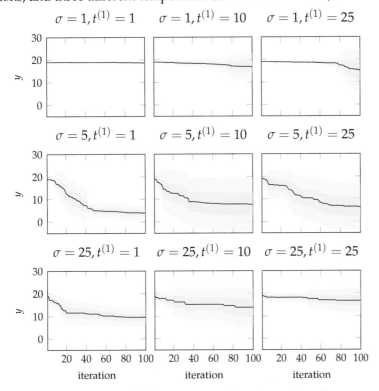

In this case, the spread of the transition distribution has the greatest impact on performance.

Example 8.2. Exploring the effect of distribution variance and temperature on the performance of simulated annealing. The blue regions indicate the 5% to 95% and 25% to 75% empirical Gaussian quantiles of the objective function value.

transition distribution, this adaptive simulated annealing method keeps track of a separate step size $\mathbf{v}$ for each coordinate direction. For a given point $\mathbf{x}$, a cycle of random moves is performed in each coordinate direction i according to:

$$\mathbf{x}' = \mathbf{x} + rv_i\mathbf{e}_i \tag{8.8}$$

where r is drawn uniformly at random from $[1, -1]$ and v_i is the maximum step size in the ith coordinate direction. Each new point is accepted according to the Metropolis criterion. The number of accepted points in each coordinate direction is stored in a vector $\mathbf{a}$.

After n_s cycles, the step sizes are adjusted with the aim to maintain an approximately equal number of accepted and rejected designs with an average acceptance rate near one-half. Rejecting too many moves is a waste of computational effort, while accepting too many moves indicates that the configuration is evolving too slowly because candidate points are too similar to the current location. The update formula used by Corana et al. is:

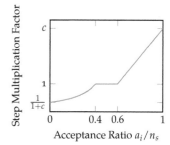

Step Multiplication Factor

Figure 8.4. The step multiplication factor as a function of acceptance rate for $c = 2$.

$$v_i = \begin{cases} v_i\left(1 + c_i\frac{a_i/n_s - 0.6}{0.4}\right) & \text{if } a_i > 0.6n_s \\ v_i\left(1 + c_i\frac{0.4 - a_i/n_s}{0.4}\right)^{-1} & \text{if } a_i < 0.4n_s \\ v_i & \text{otherwise} \end{cases} \tag{8.9}$$

The c_i parameter controls the step variation along each direction and is typically set to 2 as shown in figure 8.4. An implementation of the update is shown in algorithm 8.5.

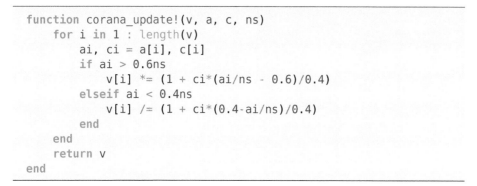

```julia
function corana_update!(v, a, c, ns)
    for i in 1 : length(v)
        ai, ci = a[i], c[i]
        if ai > 0.6ns
            v[i] *= (1 + ci*(ai/ns - 0.6)/0.4)
        elseif ai < 0.4ns
            v[i] /= (1 + ci*(0.4-ai/ns)/0.4)
        end
    end
    return v
end
```

Algorithm 8.5. The update formula used by Corana et al. in adaptive simulated annealing, where v is a vector of coordinate step sizes, a is a vector of the number of accepted steps in each coordinate direction, c is a vector of step scaling factors for each coordinate direction, and ns is the number of cycles before running the step size adjustment.

Temperature reduction occurs every n_t step adjustments, which is every $n_s \cdot n_t$ cycles. The original implementation simply multiplies the temperature by a reduction factor.

The process is terminated when the temperature sinks low enough such that improvement can no longer be expected. Termination occurs when the most recent function value is no farther than ϵ from the previous n_ϵ iterations and the best function value obtained over the course of execution. Algorithm 8.6 provides an implementation and the algorithm is visualized in figure 8.5.

8.4   Cross-Entropy Method

The *cross-entropy method*,[12] in contrast with the methods we have discussed so far in this chapter, maintains an explicit probability distribution over the design space.[13] This probability distribution, often called a *proposal distribution*, is used to propose new samples for the next iteration. At each iteration, we sample from the proposal distribution and then update the proposal distribution to fit a collection of the best samples. The aim at convergence is for the proposal distribution to focus on the global optima. Algorithm 8.7 provides an implementation.

The cross-entropy method requires choosing a family of distributions parameterized by θ. One common choice is the family of multivariate normal distributions parameterized by a mean vector and a covariance matrix. The algorithm also requires us to specify the number of *elite samples*, m_{elite}, to use when fitting the parameters for the next iteration.

Depending on the choice of distribution family, the process of fitting the distribution to the elite samples can be done analytically. In the case of the multivariate normal distribution, the parameters are updated according to the maximum likelihood estimate:

$$\mu^{(k+1)} = \frac{1}{m_{\text{elite}}} \sum_{i=1}^{m_{\text{elite}}} \mathbf{x}^{(i)} \tag{8.10}$$

$$\Sigma^{(k+1)} = \frac{1}{m_{\text{elite}}} \sum_{i=1}^{m_{\text{elite}}} \left(\mathbf{x}^{(i)} - \mu^{(k+1)}\right)\left(\mathbf{x}^{(i)} - \mu^{(k+1)}\right)^{\top} \tag{8.11}$$

Example 8.3 applies the cross-entropy method to a simple function. Figure 8.6 shows several iterations on a more complex function. Example 8.4 shows the potential limitation of using a multivariate normal distribution for fitting elite samples.

[12] R. Y. Rubinstein and D. P. Kroese, *The Cross-Entropy Method: A Unified Approach to Combinatorial Optimization, Monte-Carlo Simulation, and Machine Learning*. Springer, 2004.

[13] The name of this method comes from the fact that the process of fitting the distribution involves minimizing *cross-entropy*, which is also called the *Kullback–Leibler divergence*. Under certain conditions, minimizing the cross-entropy corresponds to finding the maximum likelihood estimate of the parameters of the distribution.

```
function adaptive_simulated_annealing(f, x, v, t, ε;
    ns=20, nε=4, nt=max(100,5length(x)),
    γ=0.85, c=fill(2,length(x)) )

    y = f(x)
    x_best, y_best = x, y
    y_arr, n, U = [], length(x), Uniform(-1.0,1.0)
    a,counts_cycles,counts_resets = zeros(n), 0, 0

    while true
        for i in 1:n
            x′ = x + basis(i,n)*rand(U)*v[i]
            y′ = f(x′)
            Δy = y′ - y
            if Δy < 0 || rand() < exp(-Δy/t)
                x, y = x′, y′
                a[i] += 1
                if y′ < y_best; x_best, y_best = x′, y′; end
            end
        end

        counts_cycles += 1
        counts_cycles ≥ ns || continue

        counts_cycles = 0
        corana_update!(v, a, c, ns)
        fill!(a, 0)
        counts_resets += 1
        counts_resets ≥ nt || continue

        t *= γ
        counts_resets = 0
        push!(y_arr, y)

        if !(length(y_arr) > nε && y_arr[end] - y_best ≤ ε &&
            all(abs(y_arr[end]-y_arr[end-u]) ≤ ε for u in 1:nε))
            x, y = x_best, y_best
        else
            break
        end
    end
    return x_best
end
```

Algorithm 8.6. The adaptive simulated annealing algorithm, where f is the multivariate objective function, x is the starting point, v is starting step vector, t is the starting temperature, and ε is the termination criterion parameter. The optional parameters are the number of cycles before running the step size adjustment ns, the number of cycles before reducing the temperature nt, the number of successive temperature reductions to test for termination nε, the temperature reduction coefficient γ, and the direction-wise varying criterion c.

Below is a flowchart for the adaptive simulated annealing algorithm as presented in the original paper.

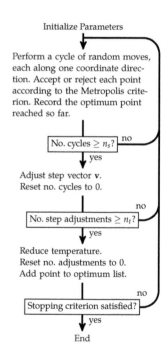

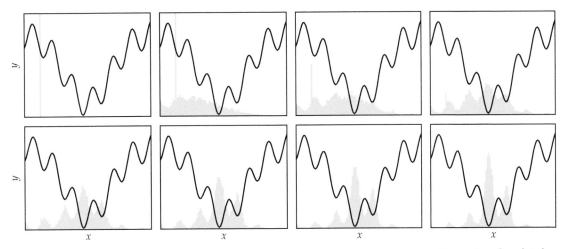

Figure 8.5. Simulated annealing with an exponentially decaying temperature, where the histograms indicate the probability of simulated annealing being at a particular position at that iteration.

```
using Distributions
function cross_entropy_method(f, P, k_max, m=100, m_elite=10)
    for k in 1 : k_max
        samples = rand(P, m)
        order = sortperm([f(samples[:,i]) for i in 1:m])
        P = fit(typeof(P), samples[:,order[1:m_elite]])
    end
    return P
end
```

Algorithm 8.7. The cross-entropy method, which takes an objective function f to be minimized, a proposal distribution P, an iteration count k_max, a sample size m, and the number of samples to use when refitting the distribution m_elite. It returns the updated distribution over where the global minimum is likely to exist.

We can use `Distributions.jl` to represent, sample from, and fit proposal distributions. The parameter vector θ is replaced by a distribution P. Calling `rand(P,m)` will produce an $n \times m$ matrix corresponding to m samples of n-dimensional samples from P, and calling `fit` will fit a new distribution of the given input type.

```
import Random: seed!
import LinearAlgebra: norm
seed!(0) # set random seed for reproducible results
f = x->norm(x)
μ = [0.5, 1.5]
Σ = [1.0 0.2; 0.2 2.0]
P = MvNormal(μ, Σ)
k_max = 10
P = cross_entropy_method(f, P, k_max)
@show P.μ
```

`P.μ = [-6.136226751273704e-7, -1.372160422208573e-6]`

Example 8.3. An example of using the cross-entropy method.

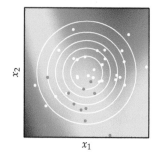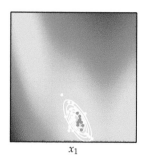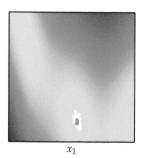

Figure 8.6.   The cross-entropy method with $m = 40$ applied to the Branin function (appendix B.3) using a multivariate Gaussian proposal distribution. The 10 elite samples in each iteration are in red.

The distribution family should be flexible enough to capture the relevant features of the objective function. Here we show the limitations of using a normal distribution on a multimodal objective function, which assigns greater density in between the two minima. A mixture model is able to center itself over each minimum.

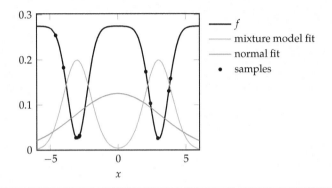

Example 8.4. The normal distribution is unable to capture multiple local minima, in contrast to mixture models which can maintain several.

8.5   Natural Evolution Strategies

Like the cross-entropy method, *natural evolution strategies*[14] optimize a proposal distribution parameterized by θ. We have to specify the proposal distribution family and the number of samples. The aim is to minimize the expectation $\mathbb{E}_{\mathbf{x} \sim p(\cdot | \theta)}[f(\mathbf{x})]$. Instead of fitting elite samples, evolution strategies apply gradient descent. The gradient is estimated from the samples:[15]

$$\nabla_\theta \, \mathbb{E}_{\mathbf{x} \sim p(\cdot | \theta)}[f(\mathbf{x})] = \int \nabla_\theta \, p(\mathbf{x} \mid \theta) f(\mathbf{x}) \, d\mathbf{x} \tag{8.12}$$

$$= \int \frac{p(\mathbf{x} \mid \theta)}{p(\mathbf{x} \mid \theta)} \nabla_\theta \, p(\mathbf{x} \mid \theta) f(\mathbf{x}) \, d\mathbf{x} \tag{8.13}$$

$$= \int p(\mathbf{x} \mid \theta) \nabla_\theta \log p(\mathbf{x} \mid \theta) f(\mathbf{x}) \, d\mathbf{x} \tag{8.14}$$

$$= \mathbb{E}_{\mathbf{x} \sim p(\cdot | \theta)}[f(\mathbf{x}) \nabla_\theta \log p(\mathbf{x} \mid \theta)] \tag{8.15}$$

$$\approx \frac{1}{m} \sum_{i=1}^{m} f(\mathbf{x}^{(i)}) \nabla_\theta \log p(\mathbf{x}^{(i)} \mid \theta) \tag{8.16}$$

Although we do not need the gradient of the objective function, we do need the gradient of the log likelihood, $\log p(\mathbf{x} \mid \theta)$. Example 8.5 shows how to compute

[14] I. Rechenberg, *Evolutionsstrategie Optimierung technischer Systeme nach Prinzipien der biologischen Evolution*. Frommann-Holzboog, 1973.

[15] This gradient estimation has recently been successfully applied to proposal distributions represented by deep neural networks. T. Salimans, J. Ho, X. Chen, and I. Sutskever, "Evolution Strategies as a Scalable Alternative to Reinforcement Learning," *ArXiv*, no. 1703.03864, 2017.

the gradient of the log likelihood for the multivariate normal distribution. The estimated gradient can be used along with any of the descent methods discussed in previous chapters to improve θ. Algorithm 8.8 uses gradient descent with a fixed step size. Figure 8.7 shows a few iterations of the algorithm.

```
using Distributions
function natural_evolution_strategies(f, θ, k_max; m=100, α=0.01)
    for k in 1 : k_max
        samples = [rand(θ) for i in 1 : m]
        θ -= α*sum(f(x)*∇logp(x, θ) for x in samples)/m
    end
    return θ
end
```

Algorithm 8.8. The natural evolution strategies method, which takes an objective function f to be minimized, an initial distribution parameter vector θ, an iteration count k_max, a sample size m, and a step factor α. An optimized parameter vector is returned. The method rand(θ) should sample from the distribution parameterized by θ, and ∇logp(x, θ) should return the log likelihood gradient.

8.6   Covariance Matrix Adaptation

Another popular method is *covariance matrix adaptation*,[16] which is also referred to as *CMA-ES* for *covariance matrix adaptation evolutionary strategy*. It has similarities with natural evolution strategies from section 8.5, but the two should not be confused. This method maintains a covariance matrix and is robust and sample efficient. Like the cross-entropy method and natural evolution strategies, a distribution is improved over time based on samples. Covariance matrix adaptation uses multivariate Gaussian distributions.[17]

Covariance matrix adaptation maintains a mean vector μ, a covariance matrix Σ, and an additional step-size scalar σ. The covariance matrix only increases or decreases in a single direction with every iteration, whereas the step-size scalar is adapted to control the overall spread of the distribution. At every iteration, m designs are sampled from the multivariate Gaussian:[18]

$$\mathbf{x} \sim \mathcal{N}(\boldsymbol{\mu}, \sigma^2 \boldsymbol{\Sigma}) \tag{8.17}$$

The designs are then sorted according to their objective function values such that $f(\mathbf{x}^{(1)}) \leq f(\mathbf{x}^{(2)}) \leq \cdots \leq f(\mathbf{x}^{(m)})$. A new mean vector $\boldsymbol{\mu}^{(k+1)}$ is formed using a weighted average of the first m_{elite} sampled designs:

$$\boldsymbol{\mu}^{(k+1)} \leftarrow \sum_{i=1}^{m_{\text{elite}}} w_i \mathbf{x}^{(i)} \tag{8.18}$$

[16] It is common to use the phrase evolution strategies to refer specifically to covariance matrix adaptation.

[17] N. Hansen, "The CMA Evolution Strategy: A Tutorial," *ArXiv*, no. 1604.00772, 2016.

[18] For optimization in $\mathbb{R}^n$, it is recommended to use at least $m = 4 + \lfloor 3 \ln n \rfloor$ samples per iteration, and $m_{\text{elite}} = \lfloor m/2 \rfloor$ elite samples.

The multivariate normal distribution $\mathcal{N}(\boldsymbol{\mu}, \boldsymbol{\Sigma})$ with mean $\boldsymbol{\mu}$ and covariance $\boldsymbol{\Sigma}$ is a popular distribution family due to having analytic solutions. The likelihood in d dimensions has the form

$$p(\mathbf{x} \mid \boldsymbol{\mu}, \boldsymbol{\Sigma}) = (2\pi)^{-\frac{d}{2}} |\boldsymbol{\Sigma}|^{-\frac{1}{2}} \exp\left(-\frac{1}{2}(\mathbf{x} - \boldsymbol{\mu})^{\top} \boldsymbol{\Sigma}^{-1} (\mathbf{x} - \boldsymbol{\mu})\right)$$

where $|\boldsymbol{\Sigma}|$ is the determinant of $\boldsymbol{\Sigma}$. The log likelihood is

$$\log p(\mathbf{x} \mid \boldsymbol{\mu}, \boldsymbol{\Sigma}) = -\frac{d}{2} \log(2\pi) - \frac{1}{2} \log |\boldsymbol{\Sigma}| - \frac{1}{2}(\mathbf{x} - \boldsymbol{\mu})^{\top} \boldsymbol{\Sigma}^{-1} (\mathbf{x} - \boldsymbol{\mu})$$

The parameters can be updated using their log likelihood gradients:

$$\nabla_{(\boldsymbol{\mu})} \log p(\mathbf{x} \mid \boldsymbol{\mu}, \boldsymbol{\Sigma}) = \boldsymbol{\Sigma}^{-1}(\mathbf{x} - \boldsymbol{\mu})$$

$$\nabla_{(\boldsymbol{\Sigma})} \log p(\mathbf{x} \mid \boldsymbol{\mu}, \boldsymbol{\Sigma}) = \frac{1}{2} \boldsymbol{\Sigma}^{-1} (\mathbf{x} - \boldsymbol{\mu})(\mathbf{x} - \boldsymbol{\mu})^{\top} \boldsymbol{\Sigma}^{-1} - \frac{1}{2} \boldsymbol{\Sigma}^{-1}$$

The term $\nabla_{(\boldsymbol{\Sigma})}$ contains the partial derivative of each entry of $\boldsymbol{\Sigma}$ with respect to the log likelihood.

Directly updating $\boldsymbol{\Sigma}$ may not result in a positive definite matrix, as is required for covariance matrices. One solution is to represent $\boldsymbol{\Sigma}$ as a product $\mathbf{A}^{\top}\mathbf{A}$, which guarantees that $\boldsymbol{\Sigma}$ remains positive semidefinite, and then update $\mathbf{A}$ instead. Replacing $\boldsymbol{\Sigma}$ by $\mathbf{A}^{\top}\mathbf{A}$ and taking the gradient with respect to $\mathbf{A}$ yields:

$$\nabla_{(\mathbf{A})} \log p(\mathbf{x} \mid \boldsymbol{\mu}, \mathbf{A}) = \mathbf{A}\left[\nabla_{(\boldsymbol{\Sigma})} \log p(\mathbf{x} \mid \boldsymbol{\mu}, \boldsymbol{\Sigma}) + \nabla_{(\boldsymbol{\Sigma})} \log p(\mathbf{x} \mid \boldsymbol{\mu}, \boldsymbol{\Sigma})^{\top}\right]$$

Example 8.5. A derivation of the log likelihood gradient equations for the multivariate Gaussian distribution. For the original derivation and several more sophisticated solutions for handling the positive definite covariance matrix, see D. Wierstra, T. Schaul, T. Glasmachers, Y. Sun, and J. Schmidhuber, "Natural Evolution Strategies," *ArXiv*, no. 1106.4487, 2011.

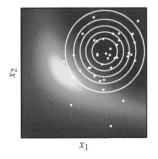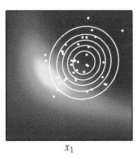

Figure 8.7. Natural evolution strategies using multivariate Gaussian distributions applied to Wheeler's Ridge, appendix B.7.

where the first m_{elite} weights sum to 1, and all the weights approximately sum to 0 and are ordered largest to smallest:[19]

$$\sum_{i=1}^{m_{\text{elite}}} w_i = 1 \qquad \sum_{i=1}^{m} w_i \approx 0 \qquad w_1 \geq w_2 \geq \cdots \geq w_m \qquad (8.19)$$

We can approximate the mean update in the cross-entropy method by setting the first m_{elite} weights to $1/m_{\text{elite}}$, and setting the remaining weights to zero. Covariance matrix adaptation instead distributes decreasing weight to all m designs, including some negative weights. The recommended weighting is obtained by normalizing

$$w_i' = \ln \frac{m+1}{2} - \ln i \ \text{ for } \ i \in \{1, \ldots, m\} \qquad (8.20)$$

to obtain $\mathbf{w}$. The positive and negative weights are normalized separately. Figure 8.8 compares the mean updates for covariance matrix adaptation and the cross-entropy method.

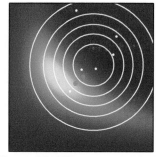

Figure 8.8. Shown is an initial proposal distribution (white contours), six samples (white dots), and the new updated means for both covariance matrix adaptation (blue dot) and the cross-entropy method (red dot) using three elite samples. Covariance matrix adaptation tends to update the mean more aggressively than the cross-entropy method (red dot), as it assigns higher weight to better sampled designs, and negative weight to worse sampled designs.

The step size is updated using a cumulative variable $\mathbf{p}_\sigma$ that tracks steps over time:

$$\mathbf{p}_\sigma^{(1)} = \mathbf{0}$$
$$\mathbf{p}_\sigma^{(k+1)} \leftarrow (1 - c_\sigma)\mathbf{p}_\sigma + \sqrt{c_\sigma(2 - c_\sigma)\mu_{\text{eff}}}(\mathbf{\Sigma}^{(k)})^{-1/2}\boldsymbol{\delta}_w \qquad (8.21)$$

where $c_\sigma < 1$ controls the rate of decay and the right hand term determines whether the step size should be increased or decreased based on the observed samples with respect to the present scale of the distribution. The variance effective selection mass μ_{eff} has the form

$$\mu_{\text{eff}} = \frac{1}{\sum_{i=1}^{m_{\text{elite}}} w_i^2} \qquad (8.22)$$

and $\boldsymbol{\delta}_w$ is computed from the sampled deviations:

$$\boldsymbol{\delta}_w = \sum_{i=1}^{m_{\text{elite}}} w_i \boldsymbol{\delta}^{(i)} \ \text{ for } \ \boldsymbol{\delta}^{(i)} = \frac{\mathbf{x}^{(i)} - \boldsymbol{\mu}^{(k)}}{\sigma^{(k)}} \qquad (8.23)$$

The new step size is obtained according to

$$\sigma^{(k+1)} \leftarrow \sigma^{(k)} \exp\left(\frac{c_\sigma}{d_\sigma}\left(\frac{\|\mathbf{p}_\sigma\|}{\mathbb{E}\|\mathcal{N}(\mathbf{0}, \mathbf{I})\|} - 1\right)\right) \qquad (8.24)$$

where

$$\mathbb{E}\|\mathcal{N}(\mathbf{0}, \mathbf{I})\| = \sqrt{2}\frac{\Gamma\left(\frac{n+1}{2}\right)}{\Gamma\left(\frac{n}{2}\right)} \approx \sqrt{n}\left(1 - \frac{1}{4n} + \frac{1}{21n^2}\right) \qquad (8.25)$$

is the expected length of a vector drawn from a Gaussian distribution. Comparing the length of $\mathbf{p}_\sigma$ to its expected length under random selection provides the mechanism by which σ is increased or decreased. The constants c_σ and d_σ have recommended values:

$$c_\sigma = (\mu_{\text{eff}} + 2)/(n + \mu_{\text{eff}} + 5)$$

$$d_\sigma = 1 + 2\max\left(0, \sqrt{(\mu_{\text{eff}} - 1)/(n+1)} - 1\right) + c_\sigma \tag{8.26}$$

The covariance matrix is also updated using a cumulative vector:

$$\mathbf{p}_\Sigma^{(1)} = \mathbf{0}$$

$$\mathbf{p}_\Sigma^{(k+1)} \leftarrow (1 - c_\Sigma)\mathbf{p}_\Sigma^{(k)} + h_\sigma \sqrt{c_\Sigma(2 - c_\Sigma)\mu_{\text{eff}}}\,\delta_w \tag{8.27}$$

where

$$h_\sigma = \begin{cases} 1 & \text{if } \dfrac{\|\mathbf{p}_\Sigma\|}{\sqrt{1-(1-c_\sigma)^{2(k+1)}}} < \left(1.4 + \frac{2}{n+1}\right) \mathbb{E}\|\mathcal{N}(\mathbf{0}, \mathbf{I})\| \\ 0 & \text{otherwise} \end{cases} \tag{8.28}$$

The h_σ stalls the update of $\mathbf{p}_\Sigma$ if $\|\mathbf{p}_\Sigma\|$ is too large, thereby preventing excessive increases in Σ when the step size is too small.

The update requires the adjusted weights $\mathbf{w}^\circ$:

$$w_i^\circ = \begin{cases} w_i & \text{if } w_i \geq 0 \\ \dfrac{nw_i}{\left\|\Sigma^{-1/2}\delta^{(i)}\right\|^2} & \text{otherwise} \end{cases} \tag{8.29}$$

The covariance update is then

$$\Sigma^{(k+1)} \leftarrow \left(1 + \underbrace{c_1c_\Sigma(1 - h_\sigma)(2 - c_\Sigma) - c_1 - c_\mu}_{\text{typically zero}}\right)\Sigma^{(k)} + \underbrace{c_1\mathbf{p}_\Sigma\mathbf{p}_\Sigma^\top}_{\text{rank-one update}} + \underbrace{c_\mu \sum_{i=1}^{\mu} w_i^\circ \delta^{(i)}\left(\delta^{(i)}\right)^\top}_{\text{rank-}\mu\text{ update}} \tag{8.30}$$

The constants c_Σ, c_1 and c_μ have recommended values

$$c_\Sigma = \frac{4 + \mu_{\text{eff}}/n}{n + 4 + 2\mu_{\text{eff}}/n}$$

$$c_1 = \frac{2}{(n + 1.3)^2 + \mu_{\text{eff}}} \tag{8.31}$$

$$c_\mu = \min\left(1 - c_1, 2\frac{\mu_{\text{eff}} - 2 + 1/\mu_{\text{eff}}}{(n + 2)^2 + \mu_{\text{eff}}}\right)$$

The covariance update consists of three components: the previous covariance matrix $\mathbf{\Sigma}^{(k)}$, a rank-one update, and a rank-μ update. The rank-one update gets its name from the fact that $\mathbf{p}_\Sigma \mathbf{p}_\Sigma^\top$ has rank one; it has only one eigenvector along $\mathbf{p}_\Sigma$. Rank-one updates using the cumulation vector allow for correlations between consecutive steps to be exploited, permitting the covariance matrix to elongate itself more quickly along a favorable axis.

The rank-μ update gets its name from the fact that $\sum_{i=1}^{\mu} w_i^\circ \boldsymbol{\delta}^{(i)} \left(\boldsymbol{\delta}^{(i)} \right)^\top$ has rank $\min(\mu, n)$. One important difference between the empirical covariance matrix update used by the cross-entropy method and the rank-μ update is that the former estimates the covariance about the new mean $\boldsymbol{\mu}^{(k+1)}$, whereas the latter estimates the covariance about the original mean $\boldsymbol{\mu}^{(k)}$. The $\boldsymbol{\delta}^{(i)}$ values thus help estimate the variances of the sampled steps rather than the variance within the sampled designs.

Covariance matrix adaptation is depicted in figure 8.9.

8.7   Summary

- Stochastic methods employ random numbers during the optimization process.

- Simulated annealing uses a temperature that controls random exploration and which is reduced over time to converge on a local minimum.

- The cross-entropy method and evolution strategies maintain proposal distributions from which they sample in order to inform updates.

- Natural evolution strategies uses gradient descent with respect to the log likelihood to update its proposal distribution.

- Covariance matrix adaptation is a robust and sample-efficient optimizer that maintains a multivariate Gaussian proposal distribution with a full covariance matrix.

8.8   Exercises

Exercise 8.1. We have shown that mixture proposal distributions can better capture multiple minima. Why might their use in the cross-entropy method be limited?

```
function covariance_matrix_adaptation(f, x, k_max;
    σ = 1.0,
    m = 4 + floor(Int, 3*log(length(x))),
    m_elite = div(m,2))

    μ, n = copy(x), length(x)
    ws = log((m+1)/2) .- log.(1:m)
    ws[1:m_elite] ./= sum(ws[1:m_elite])
    μ_eff = 1 / sum(ws[1:m_elite].^2)
    cσ = (μ_eff + 2)/(n + μ_eff + 5)
    dσ = 1 + 2max(0, sqrt((μ_eff-1)/(n+1))-1) + cσ
    cΣ = (4 + μ_eff/n)/(n + 4 + 2μ_eff/n)
    c1 = 2/((n+1.3)^2 + μ_eff)
    cμ = min(1-c1, 2*(μ_eff-2+1/μ_eff)/((n+2)^2 + μ_eff))
    ws[m_elite+1:end] .*= -(1 + c1/cμ)/sum(ws[m_elite+1:end])
    E = n^0.5*(1-1/(4n)+1/(21*n^2))
    pσ, pΣ, Σ = zeros(n), zeros(n), Matrix(1.0I, n, n)
    for k in 1 : k_max
        P = MvNormal(μ, σ^2*Σ)
        xs = [rand(P) for i in 1 : m]
        ys = [f(x) for x in xs]
        is = sortperm(ys) # best to worst

        # selection and mean update
        δs = [(x - μ)/σ for x in xs]
        δw = sum(ws[i]*δs[is[i]] for i in 1 : m_elite)
        μ += σ*δw

        # step-size control
        C = Σ^-0.5
        pσ = (1-cσ)*pσ + sqrt(cσ*(2-cσ)*μ_eff)*C*δw
        σ *= exp(cσ/dσ * (norm(pσ)/E - 1))

        # covariance adaptation
        hσ = Int(norm(pσ)/sqrt(1-(1-cσ)^(2k)) < (1.4+2/(n+1))*E)
        pΣ = (1-cΣ)*pΣ + hσ*sqrt(cΣ*(2-cΣ)*μ_eff)*δw
        w0 = [ws[i]≥0 ? ws[i] : n*ws[i]/norm(C*δs[is[i]])^2
                for i in 1:m]
        Σ = (1-c1-cμ) * Σ +
            c1*(pΣ*pΣ' + (1-hσ) * cΣ*(2-cΣ) * Σ) +
            cμ*sum(w0[i]*δs[is[i]]*δs[is[i]]' for i in 1 : m)
        Σ = triu(Σ)+triu(Σ,1)' # enforce symmetry
    end
    return μ
end
```

Algorithm 8.9. Covariance matrix adaptation, which takes an objective function f to be minimized, an initial design point x, and an iteration count k_max. One can optionally specify the step-size scalar σ, the sample size m, and the number of elite samples m_elite.

The best candidate design point is returned, which is the mean of the final sample distribution.

The covariance matrix undergoes an additional operation to ensure that it remains symmetric; otherwise small numerical inconsistencies can cause the matrix no longer to be positive definite.

This implementation uses a simplified normalization strategy for the negative weights. The original can be found in Equations 50–53 of N. Hansen, "The CMA Evolution Strategy: A Tutorial," *ArXiv*, no. 1604.00772, 2016.

Figure 8.9. Covariance matrix adaptation using multivariate Gaussian distributions applied to the flower function, appendix B.4.

Exercise 8.2. In the cross-entropy method, what is a potential effect of using an elite sample size that is very close to the total sample size?

Exercise 8.3. The log-likelihood of a value sampled from a Gaussian distribution with mean μ and variance v is:

$$\ell(x \mid \mu, v) = -\frac{1}{2} \ln 2\pi - \frac{1}{2} \ln v - \frac{(x - \mu)^2}{2v}$$

Show why evolution strategies using Gaussian distributions may encounter difficulties while applying a descent update on the variance when the mean is on the optimum, $\mu = x^*$.

Exercise 8.4. Derive the maximum likelihood estimate for the cross-entropy method using multivariate normal distributions:

$$\mu^{(k+1)} = \frac{1}{m} \sum_{i=1}^{m} \mathbf{x}^{(i)}$$

$$\Sigma^{(k+1)} = \frac{1}{m} \sum_{i=1}^{m} (\mathbf{x}^{(i)} - \mu^{(k+1)})(\mathbf{x}^{(i)} - \mu^{(k+1)})^\top$$

where the maximum likelihood estimates are the parameter values that maximize the likelihood of sampling the individuals $\left\{ \mathbf{x}^{(1)}, \cdots, \mathbf{x}^{(m)} \right\}$.

9 Population Methods

Previous chapters have focused on methods where a single design point is moved incrementally toward a minimum. This chapter presents a variety of *population methods* that involve optimization using a collection of design points, called *individuals*. Having a large number of individuals distributed throughout the design space can help the algorithm avoid becoming stuck in a local minimum. Information at different points in the design space can be shared between individuals to globally optimize the objective function. Most population methods are stochastic in nature, and it is generally easy to parallelize the computation.

9.1 Initialization

Population methods begin with an *initial population*, just as descent methods require an initial design point. The initial population should be spread over the design space to increase the chances that the samples are close to the best regions. This section presents several initialization methods, but more advanced sampling methods are discussed in detail in chapter 13.

We can often constrain the design variables to a region of interest consisting of a hyperrectangle defined by lower and upper bounds $\mathbf{a}$ and $\mathbf{b}$. Initial populations can be sampled from a uniform distribution for each coordinate:[1]

$$x_i^{(j)} \sim U(a_i, b_i) \tag{9.1}$$

where $\mathbf{x}^{(j)}$ is the jth individual in the population as seen in algorithm 9.1.

Another common approach is to use a multivariate normal distribution centered over a region of interest. The covariance matrix is typically diagonal, with diagonal entries scaled to adequately cover the search space. Algorithm 9.2 provides an implementation.

[1] Some population methods require additional information to be associated with the individual, such as velocity in the case of particle swarm optimization, discussed later. Velocity is often initialized according to a uniform or normal distribution.

```
function rand_population_uniform(m, a, b)
    d = length(a)
    return [a+rand(d).*(b-a) for i in 1:m]
end
```

Algorithm 9.1. A method for sampling an initial population of m design points over a uniform hyperrectangle with lower-bound vector a and upper-bound vector b.

```
using Distributions
function rand_population_normal(m, μ, Σ)
    D = MvNormal(μ,Σ)
    return [rand(D) for i in 1:m]
end
```

Algorithm 9.2. A method for sampling an initial population of m design points using a multivariate normal distribution with mean μ and covariance Σ.

Uniform and normal distributions limit the covered design space to a concentrated region. The *Cauchy distribution* (figure 9.1) has an unbounded variance and can cover a much broader space. Algorithm 9.3 provides an implementation. Figure 9.2, on the next page, compares example initial populations generated using different methods.

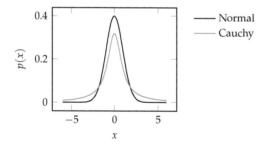

Figure 9.1. A comparison of the normal distribution with standard deviation 1 and the Cauchy distribution with scale 1. Although σ is sometimes used for the scale parameter in the Cauchy distribution, this should not be confused with the standard deviation since the standard deviation of the Cauchy distribution is undefined. The Cauchy distribution is heavy-tailed, allowing it to cover the design space more broadly.

9.2   Genetic Algorithms

Genetic algorithms (algorithm 9.4) borrow inspiration from biological evolution, where fitter individuals are more likely to pass on their genes to the next generation.[2] An individual's fitness for reproduction is inversely related to the value of the objective function at that point. The design point associated with an individual is represented as a *chromosome*. At each generation, the chromosomes of the fitter individuals are passed on to the next generation after undergoing the genetic operations of *crossover* and *mutation*.

[2] D. E. Goldberg, *Genetic Algorithms in Search, Optimization, and Machine Learning*. Addison-Wesley, 1989.

```
using Distributions
function rand_population_cauchy(m, μ, σ)
    n = length(μ)
    return [[rand(Cauchy(μ[j],σ[j])) for j in 1:n] for i in 1:m]
end
```

Algorithm 9.3. A method for sampling an initial population of m design points using a Cauchy distribution with location μ and scale σ for each dimension. The location and scale are analogous to the mean and standard deviation used in a normal distribution.

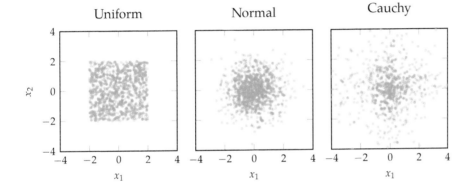

Figure 9.2. Initial populations of size 1,000 sampled using a uniform hyperrectangle with $\mathbf{a} = [-2, -2]$, $\mathbf{b} = [2, 2]$, a zero-mean normal distribution with diagonal covariance $\mathbf{\Sigma} = \mathbf{I}$, and Cauchy distributions centered at the origin with scale $\sigma = 1$.

```
function genetic_algorithm(f, population, k_max, S, C, M)
    for k in 1 : k_max
        parents = select(S, f.(population))
        children = [crossover(C,population[p[1]],population[p[2]])
                    for p in parents]
        population .= mutate.(Ref(M), children)
    end
    population[argmin(f.(population))]
end
```

Algorithm 9.4. The genetic algorithm, which takes an objective function f, an initial population, number of iterations k_max, a SelectionMethod S, a CrossoverMethod C, and a MutationMethod M.

9.2.1   Chromosomes

There are several ways to represent chromosomes. The simplest is the *binary string chromosome*, a representation that is similar to the way DNA is encoded.[3] A random binary string of length d can be generated using `bitrand(d)`. A binary string chromosome is depicted in figure 9.3.

[3] Instead of a binary representation, DNA contains four nucleobases: adenine, thymine, cytosine, and guanine, which are often abbreviated A, T, C, and G.

●●

Figure 9.3. A chromosome represented as a binary string.

Binary strings are often used due to the ease of expressing crossover and mutation. Unfortunately, the process of decoding a binary string and producing a design point is not always straightforward. Sometimes the binary string might not represent a valid point in the design space. It is often more natural to represent a chromosome using a list of real values. Such *real-valued chromosomes* are vectors in $\mathbb{R}^d$ that directly correspond to points in the design space.

9.2.2   Initialization

Genetic algorithms start with a random initial population. Binary string chromosomes are typically initialized using random bit strings as seen in algorithm 9.5. Real-valued chromosomes are typically initialized using the methods from the previous section.

```
rand_population_binary(m, n) = [bitrand(n) for i in 1:m]
```

Algorithm 9.5. A method for sampling random starting populations of m bit-string chromosomes of length n.

9.2.3   Selection

Selection is the process of choosing chromosomes to use as parents for the next generation. For a population with m chromosomes, a selection method will produce a list of m parental pairs[4] for the m children of the next generation. The selected pairs may contain duplicates.

[4] In some cases, one might use groups, should one wish to combine more than two parents to form a child.

There are several approaches for biasing the selection toward the fittest (algorithm 9.6). In *truncation selection* (figure 9.4), we choose parents from among the best k chromosomes in the population. In *tournament selection* (figure 9.5), each parent is the fittest out of k randomly chosen chromosomes of the population. In *roulette wheel selection* (figure 9.6), also known as *fitness proportionate selection*, each parent is chosen with a probability proportional to its performance relative to the population. Since we are interested in minimizing an objective function f, the fitness of the ith individual $x^{(i)}$ is inversely related to $y^{(i)} = f(x^{(i)})$. There are different ways to transform a collection $y^{(1)}, \ldots, y^{(m)}$ into fitnesses. A simple approach is to assign the fitness of individual i according to $\max\{y^{(1)}, \ldots, y^{(m)}\} - y^{(i)}$.

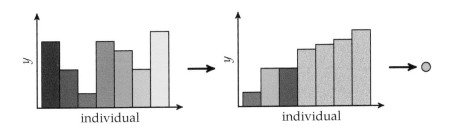

Figure 9.4. Truncation selection with a population size $m = 7$ and sample size $k = 3$. The height of a bar indicates its objective function value whereas its color indicates what individual it corresponds to.

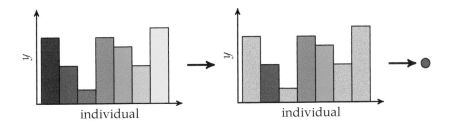

Figure 9.5. Tournament selection with a population size $m = 7$ and a sample size $k = 3$, which is run separately for each parent. The height of a bar indicates its objective function value whereas its color indicates what individual it corresponds to.

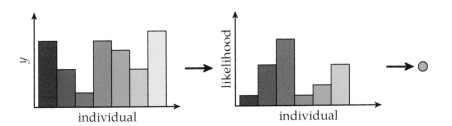

Figure 9.6. Roulette wheel selection with a population size $m = 7$, which is run separately for each parent. The approach used causes the individual with the worst objective function value to have a zero likelihood of being selected. The height of a bar indicates its objective function value (left), or its likelihood (right), whereas its color indicates what individual it corresponds to.

```
abstract type SelectionMethod end

struct TruncationSelection <: SelectionMethod
    k # top k to keep
end
function select(t::TruncationSelection, y)
    p = sortperm(y)
    return [p[rand(1:t.k, 2)] for i in y]
end


struct TournamentSelection <: SelectionMethod
    k
end
function select(t::TournamentSelection, y)
    getparent() = begin
        p = randperm(length(y))
        p[argmin(y[p[1:t.k]])]
    end
    return [[getparent(), getparent()] for i in y]
end


struct RouletteWheelSelection <: SelectionMethod end
function select(::RouletteWheelSelection, y)
    y = maximum(y) .- y
    cat = Categorical(normalize(y, 1))
    return [rand(cat, 2) for i in y]
end
```

Algorithm 9.6. Several selection methods for genetic algorithms. Calling selection with a SelectionMethod and the list of objective function values f will produce a list of parental pairs.

9.2.4 Crossover

Crossover combines the chromosomes of parents to form children. As with selection, there are several crossover schemes (algorithm 9.7).

- In *single-point crossover* (figure 9.7), the first portion of parent A's chromosome forms the first portion of the child chromosome, and the latter portion of parent B's chromosome forms the latter part of the child chromosome. The *crossover point* where the transition occurs is determined uniformly at random.

Figure 9.7. Single-point crossover.

- In *two-point crossover* (figure 9.8), we use two random crossover points.

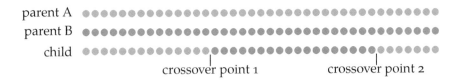

Figure 9.8. Two-point crossover.

- In *uniform crossover* (figure 9.9), each bit has a fifty percent chance of coming from either one of the two parents. This scheme is equivalent to each point having a fifty percent chance of being a crossover point.

Figure 9.9. Uniform crossover.

The previous crossover methods also work for real-valued chromosomes. We can, however, define an additional crossover routine that interpolates between real values (algorithm 9.8). Here, the real values are linearly interpolated between the parents' values $\mathbf{x}_a$ and $\mathbf{x}_b$:

$$\mathbf{x} \leftarrow (1 - \lambda)\mathbf{x}_a + \lambda\mathbf{x}_b \tag{9.2}$$

where λ is a scalar parameter typically set to one-half.

```
abstract type CrossoverMethod end
struct SinglePointCrossover <: CrossoverMethod end
function crossover(::SinglePointCrossover, a, b)
    i = rand(1:length(a))
    return vcat(a[1:i], b[i+1:end])
end

struct TwoPointCrossover <: CrossoverMethod end
function crossover(::TwoPointCrossover, a, b)
    n = length(a)
    i, j = rand(1:n, 2)
    if i > j
        (i,j) = (j,i)
    end
    return vcat(a[1:i], b[i+1:j], a[j+1:n])
end

struct UniformCrossover <: CrossoverMethod end
function crossover(::UniformCrossover, a, b)
    child = copy(a)
    for i in 1 : length(a)
        if rand() < 0.5
            child[i] = b[i]
        end
    end
    return child
end
```

Algorithm 9.7. Several crossover methods for genetic algorithms. Calling crossover with a `CrossoverMethod` and two parents a and b will produce a child chromosome that contains a mixture of the parents' genetic codes. These methods work for both binary string and real-valued chromosomes.

```
struct InterpolationCrossover <: CrossoverMethod
    λ
end
crossover(C::InterpolationCrossover, a, b) = (1-C.λ)*a + C.λ*b
```

Algorithm 9.8. A crossover method for real-valued chromosomes which performs linear interpolation between the parents.

9.2.5 Mutation

If new chromosomes were produced only through crossover, many traits that were not present in the initial random population could never occur, and the most-fit genes could saturate the population. Mutation allows new traits to spontaneously appear, allowing the genetic algorithm to explore more of the state space. Child chromosomes undergo mutation after crossover.

Each bit in a binary-valued chromosome typically has a small probability of being flipped (figure 9.10). For a chromosome with m bits, this *mutation rate* is typically set to $1/m$, yielding an average of one mutation per child chromosome. Mutation for real-valued chromosomes can be implemented using bitwise flips, but it is more common to add zero-mean Gaussian noise. Algorithm 9.9 provides implementations.

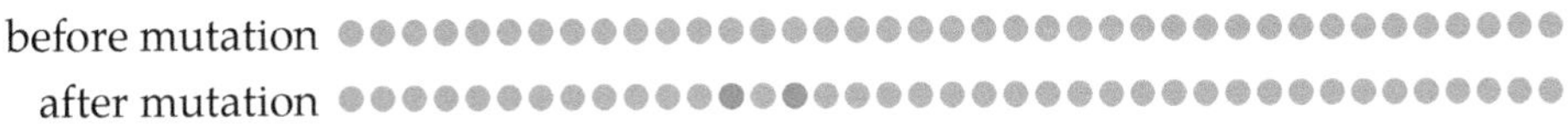

Figure 9.10. Mutation for binary string chromosomes gives each bit a small probability of flipping.

```
abstract type MutationMethod end
struct BitwiseMutation <: MutationMethod
    λ
end
function mutate(M::BitwiseMutation, child)
    return [rand() < M.λ ? !v : v for v in child]
end

struct GaussianMutation <: MutationMethod
    σ
end
function mutate(M::GaussianMutation, child)
    return child + randn(length(child))*M.σ
end
```

Algorithm 9.9. The bitwise mutation method for binary string chromosomes and the Gaussian mutation method for real-valued chromosomes. Here, λ is the mutation rate, and σ is the standard deviation.

Figure 9.11 illustrates several generations of a genetic algorithm. Example 9.1 shows how to combine selection, crossover, and mutation strategies discussed in this section.

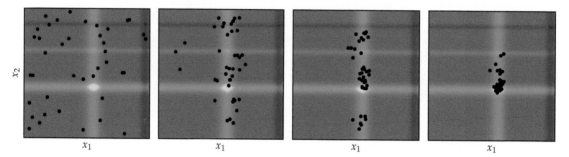

Figure 9.11. A genetic algorithm with truncation selection, single point crossover, and Gaussian mutation with $\sigma = 0.1$ applied to the Michalewicz function defined in appendix B.5.

We will demonstrate using genetic algorithms to optimize a simple function.

```julia
import Random: seed!
import LinearAlgebra: norm
seed!(0) # set random seed for reproducible results
f = x->norm(x)
m = 100 # population size
k_max = 10 # number of iterations
population = rand_population_uniform(m, [-3, 3], [3,3])
S = TruncationSelection(10) # select top 10
C = SinglePointCrossover()
M = GaussianMutation(0.5) # small mutation rate
x = genetic_algorithm(f, population, k_max, S, C, M)
@show x

x = [-0.03773045013773703, -0.03143381815543277]
```

Example 9.1. Demonstration of using a genetic algorithm for optimizing a simple function.

9.3   Differential Evolution

Differential evolution (algorithm 9.10) attempts to improve each individual in the population by recombining other individuals in the population according to a simple formula.[5] It is parameterized by a crossover probability p and a differential weight w. Typically, w is between 0.4 and 1. For each individual $\mathbf{x}$:

[5] S. Das and P. N. Suganthan, "Differential Evolution: A Survey of the State-of-the-Art," *IEEE Transactions on Evolutionary Computation*, vol. 15, no. 1, pp. 4–31, 2011.

1. Choose three random distinct individuals $\mathbf{a}$, $\mathbf{b}$, and $\mathbf{c}$.

2. Construct an interim design $\mathbf{z} = \mathbf{a} + w \cdot (\mathbf{b} - \mathbf{c})$ as shown in figure 9.12.

3. Choose a random dimension $j \in [1, \ldots, n]$ for optimization in n dimensions.

4. Construct the candidate individual $\mathbf{x}'$ using binary crossover.

$$x_i' = \begin{cases} z_i & \text{if } i = j \text{ or with probability } p \\ x_i & \text{otherwise} \end{cases} \qquad (9.3)$$

5. Insert the better design between $\mathbf{x}$ and $\mathbf{x}'$ into the next generation.

The algorithm is demonstrated in figure 9.13.

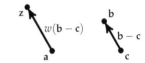

Figure 9.12. Differential evolution takes three individuals $\mathbf{a}$, $\mathbf{b}$, and $\mathbf{c}$ and combines them to form the candidate individual $\mathbf{z}$.

```julia
using StatsBase
function differential_evolution(f, population, k_max; p=0.5, w=1)
    n, m = length(population[1]), length(population)
    for k in 1 : k_max
        for (k,x) in enumerate(population)
            a, b, c = sample(population,
                Weights([j!=k for j in 1:m]), 3, replace=false)
            z = a + w*(b-c)
            j = rand(1:n)
            x′ = [i == j || rand() < p ? z[i] : x[i] for i in 1:n]
            if f(x′) < f(x)
                x[:] = x′
            end
        end
    end
    return population[argmin(f.(population))]
end
```

Algorithm 9.10. Differential evolution, which takes an objective function f, a population population, a number of iterations k_max, a crossover probability p, and a differential weight w. The best individual is returned.

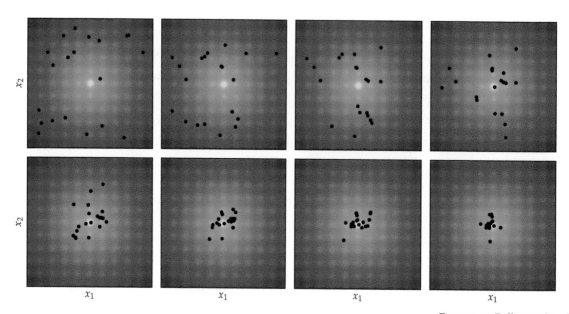

Figure 9.13. Differential evolution with $p = 0.5$ and $w = 0.2$ applied to Ackley's function, defined in appendix B.1.

9.4   Particle Swarm Optimization

Particle swarm optimization introduces momentum to accelerate convergence toward minima.[6] Each individual, or *particle*, in the population keeps track of its current position, velocity, and the best position it has seen so far (algorithm 9.11). Momentum allows an individual to accumulate speed in a favorable direction, independent of local perturbations.

[6] J. Kennedy, R. C. Eberhart, and Y. Shi, *Swarm Intelligence*. Morgan Kaufmann, 2001.

```
mutable struct Particle
    x
    v
    x_best
end
```

Algorithm 9.11. Each particle in particle swarm optimization has a position x and velocity v in design space and keeps track of the best design point found so far, x_best.

At each iteration, each individual is accelerated toward both the best position it has seen and the best position found thus far by any individual. The acceleration is weighted by a random term, with separate random numbers being generated for each acceleration. The update equations are:

$$x^{(i)} \leftarrow x^{(i)} + v^{(i)} \tag{9.4}$$

$$v^{(i)} \leftarrow wv^{(i)} + c_1 r_1 \left(x_{\text{best}}^{(i)} - x^{(i)} \right) + c_2 r_2 \left(x_{\text{best}} - x^{(i)} \right) \tag{9.5}$$

where $\mathbf{x}_{\text{best}}$ is the best location found so far over all particles; w, c_1, and c_2 are parameters; and r_1 and r_2 are random numbers drawn from $U(0,1)$.[7] Algorithm 9.12 provides an implementation. Figure 9.14 shows several iterations of the algorithm.

[7] A common strategy is to allow the inertia w to decay over time.

```
function particle_swarm_optimization(f, population, k_max;
    w=1, c1=1, c2=1)
    n = length(population[1].x)
    x_best, y_best = copy(population[1].x_best), Inf
    for P in population
        y = f(P.x)
        if y < y_best; x_best[:], y_best = P.x, y; end
    end
    for k in 1 : k_max
        for P in population
            r1, r2 = rand(n), rand(n)
            P.x += P.v
            P.v = w*P.v + c1*r1.*(P.x_best - P.x) +
                          c2*r2.*(x_best - P.x)
            y = f(P.x)
            if y < y_best; x_best[:], y_best = P.x, y; end
            if y < f(P.x_best); P.x_best[:] = P.x; end
        end
    end
    return population
end
```

Algorithm 9.12. Particle swarm optimization, which takes an objective function f, a list of particles population, a number of iterations k_max, an inertia w, and momentum coefficients c1 and c2. The default values are those used by R. Eberhart and J. Kennedy, "A New Optimizer Using Particle Swarm Theory," in *International Symposium on Micro Machine and Human Science*, 1995.

9.5 Firefly Algorithm

The *firefly algorithm* (algorithm 9.13) was inspired by the manner in which fireflies flash their lights to attract mates.[8] In the firefly algorithm, each individual in the population is a firefly and can flash to attract other fireflies. At each iteration, all fireflies are moved toward all more attractive fireflies. A firefly $\mathbf{a}$ is moved toward a firefly $\mathbf{b}$ with greater attraction according to

[8] X.-S. Yang, *Nature-Inspired Metaheuristic Algorithms*. Luniver Press, 2008. Interestingly, male fireflies flash to attract members of the opposite sex, but females sometimes flash to attract males of other species, which they then eat.

$$\mathbf{a} \leftarrow \mathbf{a} + \beta I(\|\mathbf{b} - \mathbf{a}\|)(\mathbf{b} - \mathbf{a}) + \alpha\boldsymbol{\epsilon} \tag{9.6}$$

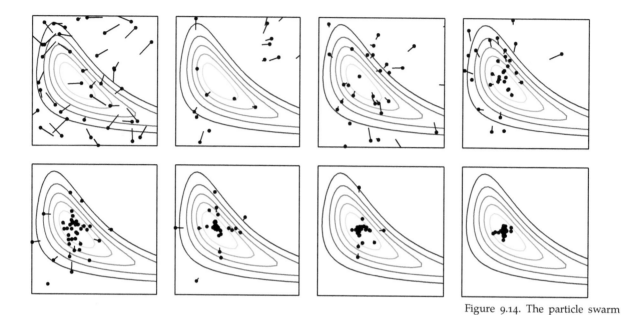

Figure 9.14. The particle swarm method with $w = 0.1$, $c_1 = 0.25$, and $c_2 = 2$ applied to Wheeler's Ridge, appendix B.7.

where I is the intensity of the attraction and β is the source intensity. A random walk component is included as well, where ϵ is drawn from a zero-mean, unit covariance multivariate Gaussian, and α scales the step size. The resulting update is a random walk biased toward brighter fireflies.[9]

The intensity I decreases as the distance r between the two fireflies increases and is defined to be 1 when $r = 0$. One approach is to model the intensity as a point source radiating into space, in which case the intensity decreases according to the inverse square law

$$I(r) = \frac{1}{r^2} \tag{9.7}$$

Alternatively, if a source is suspended in a medium that absorbs light, the intensity will decrease according to an exponential decay

$$I(r) = e^{-\gamma r} \tag{9.8}$$

where γ is a the light absorption coefficient.[10]

We generally want to avoid equation (9.7) in practice due to the singularity at $r = 0$. A combination of the inverse square law and absorption can be approximated with a Gaussian brightness drop-off:

$$I(r) = e^{-\gamma r^2} \tag{9.9}$$

[9] Yang recommends $\beta = 1$ and $\alpha \in [0, 1]$. If $\beta = 0$, the behavior is a random walk.

[10] The distance between fireflies ceases to matter as γ approaches zero.

A firefly's attraction is proportional to its performance. Attraction affects only whether one fly is attracted to another fly, whereas intensity affects how much the less attractive fly moves. Figure 9.15 shows a few iterations of the algorithm.

```
using Distributions
function firefly(f, population, k_max;
    β=1, α=0.1, brightness=r->exp(-r^2))

    m = length(population[1])
    N = MvNormal(Matrix(1.0I, m, m))
    for k in 1 : k_max
        for a in population, b in population
            if f(b) < f(a)
                r = norm(b-a)
                a[:] += β*brightness(r)*(b-a) + α*rand(N)
            end
        end
    end
    return population[argmin([f(x) for x in population])]
end
```

Algorithm 9.13. The firefly algorithm, which takes an objective function f, a population flies consisting of design points, a number of iterations k_max, a source intensity β, a random walk step size α, and an intensity function I. The best design point is returned.

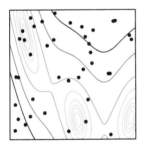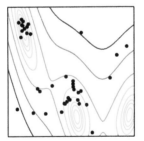

Figure 9.15. Firefly search with $\alpha = 0.5$, $\beta = 1$, and $\gamma = 0.1$ applied to the Branin function (appendix B.3).

9.6   Cuckoo Search

Cuckoo search (algorithm 9.14) is another nature-inspired algorithm named after the cuckoo bird, which engages in a form of brood parasitism.[11] Cuckoos lay their eggs in the nests of other birds, often birds of other species. When this occurs, the host bird may detect the invasive egg and then destroy it or establish a new

[11] X.-S. Yang and S. Deb, "Cuckoo Search via Lévy Flights," in *World Congress on Nature & Biologically Inspired Computing (NaBIC)*, 2009.

nest somewhere else. However, there is also a chance that the egg is accepted and raised by the host bird.[12]

In cuckoo search, each nest represents a design point. New design points can be produced using *Lévy flights* from nests, which are random walks with step-lengths from a heavy-tailed distribution. A new design point can replace a nest if it has a better objective function value, which is analogous to cuckoo eggs replacing the eggs of birds of other species.

The core rules are:

1. A cuckoo will lay an egg in a randomly chosen nest.

2. The best nests with the best eggs will survive to the next generation.

3. Cuckoo eggs have a chance of being discovered by the host bird, in which case the eggs are destroyed.

Cuckoo search relies on random flights to establish new nest locations. These flights start from an existing nest and then move randomly to a new location. While we might be tempted to use a uniform or Gaussian distribution for the walk, these restrict the search to a relatively concentrated region. Instead, cuckoo search uses a Cauchy distribution, which has a heavier tail. In addition, the Cauchy distribution has been shown to be more representative of the movements of other animals in the wild.[13] Figure 9.16 shows a few iterations of cuckoo search.

Other nature-inspired algorithms include the artificial bee colony, the gray wolf optimizer, the bat algorithm, glowworm swarm optimization, intelligent water drops, and harmony search.[14] There has been some criticism of the proliferation of methods that make analogies to nature without fundamentally contributing novel methods and understanding.[15]

9.7   Hybrid Methods

Many population methods perform well in global search, being able to avoid local minima and finding the best regions of the design space. Unfortunately, these methods do not perform as well in local search in comparison to descent methods. Several *hybrid methods*[16] have been developed to extend population methods with descent-based features to improve their performance in local search. There are two general approaches to combining population methods with local search techniques:[17]

[12] Interestingly, an instinct of newly hatched cuckoos is to knock other eggs or hatchlings (those belonging to the host bird) out of the nest.

[13] For example, a certain species of fruit fly explores its surroundings using Cauchy-like steps separated by 90° turns. A. M. Reynolds and M. A. Frye, "Free-Flight Odor Tracking in Drosophila is Consistent with an Optimal Intermittent Scale-Free Search," *PLoS ONE*, vol. 2, no. 4, e354, 2007.

[14] See, for example, D. Simon, *Evolutionary Optimization Algorithms.* Wiley, 2013.

[15] This viewpoint is expressed by K. Sörensen, "Metaheuristics—the Metaphor Exposed," *International Transactions in Operational Research*, vol. 22, no. 1, pp. 3–18, 2015.

[16] In the literature, these kinds of techniques are also referred to as *memetic algorithms* or *genetic local search*.

[17] K. W. C. Ku and M.-W. Mak, "Exploring the Effects of Lamarckian and Baldwinian Learning in Evolving Recurrent Neural Networks," in *IEEE Congress on Evolutionary Computation (CEC)*, 1997.

```julia
using Distributions
mutable struct Nest
    x # position
    y # value, f(x)
end
function cuckoo_search(f, population, k_max;
    p_a=0.1, C=Cauchy(0,1))
    m, n = length(population), length(population[1].x)
    a = round(Int, m*p_a)
    for k in 1 : k_max
        i, j = rand(1:m), rand(1:m)
        x = population[j].x + [rand(C) for k in 1 : n]
        y = f(x)
        if y < population[i].y
            population[i].x[:] = x
            population[i].y = y
        end

        p = sortperm(population, by=nest->nest.y, rev=true)
        for i in 1 : a
            j = rand(1:m-a)+a
            population[p[i]] = Nest(population[p[j]].x +
                            [rand(C) for k in 1 : n],
                            f(population[p[i]].x)
                            )
        end
    end
    return population
end
```

Algorithm 9.14. Cuckoo search, which takes an objective function f, an initial set of nests population, a number of iterations k_max, percent of nests to abandon p_a, and flight distribution C. The flight distribution is typically a centered Cauchy distribution.

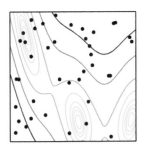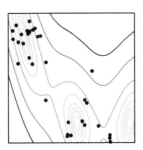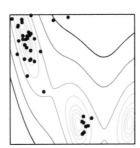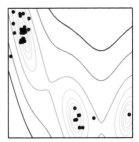

Figure 9.16. Cuckoo search applied to the Branin function (appendix B.3).

- In *Lamarckian learning*, the population method is extended with a local search method that locally improves each individual. The original individual and its objective function value are replaced by the individual's optimized counterpart and its objective function value.

- In *Baldwinian learning*, the same local search method is applied to each individual, but the results are used only to update the individual's perceived objective function value. Individuals are not replaced but are merely associated with optimized objective function values, which are not the same as their actual objective function value. Baldwinian learning can help prevent premature convergence.

The difference between these approaches is illustrated in example 9.2.

Consider optimizing $f(x) = -e^{-x^2} - 2e^{-(x-3)^2}$ using a population of individuals initialized near $x = 0$.

Example 9.2. A comparison of the Lamarckian and Baldwinian hybrid methods.

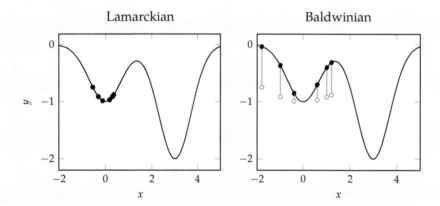

A Lamarckian local search update applied to this population would move the individuals toward the local minimum, reducing the chance that future individuals escape and find the global optimum near $x = 3$. A Baldwinian approach will compute the same update but leaves the original designs unchanged. The selection step will value each design according to its value from a local search.

9.8   Summary

- Population methods use a collection of individuals in the design space to guide progression toward an optimum.

- Genetic algorithms leverage selection, crossover, and mutations to produce better subsequent generations.

- Differential evolution, particle swarm optimization, the firefly algorithm, and cuckoo search include rules and mechanisms for attracting design points to the best individuals in the population while maintaining suitable state space exploration.

- Population methods can be extended with local search approaches to improve convergence.

9.9   Exercises

Exercise 9.1. What is the motivation behind the selection operation in genetic algorithms?

Exercise 9.2. Why does mutation play such a fundamental role in genetic algorithms? How would we choose the mutation rate if we suspect there is a better optimal solution?

Exercise 9.3. If we observe that particle swarm optimization results in fast convergence to a nonglobal minimum, how might we change the parameters of the algorithm?

10 Constraints

Previous chapters have focused on unconstrained problems where the domain of each design variable is the space of real numbers. Many problems are constrained, which forces design points to satisfy certain conditions. This chapter presents a variety of approaches for transforming problems with constraints into problems without constraints, thereby permitting the use of the optimization algorithms we have already discussed. Analytical methods are also discussed, including the concepts of duality and the necessary conditions for optimality under constrained optimization.

10.1 Constrained Optimization

Recall the core optimization problem equation (1.1):

$$\begin{aligned} \underset{\mathbf{x}}{\text{minimize}} \quad & f(\mathbf{x}) \\ \text{subject to} \quad & \mathbf{x} \in \mathcal{X} \end{aligned} \qquad (10.1)$$

In unconstrained problems, the feasible set $\mathcal{X}$ is $\mathbb{R}^n$. In constrained problems, the feasible set is some subset thereof.

Some constraints are simply upper or lower bounds on the design variables, as we have seen in bracketed line search, in which x must lie between a and b. A bracketing constraint $x \in [a, b]$ can be replaced by two inequality constraints: $a \le x$ and $x \le b$ as shown in figure 10.1. In multivariate problems, bracketing the input variables forces them to lie within a hyperrectangle as shown in figure 10.2.

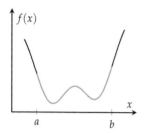

$$\underset{x}{\text{minimize}} \quad f(x)$$

$$\text{subject to} \quad x \in [a, b]$$

Figure 10.1. A simple optimization problem constrained by upper and lower bounds.

Constraints arise naturally when formulating real problems. A hedge fund manager cannot sell more stock than they have, an airplane cannot have wings with zero thickness, and the number of hours you spend per week on your homework cannot exceed 168. We include constraints in such problems to prevent the optimization algorithm from suggesting an infeasible solution.

Applying constraints to a problem can affect the solution, but this need not be the case as shown in figure 10.3.

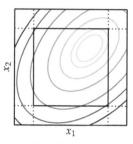

Figure 10.2. Bracketing constraints force the solution to lie within a hyperrectangle.

Unconstrained   Constrained, Same Solution   Constrained, New Solution

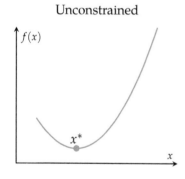 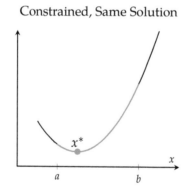 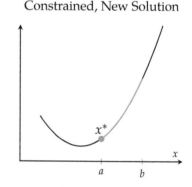

Figure 10.3. Constraints can change the solution to a problem, but do not have to.

10.2   Constraint Types

Constraints are not typically specified directly through a known feasible set $\mathcal{X}$. Instead, the feasible set is typically formed from two types of constraints:[1]

1. equality constraints, $h(\mathbf{x}) = 0$

2. inequality constraints, $g(\mathbf{x}) \le 0$

[1] We have g representing a less-than inequality constraint. Greater-than equality constraints can be translated into less-than inequality constraints by introducing a negative sign.

Any optimization problem can be rewritten using these constraints:

$$\underset{\mathbf{x}}{\text{minimize}} \quad f(\mathbf{x})$$

$$\text{subject to} \quad h_i(\mathbf{x}) = 0 \text{ for all } i \text{ in } \{1, \ldots, \ell\}$$

$$g_j(\mathbf{x}) \leq 0 \text{ for all } j \text{ in } \{1, \ldots, m\}$$

(10.2)

Of course, constraints can be constructed from a feasible set $\mathcal{X}$:

$$h(\mathbf{x}) = (\mathbf{x} \notin \mathcal{X}) \tag{10.3}$$

where Boolean expressions evaluate to 0 or 1.

We often use equality and inequality functions ($h(\mathbf{x}) = 0$, $g(\mathbf{x}) \leq 0$) to define constraints rather than set membership ($\mathbf{x} \in \mathcal{X}$) because the functions can provide information about how far a given point is from being feasible. This information helps drive solution methods toward feasibility.

Equality constraints are sometimes decomposed into two inequality constraints:

$$h(\mathbf{x}) = 0 \quad \Longleftrightarrow \quad \begin{cases} h(\mathbf{x}) \leq 0 \\ h(\mathbf{x}) \geq 0 \end{cases} \tag{10.4}$$

However, sometimes we want to handle equality constraints separately, as we will discuss later in this chapter.

10.3 Transformations to Remove Constraints

In some cases, it may be possible to transform a problem so that constraints can be removed. For example, bound constraints $a \leq x \leq b$ can be removed by passing x through a transform (figure 10.4):

$$x = t_{a,b}(\hat{x}) = \frac{b + a}{2} + \frac{b - a}{2} \left(\frac{2\hat{x}}{1 + \hat{x}^2} \right) \tag{10.5}$$

Example 10.1 demonstrates this process.

Some equality constraints can be used to solve for x_n given $x_1, \ldots, x_{n-1}$. In other words, if we know the first $n - 1$ components of $\mathbf{x}$, we can use the constraint equation to obtain x_n. In such cases, the optimization problem can be reformulated over $x_1, \ldots, x_{n-1}$ instead, removing the constraint and removing one design variable. Example 10.2 demonstrates this process.

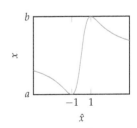

Figure 10.4. This transform ensures that x lies between a and b.

Consider the optimization problem

$$\underset{x}{\text{minimize}} \quad x\sin(x)$$

$$\text{subject to} \quad 2 \leq x \leq 6$$

Example 10.1. Removing bounds constraints using a transform on the input variable.

We can transform the problem to remove the constraints:

$$\underset{\hat{x}}{\text{minimize}} \quad t_{2,6}(\hat{x})\sin(t_{2,6}(\hat{x}))$$

$$\underset{\hat{x}}{\text{minimize}} \quad \left(4 + 2\left(\frac{2\hat{x}}{1+\hat{x}^2}\right)\right)\sin\left(4 + 2\left(\frac{2\hat{x}}{1+\hat{x}^2}\right)\right)$$

We can use the optimization method of our choice to solve the unconstrained problem. In doing so, we find two minima: $\hat{x} \approx 0.242$ and $\hat{x} \approx 4.139$, both of which have a function value of approximately -4.814.

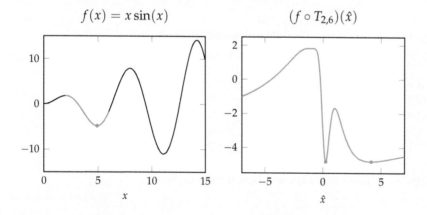

$$f(x) = x\sin(x) \qquad\qquad (f \circ T_{2,6})(\hat{x})$$

The solution for the original problem is obtained by passing $\hat{x}$ through the transform. Both values of $\hat{x}$ produce $x = t_{2,6}(\hat{x}) \approx 4.914$.

Example 10.2. Using constraint equations to eliminate design variables.

Consider the constraint:

$$h(\mathbf{x}) = c_1 x_1 + c_2 x_2 + \cdots + c_n x_n = 0$$

We can solve for x_n using the first $n-1$ variables:

$$x_n = -c_1 x_1 - c_2 x_2 - \cdots - c_{n-1} x_{n-1}$$

We can transform

$$\begin{aligned}
&\underset{\mathbf{x}}{\text{minimize}} && f(\mathbf{x}) \\
&\text{subject to} && h(\mathbf{x}) = 0
\end{aligned}$$

into

$$\underset{x_1,\ldots,x_{n-1}}{\text{minimize}} \quad f([x_1,\ldots,x_{n-1}, -c_1 x_1 - c_2 x_2 - \cdots - c_{n-1} x_{n-1}])$$

10.4 Lagrange Multipliers

The method of *Lagrange multipliers* is used to optimize a function subject to equality constraints. Consider an optimization problem with a single equality constraint:

$$\begin{aligned}
&\underset{\mathbf{x}}{\text{minimize}} && f(\mathbf{x}) \\
&\text{subject to} && h(\mathbf{x}) = 0
\end{aligned} \tag{10.6}$$

where f and h have continuous partial derivatives. Example 10.3 discusses such a problem.

The method of Lagrange multipliers is used to compute where a contour line of f is aligned with the contour line of $h(\mathbf{x}) = 0$. Since the gradient of a function at a point is perpendicular to the contour line of that function through that point, we know the gradient of h will be perpendicular to the contour line $h(\mathbf{x}) = 0$. Hence, we need to find where the gradient of f and the gradient of h are aligned.

We seek the best $\mathbf{x}$ such that the constraint

$$h(\mathbf{x}) = 0 \tag{10.7}$$

Consider the minimization problem:

$$\underset{\mathbf{x}}{\text{minimize}} \qquad -\exp\left(-\left(x_1 x_2 - \frac{3}{2}\right)^2 - \left(x_2 - \frac{3}{2}\right)^2\right)$$

$$\text{subject to} \qquad x_1 - x_2^2 = 0$$

Example 10.3. A motivating example of the method of Lagrange multipliers.

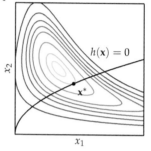

We substitute the constraint $x_1 = x_2^2$ into the objective function to obtain an unconstrained objective:

$$f_{\text{unc}} = -\exp\left(-\left(x_2^3 - \frac{3}{2}\right)^2 - \left(x_2 - \frac{3}{2}\right)^2\right)$$

whose derivative is:

$$\frac{\partial}{\partial x_2} f_{\text{unc}} = 6\exp\left(-\left(x_2^3 - \frac{3}{2}\right)^2 - \left(x_2 - \frac{3}{2}\right)^2\right)\left(x_2^5 - \frac{3}{2}x_2^2 + \frac{1}{3}x_2 - \frac{1}{2}\right)$$

Setting the derivative to zero and solving for x_2 yields $x_2 \approx 1.165$. The solution to the original optimization problem is thus $\mathbf{x}^* \approx [1.358, 1.165]$. The optimum lies where the contour line of f is aligned with h.

If the point x^* optimizes f along h, then its directional derivative at x^* along h must be zero. That is, small shifts of x^* along h cannot result in an improvement.

The contour lines of f are lines of constant f. Thus, if a contour line of f is tangent to h, then the directional derivative of h at that point, along the direction of the contour $h(\mathbf{x}) = 0$, must be zero.

is satisfied and the gradients are aligned

$$\nabla f(\mathbf{x}) = \lambda \nabla h(\mathbf{x}) \tag{10.8}$$

for some *Lagrange multiplier* λ. We need the scalar λ because the magnitudes of the gradients may not be the same.[2]

We can formulate the *Lagrangian*, which is a function of the design variables and the multiplier

$$\mathcal{L}(\mathbf{x}, \lambda) = f(\mathbf{x}) - \lambda h(\mathbf{x}) \tag{10.9}$$

Solving $\nabla \mathcal{L}(\mathbf{x}, \lambda) = \mathbf{0}$ solves equations (10.7) and (10.8). Specifically, $\nabla_{\mathbf{x}} \mathcal{L} = \mathbf{0}$ gives us the condition $\nabla f = \lambda \nabla h$, and $\nabla_\lambda \mathcal{L} = 0$ gives us $h(\mathbf{x}) = 0$. Any solution is considered a *critical point*. Critical points can be local minima, global minima, or saddle points.[3] Example 10.4 demonstrates this approach.

> [2] When ∇f is zero, the Lagrange multiplier λ equals zero, irrespective of ∇h.

> [3] The method of Lagrange multipliers gives us a first-order necessary condition to test for optimality. We will extend this method to include inequalities.

We can use the method of Lagrange multipliers to solve the problem in example 10.3. We form the Lagrangian

> Example 10.4. Using the method of Lagrange multipliers to solve the problem in example 10.3.

$$\mathcal{L}(x_1, x_2, \lambda) = -\exp\left(-\left(x_1 x_2 - \frac{3}{2}\right)^2 - \left(x_2 - \frac{3}{2}\right)^2\right) - \lambda(x_1 - x_2^2)$$

and compute the gradient

$$\frac{\partial \mathcal{L}}{\partial x_1} = 2x_2 f(\mathbf{x})\left(\frac{3}{2} - x_1 x_2\right) - \lambda$$

$$\frac{\partial \mathcal{L}}{\partial x_2} = 2\lambda x_2 + f(\mathbf{x})\left(-2x_1(x_1 x_2 - \frac{3}{2}) - 2(x_2 - \frac{3}{2})\right)$$

$$\frac{\partial \mathcal{L}}{\partial \lambda} = x_2^2 - x_1$$

Setting these derivatives to zero and solving yields $x_1 \approx 1.358$, $x_2 \approx 1.165$, and $\lambda \approx 0.170$.

The method of Lagrange multipliers can be extended to multiple equality constraints. Consider a problem with two equality constraints:

$$\begin{aligned}
\underset{\mathbf{x}}{\text{minimize}} \quad & f(\mathbf{x}) \\
\text{subject to} \quad & h_1(\mathbf{x}) = 0 \\
& h_2(\mathbf{x}) = 0
\end{aligned} \tag{10.10}$$

We can collapse these constraints into a single constraint. The new constraint is satisfied by exactly the same points as before, so the solution is unchanged.

$$\underset{\mathbf{x}}{\text{minimize}} \qquad f(\mathbf{x})$$
$$\text{subject to} \quad h_{\text{comb}}(\mathbf{x}) = h_1(\mathbf{x})^2 + h_2(\mathbf{x})^2 = 0 \tag{10.11}$$

We can now apply the method of Lagrange multipliers as we did before. In particular, we compute the gradient condition

$$\nabla f - \lambda \nabla h_{\text{comb}} = \mathbf{0} \tag{10.12}$$
$$\nabla f - 2\lambda(h_1 \nabla h_1 + h_2 \nabla h_2) = \mathbf{0} \tag{10.13}$$

Our choice for h_{comb} was somewhat arbitrary. We could have used

$$h_{\text{comb}}(\mathbf{x}) = h_1(\mathbf{x})^2 + c \cdot h_2(\mathbf{x})^2 \tag{10.14}$$

for some constant $c > 0$.

With this more general formulation, we get

$$\mathbf{0} = \nabla f - \lambda \nabla h_{\text{comb}} \tag{10.15}$$
$$= \nabla f - 2\lambda h_1 \nabla h_1 - 2c\lambda h_2 \nabla h_2 \tag{10.16}$$
$$= \nabla f - \lambda_1 \nabla h_1 - \lambda_2 \nabla h_2 \tag{10.17}$$

We can thus define a Lagrangian with ℓ Lagrange multipliers for problems with ℓ equality constraints

$$\mathcal{L}(\mathbf{x}, \boldsymbol{\lambda}) = f(\mathbf{x}) - \sum_{i=1}^{\ell} \lambda_i h_i(\mathbf{x}) = f(\mathbf{x}) - \boldsymbol{\lambda}^\top \mathbf{h}(\mathbf{x}) \tag{10.18}$$

10.5   Inequality Constraints

Consider a problem with a single inequality constraint:

$$\underset{\mathbf{x}}{\text{minimize}} \qquad f(\mathbf{x})$$
$$\text{subject to} \quad g(\mathbf{x}) \leq 0 \tag{10.19}$$

We know that if the solution lies at the constraint boundary, then the Lagrange condition holds

$$\nabla f + \mu \nabla g = \mathbf{0} \tag{10.20}$$

for some non-negative constant μ. When this occurs, the constraint is considered *active*, and the gradient of the objective function is limited exactly as it was with equality constraints. Figure 10.5 shows an example.

If the solution to the problem does not lie at the constraint boundary, then the constraint is considered *inactive*. Solutions of f will simply lie where the gradient of f is zero, as with unconstrained optimization. In this case, equation (10.20) will hold by setting μ to zero. Figure 10.6 shows an example.

We could optimize a problem with an inequality constraint by introducing an infinite step penalty for infeasible points:

$$f_{\infty\text{-step}}(\mathbf{x}) = \begin{cases} f(\mathbf{x}) & \text{if } g(\mathbf{x}) \leq 0 \\ \infty & \text{otherwise} \end{cases} \tag{10.21}$$

$$= f(\mathbf{x}) + \infty \cdot (g(\mathbf{x}) > 0) \tag{10.22}$$

Unfortunately, $f_{\infty\text{-step}}$ is inconvenient to optimize.[4] It is discontinuous and nondifferentiable. Search routines obtain no directional information to steer themselves toward feasibility.

We can instead use a linear penalty $\mu g(\mathbf{x})$, which forms a lower bound on $\infty \cdot (g(\mathbf{x}) > 0)$ and penalizes the objective toward feasibility as long as $\mu > 0$. This linear penalty is visualized in figure 10.7.

We can use this linear penalty to construct a Lagrangian for inequality constraints

$$\mathcal{L}(\mathbf{x}, \mu) = f(\mathbf{x}) + \mu g(\mathbf{x}) \tag{10.23}$$

We can recover $f_{\infty\text{-step}}$ by maximizing with respect to μ

$$f_{\infty\text{-step}} = \underset{\mu \geq 0}{\text{maximize}} \, \mathcal{L}(\mathbf{x}, \mu) \tag{10.24}$$

For any infeasible $\mathbf{x}$ we get infinity and for any feasible $\mathbf{x}$ we get $f(\mathbf{x})$.

The new optimization problem is thus

$$\underset{\mathbf{x}}{\text{minimize}} \, \underset{\mu \geq 0}{\text{maximize}} \, \mathcal{L}(\mathbf{x}, \mu) \tag{10.25}$$

This reformulation is known as the *primal* problem.

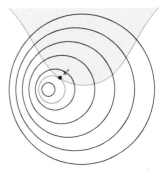

Figure 10.5. An active inequality constraint. The corresponding contour line is shown in red.

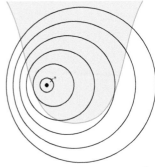

Figure 10.6. An inactive inequality constraint.

[4] Such a problem formulation can be optimized using direct methods such as the mesh adaptive direct search (section 8.2).

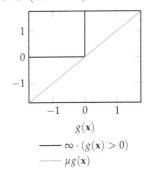

— $\infty \cdot (g(\mathbf{x}) > 0)$
— $\mu g(\mathbf{x})$

Figure 10.7. The linear function $\mu g(\mathbf{x})$ is a lower bound to the infinite step penalty so long as $\mu \geq 0$.

Optimizing the primal problem requires finding critical points $\mathbf{x}^*$ such that:

1. $g(\mathbf{x}^*) \leq 0$
   The point is feasible.

2. $\mu \geq 0$
   The penalty must point in the right direction. This requirement is sometimes called *dual feasibility*.

3. $\mu g(\mathbf{x}^*) = 0$
   A feasible point on the boundary will have $g(\mathbf{x}) = 0$, whereas a feasible point with $g(\mathbf{x}) < 0$ will have $\mu = 0$ to recover $f(\mathbf{x}^*)$ from the Lagrangian.

4. $\nabla f(\mathbf{x}^*) + \mu \nabla g(\mathbf{x}^*) = \mathbf{0}$
   When the constraint is active, we require that the contour lines of f and g be aligned, which is equivalent to saying that their gradients be aligned. When the constraint is inactive, our optimum will have $\nabla f(\mathbf{x}^*) = \mathbf{0}$ and $\mu = 0$.

These four requirements can be generalized to optimization problems with any number of equality and inequality constraints:[5]

$$\begin{aligned} \underset{\mathbf{x}}{\text{minimize}} \quad & f(\mathbf{x}) \\ \text{subject to} \quad & \mathbf{g}(\mathbf{x}) \leq \mathbf{0} \\ & \mathbf{h}(\mathbf{x}) = \mathbf{0} \end{aligned} \qquad (10.26)$$

where each component of $\mathbf{g}$ is an inequality constraint and each component of $\mathbf{h}$ is an equality constraint. The four conditions are called the *KKT conditions*.[6]

1. **Feasibility:** The constraints are all satisfied.

$$\mathbf{g}(\mathbf{x}^*) \leq \mathbf{0} \qquad (10.27)$$
$$\mathbf{h}(\mathbf{x}^*) = \mathbf{0} \qquad (10.28)$$

2. **Dual feasibility:** Penalization is toward feasibility.

$$\mu \geq \mathbf{0} \qquad (10.29)$$

3. **Complementary slackness:** The Lagrange multipliers takes up the slack. Either μ_i is zero or $g_i(\mathbf{x}^*)$ is zero.[7]

$$\mu \odot \mathbf{g} = \mathbf{0} \qquad (10.30)$$

[5] If $\mathbf{u}$ and $\mathbf{v}$ are vectors of the same length, then we say $\mathbf{u} \leq \mathbf{v}$ when $u_i \leq v_i$ for all i. We define $\geq$, $<$, and $>$ similarly for vectors.

[6] Named after Harold W. Kuhn and Albert W. Tucker who published the conditions in 1951. It was later discovered that William Karush studied these conditions in an unpublished master's thesis in 1939. A historical prospective is provided by T. H. Kjeldsen, "A Contextualized Historical Analysis of the Kuhn-Tucker Theorem in Nonlinear Programming: The Impact of World War II," *Historia Mathematica*, vol. 27, no. 4, pp. 331–361, 2000.

[7] The operation $\mathbf{a} \odot \mathbf{b}$ indicates the element-wise product between vectors $\mathbf{a}$ and $\mathbf{b}$.

4. **Stationarity:** The objective function contour is tangent to each active constraint.[8]

$$\nabla f(\mathbf{x}^*) + \sum_i \mu_i \nabla g_i(\mathbf{x}^*) + \sum_j \lambda_j \nabla h_j(\mathbf{x}^*) = \mathbf{0} \qquad (10.31)$$

[8] Since the sign of λ is not restricted, we can reverse the sign for the equality constraints from the method of Lagrange multipliers.

These four conditions are first-order necessary conditions for optimality and are thus FONCs for problems with smooth constraints. Just as with the FONCs for unconstrained optimization, special care must be taken to ensure that identified critical points are actually local minima.

10.6   Duality

In deriving the FONCs for constrained optimization, we also find a more general form for the Lagrangian. This *generalized Lagrangian* is

$$\mathcal{L}(\mathbf{x}, \boldsymbol{\mu}, \boldsymbol{\lambda}) = f(\mathbf{x}) + \sum_i \mu_i g_i(\mathbf{x}) + \sum_j \lambda_j h_j(\mathbf{x}) \qquad (10.32)$$

The *primal form* of the optimization problem is the original optimization problem formulated using the generalized Lagrangian

$$\underset{\mathbf{x}}{\text{minimize}} \, \underset{\boldsymbol{\mu} \geq 0, \boldsymbol{\lambda}}{\text{maximize}} \, \mathcal{L}(\mathbf{x}, \boldsymbol{\mu}, \boldsymbol{\lambda}) \qquad (10.33)$$

The primal problem is identical to the original problem and is just as difficult to optimize.

The *dual* form of the optimization problem reverses the order of the minimization and maximization in equation (10.33):

$$\underset{\boldsymbol{\mu} \geq 0, \boldsymbol{\lambda}}{\text{maximize}} \, \underset{\mathbf{x}}{\text{minimize}} \, \mathcal{L}(\mathbf{x}, \boldsymbol{\mu}, \boldsymbol{\lambda}) \qquad (10.34)$$

The *max-min inequality* states that for any function $f(\mathbf{a}, \mathbf{b})$:

$$\underset{\mathbf{a}}{\text{maximize}} \, \underset{\mathbf{b}}{\text{minimize}} \, f(\mathbf{a}, \mathbf{b}) \leq \underset{\mathbf{b}}{\text{minimize}} \, \underset{\mathbf{a}}{\text{maximize}} \, f(\mathbf{a}, \mathbf{b}) \qquad (10.35)$$

The solution to the dual problem is thus a lower bound to the solution of the primal problem. That is, $d^* \leq p^*$, where d^* is the *dual value* and p^* is the *primal value*.

The inner minimization in the dual problem is often folded into a *dual function*,

$$\mathcal{D}(\boldsymbol{\mu} \geq \mathbf{0}, \boldsymbol{\lambda}) = \underset{\mathbf{x}}{\text{minimize}} \, \mathcal{L}(\mathbf{x}, \boldsymbol{\mu}, \boldsymbol{\lambda}) \qquad (10.36)$$

for notational convenience. The dual function is concave.[9] Gradient ascent on a concave function always converges to the global maximum. Optimizing the dual problem is easy whenever minimizing the Lagrangian with respect to $\mathbf{x}$ is easy.

We know that maximize$_{\mu \geq 0, \lambda}\, \mathcal{D}(\mu, \lambda) \leq p^*$. It follows that the dual function is always a lower bound on the primal problem (see example 10.5). For any $\mu \geq 0$ and any λ, we have

$$\mathcal{D}(\mu \geq \mathbf{0}, \lambda) \leq p^* \tag{10.37}$$

The difference $p^* - d^*$ between the dual and primal values is called the *duality gap*. In some cases, the dual problem is guaranteed to have the same solution as the original problem—the duality gap is zero.[10] In such cases, duality provides an alternative approach for optimizing our problem. Example 10.6 demonstrates this approach.

10.7   *Penalty Methods*

We can use *penalty methods* to convert constrained optimization problems into unconstrained optimization problems by adding penalty terms to the objective function, allowing us to use the methods developed in previous chapters.

Consider a general optimization problem:

$$\begin{aligned} \underset{\mathbf{x}}{\text{minimize}} \quad & f(\mathbf{x}) \\ \text{subject to} \quad & \mathbf{g}(\mathbf{x}) \leq \mathbf{0} \\ & \mathbf{h}(\mathbf{x}) = \mathbf{0} \end{aligned} \tag{10.38}$$

A simple penalty method counts the number of constraint equations that are violated:

$$p_{\text{count}}(\mathbf{x}) = \sum_i (g_i(\mathbf{x}) > 0) + \sum_j (h_j(\mathbf{x}) \neq 0) \tag{10.39}$$

which results in the unconstrained optimization problem that penalizes infeasibility

$$\underset{\mathbf{x}}{\text{minimize}}\, f(\mathbf{x}) + \rho \cdot p_{\text{count}}(\mathbf{x}) \tag{10.40}$$

where $\rho > 0$ adjusts the penalty magnitude. Figure 10.8 shows an example.

[9] For a detailed overview, see S. Nash and A. Sofer, *Linear and Nonlinear Programming*. McGraw-Hill, 1996.

[10] Conditions that guarantee a zero duality gap are discussed in S. Boyd and L. Vandenberghe, *Convex Optimization*. Cambridge University Press, 2004.

Consider the optimization problem:

Example 10.5. The dual function is a lower bound of the primal problem.

$$\underset{x}{\text{minimize}} \quad \sin(x)$$

$$\text{subject to} \quad x^2 \leq 1$$

The generalized Lagrangian is $\mathcal{L}(x, \mu) = \sin(x) + \mu(x^2 - 1)$, making the primal problem:

$$\underset{x}{\text{minimize}} \, \underset{\mu \geq 0}{\text{maximize}} \, \sin(x) + \mu(x^2 - 1)$$

and the dual problem:

$$\underset{\mu \geq 0}{\text{maximize}} \, \underset{x}{\text{minimize}} \, \sin(x) + \mu(x^2 - 1)$$

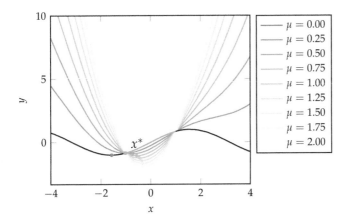

The objective function is plotted in black, with the feasible region traced over in blue. The minimum is at $x^* = -1$ with $p^* \approx -0.841$. The purple lines are the Lagrangian $\mathcal{L}(x, \mu)$ for different values of μ, each of which has a minimum lower than p^*.

Consider the problem:

$$\operatorname*{minimize}_{x} \quad x_1 + x_2 + x_1 x_2$$

$$\text{subject to} \quad x_1^2 + x_2^2 = 1$$

The Lagrangian is $\mathcal{L}(x_1, x_2, \lambda) = x_1 + x_2 + x_1 x_2 + \lambda(x_1^2 + x_2^2 - 1)$.

We apply the method of Lagrange multipliers:

$$\frac{\partial \mathcal{L}}{\partial x_1} = 1 + x_2 + 2\lambda x_1 = 0$$

$$\frac{\partial \mathcal{L}}{\partial x_2} = 1 + x_1 + 2\lambda x_2 = 0$$

$$\frac{\partial \mathcal{L}}{\partial \lambda} = x_1^2 + x_2^2 - 1 = 0$$

Solving yields four potential solutions, and thus four critical points:

x_1	x_2	λ	$x_1 + x_2 + x_1 x_2$
-1	0	$1/2$	-1
0	-1	$1/2$	-1
$\frac{\sqrt{2}+1}{\sqrt{2}+2}$	$\frac{\sqrt{2}+1}{\sqrt{2}+2}$	$\frac{1}{2}\left(-1-\sqrt{2}\right)$	$\frac{1}{2}+\sqrt{2} \approx 1.914$
$\frac{\sqrt{2}-1}{\sqrt{2}-2}$	$\frac{\sqrt{2}-1}{\sqrt{2}-2}$	$\frac{1}{2}\left(-1+\sqrt{2}\right)$	$\frac{1}{2}-\sqrt{2} \approx -0.914$

We find that the two optimal solutions are $[-1, 0]$ and $[0, -1]$.
The dual function has the form

$$\mathcal{D}(\lambda) = \operatorname*{minimize}_{x_1, x_2} x_1 + x_2 + x_1 x_2 + \lambda(x_1^2 + x_2^2 - 1)$$

The dual function is unbounded below when λ is less than $1/2$ (consider $x_1 \to \infty$ and $x_2 \to -\infty$). Setting the gradient to $\mathbf{0}$ and solving yields $x_2 = -1 - 2\lambda x_1$ and $x_1 = (2\lambda - 1)/(1 - 4\lambda^2)$ for $\lambda \neq \pm 1/2$. When $\lambda = 1/2$, $x_1 = -1 - x_2$ and $\mathcal{D}(1/2) = -1$. Substituting these into the dual function yields:

$$\mathcal{D}(\lambda) = \begin{cases} -\lambda - \frac{1}{2\lambda+1} & \lambda \geq \frac{1}{2} \\ -\infty & \text{otherwise} \end{cases}$$

The dual problem $\operatorname{maximize}_\lambda \mathcal{D}(\lambda)$ is maximized at $\lambda = 1/2$.

Example 10.6. An example of Lagrangian duality applied to a problem with an equality constraint. The top figure shows the objective function contour and the constraint with the four critical points marked by scatter points. The bottom figure shows the dual function.

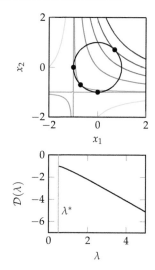

Penalty methods start with an initial point **x** and a small value for ρ. The unconstrained optimization problem equation (10.40) is then solved. The resulting design point is then used as the starting point for another optimization with an increased penalty. We continue with this procedure until the resulting point is feasible, or a maximum number of iterations has been reached. Algorithm 10.1 provides an implementation.

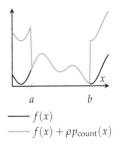

$f(x)$

$f(x) + \rho p_{\text{count}}(x)$

Figure 10.8. The original and count-penalized objective functions for minimizing f subject to $x \in [a,b]$.

```
function penalty_method(f, p, x, k_max; ρ=1, γ=2)
    for k in 1 : k_max
        x = minimize(x -> f(x) + ρ*p(x), x)
        ρ *= γ
        if p(x) == 0
            return x
        end
    end
    return x
end
```

Algorithm 10.1. The penalty method for objective function f, penalty function p, initial point x, number of iterations k_max, initial penalty ρ > 0, and penalty multiplier γ > 1. The method minimize should be replaced with a suitable unconstrained minimization method.

This penalty will preserve the problem solution for large values of ρ, but it introduces a sharp discontinuity. Points not inside the feasible set lack gradient information to guide the search towards feasibility.

We can use *quadratic penalties* to produce a smooth objective function (figure 10.9):

$$p_{\text{quadratic}}(\mathbf{x}) = \sum_i \max(g_i(\mathbf{x}), 0)^2 + \sum_j h_j(\mathbf{x})^2 \tag{10.41}$$

Quadratic penalties close to the constraint boundary are very small and may require ρ to approach infinity before the solution ceases to violate the constraints.

It is also possible to mix a count and a quadratic penalty function (figure 10.10):

$$p_{\text{mixed}}(\mathbf{x}) = \rho_1 p_{\text{count}}(\mathbf{x}) + \rho_2 p_{\text{quadratic}}(\mathbf{x}) \tag{10.42}$$

Such a penalty mixture provides a clear boundary between the feasible region and the infeasible region while providing gradient information to the solver.

Figure 10.11 shows the progress of the penalty function as ρ is increased. Quadratic penalty functions cannot ensure feasibility as discussed in example 10.7.

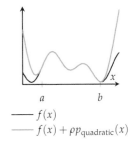

$f(x)$

$f(x) + \rho p_{\text{quadratic}}(x)$

Figure 10.9. Using a quadratic penalty function for minimizing f subject to $x \in [a,b]$.

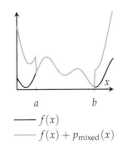

$f(x)$

$f(x) + p_{\text{mixed}}(x)$

Figure 10.10. Using both a quadratic and discrete penalty function for minimizing f subject to $x \in [a,b]$.

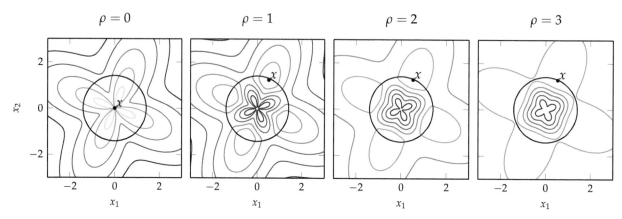

Figure 10.11. The penalty method applied to the flower function, appendix B.4, and the circular constraint

$$x_1^2 + x_2^2 \geq 2$$

Example 10.7. An example showing how quadratic penalties cannot ensure feasibility.

Consider the problem

$$\underset{x}{\text{minimize}} \quad x$$

$$\text{subject to} \quad x \geq 5$$

using a quadratic penalty function.

The unconstrained objective function is

$$f(x) = x + \rho \max(5 - x, 0)^2$$

The minimum of the unconstrained objective function is

$$x^* = 5 - \frac{1}{2\rho}$$

While the minimum of the constrained optimization problem is clearly $x = 5$, the minimum of the penalized optimization problem merely approaches $x = 5$, requiring an infinite penalty to achieve feasibility.

10.8   Augmented Lagrange Method

The *augmented Lagrange method*[11] is an adaptation of the penalty method specifically for equality constraints. Unlike the penalty method, where ρ must sometimes approach infinity before a feasible solution is found, the augmented Lagrange method will work with smaller values of ρ. It uses both a quadratic and a linear penalty for each constraint.

For an optimization problem with equality constraints $\mathbf{h}(\mathbf{x}) = \mathbf{0}$, the penalty function is:

$$p_{\text{Lagrange}}(\mathbf{x}) = \frac{1}{2}\rho \sum_i (h_i(\mathbf{x}))^2 - \sum_i \lambda_i h_i(\mathbf{x}) \tag{10.43}$$

where λ converges toward the Lagrange multiplier.

In addition to increasing ρ with each iteration, the linear penalty vector is updated according to:

$$\boldsymbol{\lambda}^{(k+1)} = \boldsymbol{\lambda}^{(k)} - \rho\mathbf{h}(\mathbf{x}) \tag{10.44}$$

Algorithm 10.2 provides an implementation.

```
function augmented_lagrange_method(f, h, x, k_max; ρ=1, γ=2)
    λ = zeros(length(h(x)))
    for k in 1 : k_max
        p = x -> ρ/2*sum(h(x).^2) - λ·h(x)
        x = minimize(x -> f(x) + p(x), x)
        λ -= ρ*h(x)
        ρ *= γ
    end
    return x
end
```

Algorithm 10.2. The augmented Lagrange method for objective function f, equality constraint function h, initial point x, number of iterations k_max, initial penalty scalar ρ > 0, and penalty multiplier γ > 1. The function minimize should be replaced with the minimization method of your choice.

10.9   Interior Point Methods

Interior point methods (algorithm 10.3), sometimes referred to as *barrier methods*, are optimization methods that ensure that the search points always remain feasible.[12] Interior point methods use a barrier function that approaches infinity as one approaches a constraint boundary. This barrier function, $p_{\text{barrier}}(\mathbf{x})$, must satisfy several properties:

1. $p_{\text{barrier}}(\mathbf{x})$ is continuous

2. $p_{\text{barrier}}(\mathbf{x})$ is nonnegative $(p_{\text{barrier}}(\mathbf{x}) \geq 0)$

3. $p_{\text{barrier}}(\mathbf{x})$ approaches infinity as $\mathbf{x}$ approaches any constraint boundary

Some examples of barrier functions are:

Inverse Barrier:

$$p_{\text{barrier}}(\mathbf{x}) = -\sum_i \frac{1}{g_i(\mathbf{x})} \qquad (10.45)$$

Log Barrier:

$$p_{\text{barrier}}(\mathbf{x}) = -\sum_i \begin{cases} \log(-g_i(\mathbf{x})) & \text{if } g_i(\mathbf{x}) \geq -1 \\ 0 & \text{otherwise} \end{cases} \qquad (10.46)$$

A problem with inequality constraints can be transformed into an unconstrained optimization problem

$$\underset{\mathbf{x}}{\text{minimize}} \; f(\mathbf{x}) + \frac{1}{\rho} p_{\text{barrier}}(\mathbf{x}) \qquad (10.47)$$

When ρ is increased, the penalty for approaching the boundary decreases (figure 10.12).

Special care must be taken such that line searches do not leave the feasible region. Line searches $f(\mathbf{x} + \alpha\mathbf{d})$ are constrained to the interval $\alpha = [0, \alpha_u]$, where α_u is the step to the nearest boundary. In practice, α_u is chosen such that $\mathbf{x} + \alpha\mathbf{d}$ is just inside the boundary to avoid the boundary singularity.

Like the penalty method, the interior point method begins with a low value for ρ and slowly increases it until convergence. The interior point method is typically terminated when the difference between subsequent points is less than a certain threshold. Figure 10.13 shows an example of the effect of incrementally increasing ρ.

The interior point method requires a feasible point from which to start the search. One convenient method for finding a feasible point is to optimize the quadratic penalty function

$$\underset{\mathbf{x}}{\text{minimize}} \; p_{\text{quadratic}}(\mathbf{x}) \qquad (10.48)$$

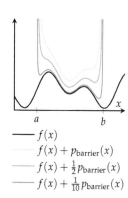

$$\text{———} \; f(x)$$
$$\text{———} \; f(x) + p_{\text{barrier}}(x)$$
$$\text{———} \; f(x) + \tfrac{1}{2} p_{\text{barrier}}(x)$$
$$\text{———} \; f(x) + \tfrac{1}{10} p_{\text{barrier}}(x)$$

Figure 10.12. Applying the interior point method with an inverse barrier for minimizing f subject to $x \in [a, b]$.

```
function interior_point_method(f, p, x; ρ=1, γ=2, ϵ=0.001)
    delta = Inf
    while delta > ϵ
        x′ = minimize(x -> f(x) + p(x)/ρ, x)
        delta = norm(x′ - x)
        x = x′
        ρ *= γ
    end
    return x
end
```

Algorithm 10.3. The interior point method for objective function f, barrier function p, initial point x, initial penalty ρ > 0, penalty multiplier γ > 1, and stopping tolerance ϵ > 0.

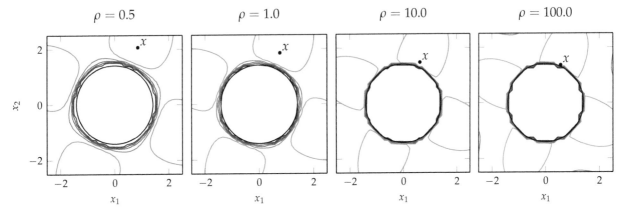

Figure 10.13. The interior point method with the inverse barrier applied to the flower function, appendix B.4, and the constraint

$$x_1^2 + x_2^2 \geq 2$$

10.10   Summary

- Constraints are requirements on the design points that a solution must satisfy.

- Some constraints can be transformed or substituted into the problem to result in an unconstrained optimization problem.

- Analytical methods using Lagrange multipliers yield the generalized Lagrangian and the necessary conditions for optimality under constraints.

- A constrained optimization problem has a dual problem formulation that is easier to solve and whose solution is a lower bound of the solution to the original problem.

- Penalty methods penalize infeasible solutions and often provide gradient information to the optimizer to guide infeasible points toward feasibility.

- Interior point methods maintain feasibility but use barrier functions to avoid leaving the feasible set.

10.11   Exercises

Exercise 10.1. Solve

$$\begin{array}{ll} \underset{x}{\text{minimize}} & x \\ \text{subject to} & x \geq 0 \end{array} \tag{10.49}$$

using the quadratic penalty method with $\rho > 0$. Solve the problem in closed form.

Exercise 10.2. Solve the problem above using the count penalty method with $\rho > 1$ and compare it to the quadratic penalty method.

Exercise 10.3. Suppose that you are solving a constrained problem with the penalty method. You notice that the iterates remain infeasible and you decide to stop the algorithm. What can you do to be more successful in your next attempt?

Exercise 10.4. Consider a simple univariate minimization problem where you minimize a function $f(x)$ subject to $x \geq 0$. Assume that you know that the constraint is active, that is, $x^* = 0$ where x^* is the minimizer and $f'(x^*) > 0$ from the optimality conditions. Show that solving the same problem with the penalty method

$$f(x) + (\min(x, 0))^2 \tag{10.50}$$

yields an infeasible solution with respect to the original problem.

Exercise 10.5. What is the advantage of the augmented Lagrange method compared to the quadratic penalty method?

Exercise 10.6. When would you use the barrier method in place of the penalty method?

Exercise 10.7. Give an example of a smooth optimization problem, such that, for any penalty parameter $\rho > 0$, there exists a starting point $x^{(1)}$ for which the steepest descent method diverges.

Exercise 10.8. Suppose you have an optimization problem

$$
\begin{aligned}
& \underset{x}{\text{minimize}} && f(\mathbf{x}) \\
& \text{subject to} && \mathbf{h}(\mathbf{x}) = \mathbf{0} \\
& && \mathbf{g}(\mathbf{x}) \le \mathbf{0}
\end{aligned}
\tag{10.51}
$$

but do not have an initial feasible design. How would you find a feasible point with respect to the constraints, provided that one exists?

Exercise 10.9. Solve the constrained optimization problem

$$
\begin{aligned}
& \underset{x}{\text{minimize}} && \sin\left(\frac{4}{x}\right) \\
& \text{subject to} && x \in [1, 10]
\end{aligned}
\tag{10.52}
$$

using both the transform $x = t_{a,b}(\hat{x})$ and a sigmoid transform for constraint bounds $x \in [a, b]$:

$$
x = s(\hat{x}) = a + \frac{(b-a)}{1 + e^{-\hat{x}}}
\tag{10.53}
$$

Why is the t transform better than the sigmoid transform?

Exercise 10.10. Give an example of a quadratic objective function involving two design variables where the addition of a linear constraint results in a different optimum.

Exercise 10.11. Suppose we want to minimize $x_1^3 + x_2^2 + x_3$ subject to the constraint that $x_1 + 2x_2 + 3x_3 = 6$. How might we transform this into an unconstrained problem with the same minimizer?

Exercise 10.12. Suppose we want to minimize $-x_1 - 2x_2$ subject to the constraints $ax_1 + x_2 \leq 5$ and $x_1, x_2 \geq 0$. If a is a bounded constant, what range of values of a will result in an infinite number of optimal solutions?

Exercise 10.13. Consider using a penalty method to optimize

$$
\begin{aligned}
\underset{x}{\text{minimize}} \quad & 1 - x^2 \\
\text{subject to} \quad & |x| \leq 2
\end{aligned}
\tag{10.54}
$$

Optimization with the penalty method typically involves running several optimizations with increasing penalty weights. Impatient engineers may wish to optimize once using a very large penalty weight. Explain what issues are encountered for both the count penalty method and the quadratic penalty method.

11 *Linear Constrained Optimization*

Linear programming involves solving optimization problems with linear objective functions and linear constraints. Many problems are naturally described by linear programs, including problems from fields as diverse as transportation, communication networks, manufacturing, economics, and operations research. Many problems that are not naturally linear can often be approximated by linear programs. Several methods have been developed for exploiting the linear structure. Modern techniques and hardware can globally minimize problems with millions of variables and millions of constraints.[1]

11.1 *Problem Formulation*

A linear programming problem, called a *linear program*,[2] can be expressed in several forms. Each linear program consists of a linear objective function and a set of linear constraints:

$$
\begin{aligned}
\underset{\mathbf{x}}{\text{minimize}} \quad & \mathbf{c}^\top \mathbf{x} \\
\text{subject to} \quad & \mathbf{w}_{\text{LE}}^{(i)\top} \mathbf{x} \le b_i \ \text{ for } \ i \ \in \ \{1,2,\ldots\} \\
& \mathbf{w}_{\text{GE}}^{(j)\top} \mathbf{x} \ge b_j \ \text{ for } \ j \ \in \ \{1,2,\ldots\} \\
& \mathbf{w}_{\text{EQ}}^{(k)\top} \mathbf{x} = b_k \ \text{ for } \ k \ \in \ \{1,2,\ldots\}
\end{aligned}
\tag{11.1}
$$

where i, j, and k vary over finite sets of constraints. Such an optimization problem is given in example 11.1. Transforming real problems into this mathematical form is often nontrivial. This text focuses on the algorithms for obtaining solutions, but other texts discuss how to go about modeling real problems.[3] Several interesting conversions are given in example 11.2.

[1] This chapter is a short introduction to linear programs and one variation of the simplex algorithm used to solve them. Several textbooks are dedicated entirely to linear programs, including R. J. Vanderbei, *Linear Programming: Foundations and Extensions*, 4th ed. Springer, 2014. There are a variety of packages for solving linear programs, such as Convex.jl and JuMP.jl, both of which include interfaces to open-source and commercial solvers.

[2] A *quadratic program* is a generalization of a linear program, where the objective function is quadratic and the constraints are linear. Common approaches for solving such problems include some of the algorithms discussed in earlier chapters, including the interior point method, augmented Lagrange method, and conjugate gradient. The simplex method, covered in this chapter, has also been adapted for optimizing quadratic programs. J. Nocedal and S. J. Wright, *Numerical Optimization*, 2nd ed. Springer, 2006.

[3] See, for example, H. P. Williams, *Model Building in Mathematical Programming*, 5th ed. Wiley, 2013.

The following problem has a linear objective and linear constraints, making it a linear program.

$$\begin{array}{rl} \underset{x_1,x_2,x_3}{\text{minimize}} & 2x_1 - 3x_2 + 7x_3 \\ \text{subject to} & 2x_1 + 3x_2 - 8x_3 \leq 5 \\ & 4x_1 + x_2 + 3x_3 \leq 9 \\ & x_1 - 5x_2 - 3x_3 \geq -4 \\ & x_1 + x_2 + 2x_3 = 1 \end{array}$$

Example 11.1. An example linear program.

Many problems can be converted into linear programs that have the same solution. Two examples are L_1 and L_∞ minimization problems:

$$\text{minimize}\|\mathbf{Ax} - \mathbf{b}\|_1 \qquad \text{minimize}\|\mathbf{Ax} - \mathbf{b}\|_\infty$$

The first problem is equivalent to solving

$$\begin{array}{rl} \underset{\mathbf{x},\mathbf{s}}{\text{minimize}} & \mathbf{1}^\top \mathbf{s} \\ \text{subject to} & \mathbf{Ax} - \mathbf{b} \leq \mathbf{s} \\ & \mathbf{Ax} - \mathbf{b} \geq -\mathbf{s} \end{array}$$

with the additional variables $\mathbf{s}$.

The second problem is equivalent to solving

$$\begin{array}{rl} \underset{\mathbf{x},t}{\text{minimize}} & t \\ \text{subject to} & \mathbf{Ax} - \mathbf{b} \leq t\mathbf{1} \\ & \mathbf{Ax} - \mathbf{b} \geq -t\mathbf{1} \end{array}$$

with the additional variable t.

Example 11.2. Common norm minimization problems that can be converted into linear programs.

11.1.1 General Form

We can write linear programs more compactly using matrices and arrive at the general form:[4]

$$\begin{aligned} \underset{\mathbf{x}}{\text{minimize}} \quad & \mathbf{c}^\top \mathbf{x} \\ \text{subject to} \quad & \mathbf{A}_{LE}\mathbf{x} \leq \mathbf{b}_{LE} \\ & \mathbf{A}_{GE}\mathbf{x} \geq \mathbf{b}_{GE} \\ & \mathbf{A}_{EQ}\mathbf{x} = \mathbf{b}_{EQ} \end{aligned} \tag{11.2}$$

[4] Here, each constraint is element-wise. For example, in writing

$$\mathbf{a} \leq \mathbf{b},$$

we mean $a_i \leq b_i$ for all i.

11.1.2 Standard Form

The general linear program given in equation (11.2) can be converted into *standard form* where all constraints are less-than inequalities and the design variables are nonnegative

$$\begin{aligned} \underset{\mathbf{x}}{\text{minimize}} \quad & \mathbf{c}^\top \mathbf{x} \\ \text{subject to} \quad & \mathbf{A}\mathbf{x} \leq \mathbf{b} \\ & \mathbf{x} \geq \mathbf{0} \end{aligned} \tag{11.3}$$

Greater-than inequalities are inverted, and equality constraints are split in two

$$\begin{aligned} \mathbf{A}_{GE}\mathbf{x} \geq \mathbf{b}_{GE} \quad &\rightarrow \quad -\mathbf{A}_{GE}\mathbf{x} \leq -\mathbf{b}_{GE} \\ \mathbf{A}_{EQ}\mathbf{x} = \mathbf{b}_{EQ} \quad &\rightarrow \quad \begin{cases} \mathbf{A}_{EQ}\mathbf{x} \leq \ \ \mathbf{b}_{EQ} \\ -\mathbf{A}_{EQ}\mathbf{x} \leq -\mathbf{b}_{EQ} \end{cases} \end{aligned} \tag{11.4}$$

We must ensure that all $\mathbf{x}$ entries are nonnegative as well. Suppose we start with a linear program where $\mathbf{x}$ is not constrained to be nonnegative:

$$\begin{aligned} \underset{\mathbf{x}}{\text{minimize}} \quad & \mathbf{c}^\top \mathbf{x} \\ \text{subject to} \quad & \mathbf{A}\mathbf{x} \leq \mathbf{b} \end{aligned} \tag{11.5}$$

We replace $\mathbf{x}$ with $\mathbf{x}^+ - \mathbf{x}^-$ and constrain $\mathbf{x}^+ \geq 0$ and $\mathbf{x}^- \geq 0$:

$$
\begin{aligned}
\underset{\mathbf{x}^+, \mathbf{x}^-}{\text{minimize}} \quad & \begin{bmatrix} \mathbf{c}^\top & -\mathbf{c}^\top \end{bmatrix} \begin{bmatrix} \mathbf{x}^+ \\ \mathbf{x}^- \end{bmatrix} \\
\text{subject to} \quad & \begin{bmatrix} \mathbf{A} & -\mathbf{A} \end{bmatrix} \begin{bmatrix} \mathbf{x}^+ \\ \mathbf{x}^- \end{bmatrix} \leq \mathbf{b} \\
& \begin{bmatrix} \mathbf{x}^+ \\ \mathbf{x}^- \end{bmatrix} \geq 0
\end{aligned}
\tag{11.6}
$$

The linear objective function $\mathbf{c}^\top \mathbf{x}$ forms a flat ramp. The function increases in the direction of $\mathbf{c}$, and, as a result, all contour lines are perpendicular to $\mathbf{c}$ and parallel to one another as shown in figure 11.1.

A single inequality constraint $\mathbf{w}^\top \mathbf{x} \leq b$ forms a *half-space*, or a region on one side of a hyperplane. The hyperplane is perpendicular to $\mathbf{w}$ and is defined by $\mathbf{w}^\top \mathbf{x} = b$ as shown in figure 11.2. The region $\mathbf{w}^\top \mathbf{x} > b$ is on the $+\mathbf{w}$ side of the hyperplane, whereas $\mathbf{w}^\top \mathbf{x} < b$ is on the $-\mathbf{w}$ side of the hyperplane.

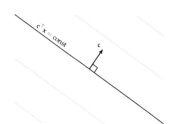

Figure 11.1. The contours of a linear objective function $\mathbf{c}^\top \mathbf{x}$, which increase in the direction of $\mathbf{c}$.

Figure 11.2. A linear constraint.

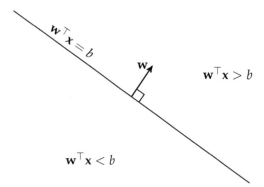

Half-spaces are convex sets (see appendix C.3), and the intersection of convex sets is convex, as shown in figure 11.3. Thus, the feasible set of a linear program will always form a convex set. Convexity of the feasible set, along with convexity of the objective function, implies that if we find a local feasible minimum, it is also a global feasible minimum.

The feasible set is a convex region enclosed by flat faces. Depending on the region's configuration, the solution can lie at a vertex, on an edge, or on an entire face. If the problem is not properly constrained, the solution can be unbounded,

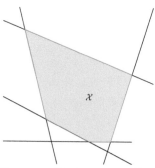

Figure 11.3. The intersection of linear constraints is a convex set.

and, if the system is *over-constrained*, there is no feasible solution. Several such cases are shown in figure 11.4.

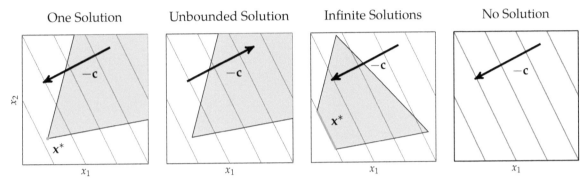

Figure 11.4. Several different linear problem forms with different solutions.

11.1.3   Equality Form

Linear programs are often solved in a third form, the *equality form*

$$\begin{aligned}
\underset{\mathbf{x}}{\text{minimize}} \quad & \mathbf{c}^\top \mathbf{x} \\
\text{subject to} \quad & \mathbf{A}\mathbf{x} = \mathbf{b} \\
& \mathbf{x} \geq \mathbf{0}
\end{aligned} \quad (11.7)$$

where $\mathbf{x}$ and $\mathbf{c}$ each have n components, $\mathbf{A}$ is an $m \times n$ matrix, and $\mathbf{b}$ has m components. In other words, we have n nonnegative design variables and a system of m equations defining equality constraints.

The equality form has constraints in two parts. The first, $\mathbf{A}\mathbf{x} = \mathbf{b}$, forces the solution to lie in an affine subspace.[5] Such a constraint is convenient because search techniques can constrain themselves to the constrained affine subspace to remain feasible. The second part of the constraints requires $\mathbf{x} \geq \mathbf{0}$, which forces the solution to lie in the positive quadrant. The feasible set is thus the nonnegative portion of an affine subspace. Example 11.3 provides a visualization for a simple linear program.

Any linear program in standard form can be transformed to equality form. The constraints are converted to:

$$\mathbf{A}\mathbf{x} \leq \mathbf{b} \quad \rightarrow \quad \mathbf{A}\mathbf{x} + \mathbf{s} = \mathbf{b}, \quad \mathbf{s} \geq \mathbf{0} \quad (11.8)$$

[5] Informally, an *affine subspace* is a vector space that has been translated such that its origin in a higher-dimensional space is not necessarily 0.

Consider the standard-form linear program:

$$\underset{x}{\text{minimize}} \quad x$$

$$\text{subject to} \quad x \geq 1$$

When we convert this to equality form, we get

$$\underset{x,s}{\text{minimize}} \quad x$$

$$\text{subject to} \quad x - s = 1$$

$$x, s \geq 0$$

The equality constraint requires that feasible points fall on the line $x - s = 1$. That line is a one-dimensional affine subspace of the two-dimensional Euclidean space.

Example 11.3.   Feasible sets for the equality form are hyperplanes.

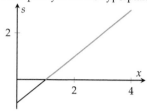

by introducing *slack variables* **s**. These variables take up the extra slack to enforce equality.

Starting with a linear program:

$$\underset{x}{\text{minimize}} \quad c^\top x$$

$$\text{subject to} \quad Ax \leq b \tag{11.9}$$

$$x \geq 0$$

We introduce the slack variables:

$$\underset{x,s}{\text{minimize}} \quad \begin{bmatrix} c^\top & 0^\top \end{bmatrix} \begin{bmatrix} x \\ s \end{bmatrix}$$

$$\text{subject to} \quad \begin{bmatrix} A & I \end{bmatrix} \begin{bmatrix} x \\ s \end{bmatrix} = b \tag{11.10}$$

$$\begin{bmatrix} x \\ s \end{bmatrix} \geq 0$$

Example 11.4 demonstrates converting from standard to equality form.

Consider the linear program

$$\underset{\mathbf{x}}{\text{minimize}} \quad 5x_1 + 4x_2$$

$$\text{subject to} \quad 2x_1 + 3x_2 \leq 5$$

$$4x_1 + x_2 \leq 11$$

To convert to equality form, we first introduce two slack variables:

$$\underset{\mathbf{x},\mathbf{s}}{\text{minimize}} \quad 5x_1 + 4x_2$$

$$\text{subject to} \quad 2x_1 + 3x_2 + s_1 = 5$$

$$4x_1 + x_2 + s_2 = 11$$

$$s_1, s_2 \geq 0$$

We then split $\mathbf{x}$:

$$\underset{\mathbf{x}^+,\mathbf{x}^-,\mathbf{s}}{\text{minimize}} \quad 5(x_1^+ - x_1^-) + 4(x_2^+ - x_2^-)$$

$$\text{subject to} \quad 2(x_1^+ - x_1^-) + 3(x_2^+ - x_2^-) + s_1 = 5$$

$$4(x_1^+ - x_1^-) + (x_2^+ - x_2^-) + s_2 = 11$$

$$x_1^+, x_1^-, x_2^+, x_2^-, s_1, s_2 \geq 0$$

Example 11.4. Converting a linear program to equality form.

11.2 Simplex Algorithm

The *simplex algorithm* solves linear programs by moving from vertex to vertex of the feasible set.[6] The method is guaranteed to arrive at an optimal solution so long as the linear program is feasible and bounded.

The simplex algorithm operates on equality-form linear programs ($\mathbf{A}\mathbf{x} = \mathbf{b}$, $\mathbf{x} \geq \mathbf{0}$). We assume that the rows of $\mathbf{A}$ are linearly independent.[7] We also assume that the problem has no more equality constraints than it has design variables ($m \leq n$), which ensures that the problem is not over constrained. A preprocessing phase guarantees that $\mathbf{A}$ satisfies these conditions.

[6] The simplex algorithm was originally developed in the 1940s by George Dantzig. A history of the development can be found here: G. B. Dantzig, "Origins of the Simplex Method," in *A History of Scientific Computing*, S. G. Nash, ed., ACM, 1990, pp. 141–151.

[7] A matrix whose rows are linearly independent is said to have full row rank. Linear independence is achieved by removing redundant equality constraints.

11.2.1   Vertices

Linear programs in equality form have feasible sets in the form of convex *polytopes*, which are geometric objects with flat faces. These polytopes are formed by the intersection of the equality constraints with the positive quadrant. Associated with a polytope are *vertices*, which are points in the feasible set that do not lie between any other points in the feasible set.

The feasible set consists of several different types of design points. Points on the interior are never optimal because they can be improved by moving along $-\mathbf{c}$. Points on faces can be optimal only if the face is perpendicular to $\mathbf{c}$. Points on faces not perpendicular to $\mathbf{c}$ can be improved by sliding along the face in the direction of the projection of $-\mathbf{c}$ onto the face. Similarly, points on edges can be optimal only if the edge is perpendicular to $\mathbf{c}$, and can otherwise be improved by sliding along the projection of $-\mathbf{c}$ onto the edge. Finally, vertices can also be optimal.

The simplex algorithm produces an optimal vertex. If a linear program has a bounded solution, then it also contains at least one vertex. Furthermore, at least one solution must lie at a vertex. In the case where an entire edge or face is optimal, a vertex-solution is just as good as any other.

Every vertex for a linear program in equality form can be uniquely defined by $n - m$ components of $\mathbf{x}$ that equal zero. These components are actively constrained by $x_i \geq 0$. Figure 11.5 visualizes the active constraints needed to identify a vertex.

The equality constraint $\mathbf{A}\mathbf{x} = \mathbf{b}$ has a unique solution when $\mathbf{A}$ is square. We have assumed that $m \leq n$, so choosing m design variables and setting the remaining variables to zero effectively removes $n - m$ columns of $\mathbf{A}$, yielding an $m \times m$ constraint matrix (see example 11.5).

The indices into the components $\{1, \dots, n\}$ of any vertex can be partitioned into two sets, $\mathcal{B}$ and $\mathcal{V}$, such that:

- The design values associated with indices in $\mathcal{V}$ are zero:

$$i \in \mathcal{V} \implies x_i = 0 \tag{11.11}$$

- The design values associated with indices in $\mathcal{B}$ may or may not be zero:

$$i \in \mathcal{B} \implies x_i \geq 0 \tag{11.12}$$

- $\mathcal{B}$ has exactly m elements and $\mathcal{V}$ has exactly $n - m$ elements.

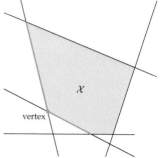

Figure 11.5. In two dimensions, any vertex will have at least two active constraints.

For a problem with 5 design variables and 3 constraints, setting 2 variables to zero uniquely defines a point.

$$\begin{bmatrix} a_{11} & a_{12} & a_{13} & a_{14} & a_{15} \\ a_{21} & a_{22} & a_{23} & a_{24} & a_{25} \\ a_{31} & a_{32} & a_{33} & a_{34} & a_{35} \end{bmatrix} \begin{bmatrix} x_1 \\ 0 \\ x_3 \\ x_4 \\ 0 \end{bmatrix} = \begin{bmatrix} a_{11} & a_{13} & a_{14} \\ a_{21} & a_{23} & a_{24} \\ a_{31} & a_{33} & a_{34} \end{bmatrix} \begin{bmatrix} x_1 \\ x_3 \\ x_4 \end{bmatrix} = \begin{bmatrix} b_1 \\ b_2 \\ b_3 \end{bmatrix}$$

Example 11.5. Setting $n - m$ components of $\mathbf{x}$ to zero can uniquely define a point.

We use $\mathbf{x}_\mathcal{B}$ to refer to the vector consisting of the components of $\mathbf{x}$ that are in $\mathcal{B}$ and $\mathbf{x}_\mathcal{V}$ to refer to the vector consisting of the components of $\mathbf{x}$ that are in $\mathcal{V}$. Note that $\mathbf{x}_\mathcal{V} = \mathbf{0}$.

The vertex associated with a partition $(\mathcal{B}, \mathcal{V})$ can be obtained using the $m \times m$ matrix $\mathbf{A}_\mathcal{B}$ formed by the m columns of $\mathbf{A}$ selected by $\mathcal{B}$:[8]

$$\mathbf{A}\mathbf{x} = \mathbf{A}_\mathcal{B}\mathbf{x}_\mathcal{B} = \mathbf{b} \quad \rightarrow \quad \mathbf{x}_\mathcal{B} = \mathbf{A}_\mathcal{B}^{-1}\mathbf{b} \tag{11.13}$$

Knowing $\mathbf{x}_\mathcal{B}$ is sufficient to construct $\mathbf{x}$; the remaining design variables are zero. Algorithm 11.1 implements this procedure, and example 11.6 demonstrates verifying that a given design point is a vertex.

[8] If $\mathcal{B}$ and $\mathcal{V}$ identify a vertex, then the columns of $\mathbf{A}_\mathcal{B}$ must be linearly independent because $\mathbf{A}\mathbf{x} = \mathbf{b}$ must have exactly one solution. Hence, $\mathbf{A}_\mathcal{B}\mathbf{x}_\mathcal{B} = \mathbf{b}$ must have exactly one solution. This linear independence guarantees that $\mathbf{A}_\mathcal{B}$ is invertible.

```
mutable struct LinearProgram
    A
    b
    c
end
function get_vertex(B, LP)
    A, b, c = LP.A, LP.b, LP.c
    b_inds = sort!(collect(B))
    AB = A[:,b_inds]
    xB = AB\b
    x = zeros(length(c))
    x[b_inds] = xB
    return x
end
```

Algorithm 11.1. A method for extracting the vertex associated with a partition B and a linear program LP in equality form. We introduce the special type LinearProgram for linear programs.

While every vertex has an associated partition $(\mathcal{B}, \mathcal{V})$, not every partition corresponds to a vertex. A partition corresponds to a vertex only if $\mathbf{A}_\mathcal{B}$ is nonsingular and the design obtained by applying equation (11.13) is feasible.[9] Identifying partitions that correspond to vertices is nontrivial, and we show in section 11.2.4 that finding such a partition involves solving a linear program! The simplex algorithm operates in two phases—an *initialization phase* that identifies a vertex partition and an *optimization phase* that transitions between vertex partitions toward a partition corresponding to an optimal vertex. We will discuss both of these phases later in this section.

[9] For example, $\mathcal{B} = \{1, 2\}$ for the constraints

$$\begin{bmatrix} 1 & 2 & 0 \\ 1 & 2 & 1 \end{bmatrix} \begin{bmatrix} x_1 \\ x_2 \\ x_3 \end{bmatrix} = \begin{bmatrix} 1 \\ 1 \end{bmatrix}$$

corresponds to

$$\begin{bmatrix} 1 & 2 \\ 1 & 2 \end{bmatrix} \begin{bmatrix} x_1 \\ x_2 \end{bmatrix} = \begin{bmatrix} 1 \\ 1 \end{bmatrix}$$

which does not produce an invertible $\mathbf{A}_\mathcal{B}$ and does not have a unique solution.

Consider the constraints:

$$\begin{bmatrix} 1 & 1 & 1 & 1 \\ 0 & -1 & 2 & 3 \\ 2 & 1 & 2 & -1 \end{bmatrix} \mathbf{x} = \begin{bmatrix} 2 \\ -1 \\ 3 \end{bmatrix}, \quad \mathbf{x} \geq \mathbf{0}$$

Consider the design point $\mathbf{x} = [1, 1, 0, 0]$. We can verify that $\mathbf{x}$ is feasible and that it has no more than three nonzero components. We can choose either $\mathcal{B} = \{1, 2, 3\}$ or $\mathcal{B} = \{1, 2, 4\}$. Both

$$\mathbf{A}_{\{1,2,3\}} = \begin{bmatrix} 1 & 1 & 1 \\ 0 & -1 & 2 \\ 2 & 1 & 2 \end{bmatrix}$$

and

$$\mathbf{A}_{\{1,2,4\}} = \begin{bmatrix} 1 & 1 & 1 \\ 0 & -1 & 3 \\ 2 & 1 & -1 \end{bmatrix}$$

are invertible. Thus, $\mathbf{x}$ is a vertex of the feasible set polytope.

Example 11.6. Verifying that a design point is a vertex for constraints in equality form.

11.2.2   First-Order Necessary Conditions

The first-order necessary conditions (FONCs) for optimality are used to determine when a vertex is optimal and to inform how to transition to a more favorable vertex. We construct a Lagrangian for the equality form of the linear program:[10]

$$\mathcal{L}(\mathbf{x}, \boldsymbol{\mu}, \boldsymbol{\lambda}) = \mathbf{c}^\top \mathbf{x} - \boldsymbol{\mu}^\top \mathbf{x} - \boldsymbol{\lambda}^\top (\mathbf{A}\mathbf{x} - \mathbf{b}) \tag{11.14}$$

[10] Note that in $\mathbf{x} \geq \mathbf{0}$ the polarity of the inequality must be inverted by multiplying both sides by -1, yielding the negative sign in front of μ. The Lagrangian can be defined with either positive or negative λ.

with the following FONCs:

1. **feasibility:** $\mathbf{Ax} = \mathbf{b}, \mathbf{x} \geq \mathbf{0}$

2. **dual feasibility:** $\mathbf{\mu} \geq \mathbf{0}$

3. **complementary slackness:** $\mathbf{\mu} \odot \mathbf{x} = \mathbf{0}$

4. **stationarity:** $\mathbf{A}^\top \mathbf{\lambda} + \mathbf{\mu} = \mathbf{c}$

The FONCs are sufficient conditions for optimality for linear programs. Thus, if $\mathbf{\mu}$ and $\mathbf{\lambda}$ can be computed for a given vertex and all four FONC equations are satisfied, then the vertex is optimal.

We can decompose the stationarity condition into $\mathcal{B}$ and $\mathcal{V}$ components:

$$\mathbf{A}^\top \mathbf{\lambda} + \mathbf{\mu} = \mathbf{c} \quad \rightarrow \quad \begin{cases} \mathbf{A}_{\mathcal{B}}^\top \mathbf{\lambda} + \mathbf{\mu}_{\mathcal{B}} = \mathbf{c}_{\mathcal{B}} \\ \mathbf{A}_{\mathcal{V}}^\top \mathbf{\lambda} + \mathbf{\mu}_{\mathcal{V}} = \mathbf{c}_{\mathcal{V}} \end{cases} \qquad (11.15)$$

We can choose $\mathbf{\mu}_{\mathcal{B}} = \mathbf{0}$ to satisfy complementary slackness. The value of $\mathbf{\lambda}$ can be computed from $\mathcal{B}$:[11]

$$\mathbf{A}_{\mathcal{B}}^\top \mathbf{\lambda} + \underbrace{\mathbf{\mu}_{\mathcal{B}}}_{=\mathbf{0}} = \mathbf{c}_{\mathcal{B}}$$

$$\mathbf{\lambda} = \mathbf{A}_{\mathcal{B}}^{-\top} \mathbf{c}_{\mathcal{B}}$$

$\qquad (11.16)$

We can use this to obtain

$$\mathbf{A}_{\mathcal{V}}^\top \mathbf{\lambda} + \mathbf{\mu}_{\mathcal{V}} = \mathbf{c}_{\mathcal{V}}$$

$$\mathbf{\mu}_{\mathcal{V}} = \mathbf{c}_{\mathcal{V}} - \mathbf{A}_{\mathcal{V}}^\top \mathbf{\lambda}$$

$$\mathbf{\mu}_{\mathcal{V}} = \mathbf{c}_{\mathcal{V}} - \left(\mathbf{A}_{\mathcal{B}}^{-1} \mathbf{A}_{\mathcal{V}} \right)^\top \mathbf{c}_{\mathcal{B}}$$

$\qquad (11.17)$

[11] We use $\mathbf{A}^{-\top}$ to refer to the transpose of the inverse of $\mathbf{A}$:

$$\mathbf{A}^{-\top} = \left(\mathbf{A}^{-1} \right)^\top = \left(\mathbf{A}^\top \right)^{-1}$$

Knowing $\mathbf{\mu}_{\mathcal{V}}$ allows us to assess the optimality of the vertices. If $\mathbf{\mu}_{\mathcal{V}}$ contains negative components, then dual feasibility is not satisfied and the vertex is suboptimal.

11.2.3   Optimization Phase

The simplex algorithm maintains a partition $(\mathcal{B}, \mathcal{V})$, which corresponds to a vertex of the feasible set polytope. The partition can be updated by swapping indices between $\mathcal{B}$ and $\mathcal{V}$. Such a swap equates to moving from one vertex along an edge of the feasible set polytope to another vertex. If the initial partition corresponds to a vertex and the problem is bounded, the simplex algorithm is guaranteed to converge to an optimum.

A transition $\mathbf{x} \rightarrow \mathbf{x}'$ between vertices must satisfy $\mathbf{Ax}' = \mathbf{b}$. Starting with a partition defined by $\mathcal{B}$, we choose an *entering index* $q \in \mathcal{V}$ that is to enter $\mathcal{B}$ using one of the heuristics described near the end of this section. The new vertex $\mathbf{x}'$ must satisfy:

$$\mathbf{Ax}' = \mathbf{A}_{\mathcal{B}}\mathbf{x}'_{\mathcal{B}} + \mathbf{A}_{\{q\}}x'_q = \mathbf{A}_{\mathcal{B}}\mathbf{x}_{\mathcal{B}} = \mathbf{Ax} = \mathbf{b} \tag{11.18}$$

One *leaving index* $p \in \mathcal{B}$ in $\mathbf{x}'_{\mathcal{B}}$ becomes zero during the transition and is replaced by the column of $\mathbf{A}$ corresponding to index q. This action is referred to as *pivoting*.

We can solve for the new design point

$$\mathbf{x}'_{\mathcal{B}} = \mathbf{x}_{\mathcal{B}} - \mathbf{A}_{\mathcal{B}}^{-1}\mathbf{A}_{\{q\}}x'_q \tag{11.19}$$

A particular leaving index $p \in \mathcal{B}$ becomes active when:

$$\left(\mathbf{x}'_{\mathcal{B}}\right)_p = 0 = (\mathbf{x}_{\mathcal{B}})_p - \left(\mathbf{A}_{\mathcal{B}}^{-1}\mathbf{A}_{\{q\}}\right)_p x'_q \tag{11.20}$$

and is thus obtained by increasing $x_q = 0$ to x'_q with:

$$x'_q = \frac{(\mathbf{x}_{\mathcal{B}})_p}{\left(\mathbf{A}_{\mathcal{B}}^{-1}\mathbf{A}_{\{q\}}\right)_p} \tag{11.21}$$

The leaving index is obtained using the *minimum ratio test*, which computes for each potential leaving index and selects the one with minimum x'_q. We then swap p and q between $\mathcal{B}$ and $\mathcal{V}$. The edge transition is implemented in algorithm 11.2.

```
function edge_transition(LP, B, q)
    A, b, c = LP.A, LP.b, LP.c
    n = size(A, 2)
    b_inds = sort(B)
    n_inds = sort!(setdiff(1:n, B))
    AB = A[:,b_inds]
    d, xB = AB\A[:,n_inds[q]], AB\b

    p, xq´ = 0, Inf
    for i in 1 : length(d)
        if d[i] > 0
            v = xB[i] / d[i]
            if v < xq´
                p, xq´ = i, v
            end
        end
    end
    return (p, xq´)
end
```

Algorithm 11.2. A method for computing the index p and the new coordinate value x_q' obtained by increasing index q of the vertex defined by the partition B in the equality-form linear program LP.

The effect that a transition has on the objective function can be computed using x_q'. The objective function value at the new vertex is[12]

[12] Here, we used the fact that $\lambda = \mathbf{A}_\mathcal{B}^{-\top} \mathbf{c}_\mathcal{B}$ and that $\mathbf{A}_{\{q\}}^\top \lambda = c_q - \mu_q$.

$$\mathbf{c}^\top \mathbf{x}' = \mathbf{c}_\mathcal{B}^\top \mathbf{x}_\mathcal{B}' + c_q x_q' \tag{11.22}$$

$$= \mathbf{c}_\mathcal{B}^\top \left(\mathbf{x}_\mathcal{B} - \mathbf{A}_\mathcal{B}^{-1} \mathbf{A}_{\{q\}} x_q' \right) + c_q x_q' \tag{11.23}$$

$$= \mathbf{c}_\mathcal{B}^\top \mathbf{x}_\mathcal{B} - \mathbf{c}_\mathcal{B}^\top \mathbf{A}_\mathcal{B}^{-1} \mathbf{A}_{\{q\}} x_q' + c_q x_q' \tag{11.24}$$

$$= \mathbf{c}_\mathcal{B}^\top \mathbf{x}_\mathcal{B} - \left(c_q - \mu_q \right) x_q' + c_q x_q' \tag{11.25}$$

$$= \mathbf{c}^\top \mathbf{x} + \mu_q x_q' \tag{11.26}$$

Choosing an entering index q decreases the objective function value by

$$\mathbf{c}^\top \mathbf{x}' - \mathbf{c}^\top \mathbf{x} = \mu_q x_q' \tag{11.27}$$

The objective function decreases only if μ_q is negative. In order to progress toward optimality, we must choose an index q in $\mathcal{V}$ such that μ_q is negative. If all components of $\mu_\mathcal{V}$ are non-negative, we have found a global optimum.

Since there can be multiple negative entries in $\mu_\mathcal{V}$, different heuristics can be used to select an entering index:[13]

[13] Modern implementations use more sophisticated rules. For example, see J. J. Forrest and D. Goldfarb, "Steepest-Edge Simplex Algorithms for Linear Programming," *Mathematical Programming*, vol. 57, no. 1, pp. 341–374, 1992.

- *Greedy heuristic*, which chooses a q that maximally reduces $\mathbf{c}^\top \mathbf{x}$.

- *Dantzig's rule*, which chooses the q with the most negative entry in $\boldsymbol{\mu}$. This rule is easy to calculate, but it does not guarantee the maximum reduction in $\mathbf{c}^\top \mathbf{x}$. It is also sensitive to scaling of the constraints.[14]

- *Bland's rule*, which chooses the first q with a negative entry in $\boldsymbol{\mu}$. When used on its own, Bland's rule tends to result in poor performance in practice. However, this rule can help us prevent *cycles*, which occur when we return to a vertex we have visited before without decreasing the objective function. This rule is usually used only after no improvements have been made for several iterations of a different rule to break out of a cycle and ensure convergence.

One iteration of the simplex method's optimization phase moves a vertex partition to a neighboring vertex based on a heuristic for the entering index. Algorithm 11.3 implements such an iteration with the greedy heuristic. Example 11.7 demonstrates using the simplex algorithm starting from a known vertex partition to solve a linear program.

11.2.4   Initialization Phase

The *optimization phase* of the simplex algorithm is implemented in algorithm 11.4. Unfortunately, algorithm 11.4 requires an initial partition that corresponds to a vertex. If we do not have such a partition, we must solve an *auxiliary linear program* to obtain this partition as part of an *initialization phase*.

The auxiliary linear program to be solved in the initialization phase includes extra variables $\mathbf{z} \in \mathbb{R}^m$, which we seek to zero out:[15]

$$
\begin{aligned}
\underset{\mathbf{x},\mathbf{z}}{\text{minimize}} \quad & \begin{bmatrix} \mathbf{0}^\top & \mathbf{1}^\top \end{bmatrix} \begin{bmatrix} \mathbf{x} \\ \mathbf{z} \end{bmatrix} \\
\text{subject to} \quad & \begin{bmatrix} \mathbf{A} & \mathbf{Z} \end{bmatrix} \begin{bmatrix} \mathbf{x} \\ \mathbf{z} \end{bmatrix} = \mathbf{b} \\
& \begin{bmatrix} \mathbf{x} \\ \mathbf{z} \end{bmatrix} \geq \mathbf{0}
\end{aligned}
\tag{11.28}
$$

[14] For constraint $\mathbf{A}^\top \mathbf{x} = \mathbf{b} \rightarrow \alpha \mathbf{A}^\top \mathbf{x} = \alpha \mathbf{b}$, $\alpha > 0$, we do not change the solution but the Lagrange multipliers are scaled, $\lambda \rightarrow \alpha^{-1} \lambda$.

[15] The values for $\mathbf{z}$ represent the amount by which $\mathbf{A}\mathbf{x} = \mathbf{b}$ is violated. By zeroing out $\mathbf{z}$, we find a feasible point. If, in solving the auxiliary problem, we do not find a vertex with a zeroed-out $\mathbf{z}$, then we can conclude that the problem is infeasible. Furthermore, it is not always necessary to add all m extra variables, especially when slack variables are included when transforming between standard and equality form.

Consider the equality-form linear program with

Example 11.7. Solving a linear program with the simplex algorithm.

$$A = \begin{bmatrix} 1 & 1 & 1 & 0 \\ -4 & 2 & 0 & 1 \end{bmatrix}, \quad b = \begin{bmatrix} 9 \\ 2 \end{bmatrix}, \quad c = \begin{bmatrix} 3 \\ -1 \\ 0 \\ 0 \end{bmatrix}$$

and the initial vertex defined by $\mathcal{B} = \{3, 4\}$. After verifying that $\mathcal{B}$ defines a feasible vertex, we can begin one iteration of the simplex algorithm.

We extract $x_{\mathcal{B}}$:

$$x_{\mathcal{B}} = A_{\mathcal{B}}^{-1}b = \begin{bmatrix} 1 & 0 \\ 0 & 1 \end{bmatrix}^{-1} \begin{bmatrix} 9 \\ 2 \end{bmatrix} = \begin{bmatrix} 9 \\ 2 \end{bmatrix}$$

We then compute λ:

$$\lambda = A_{\mathcal{B}}^{-\top}c_{\mathcal{B}} = \begin{bmatrix} 1 & 0 \\ 0 & 1 \end{bmatrix}^{-\top} \begin{bmatrix} 0 \\ 0 \end{bmatrix} = 0$$

and $\mu_{\mathcal{V}}$:

$$\mu_{\mathcal{V}} = c_{\mathcal{V}} - \left(A_{\mathcal{B}}^{-1}A_{\mathcal{V}}\right)^{\top}c_{\mathcal{B}} = \begin{bmatrix} 3 \\ -1 \end{bmatrix} - \left(\begin{bmatrix} 1 & 0 \\ 0 & 1 \end{bmatrix}^{-1} \begin{bmatrix} 1 & 1 \\ -4 & 2 \end{bmatrix}\right)^{\top} \begin{bmatrix} 0 \\ 0 \end{bmatrix} = \begin{bmatrix} 3 \\ -1 \end{bmatrix}$$

We find that $\mu_{\mathcal{V}}$ contains negative elements, so our current $\mathcal{B}$ is suboptimal. We will pivot on the index of the only negative element, $q = 2$. An edge transition is run from $x_{\mathcal{B}}$ in the direction $-A_{\mathcal{B}}^{-1}A_{\{q\}} = [1, 2]$.

Using equation (11.19), we increase x_q' until a new constraint becomes active. In this case, $x_q' = 1$ causes x_4 to become zero. We update our set of basic indices to $\mathcal{B} = \{2, 3\}$.

In the second iteration, we find:

$$x_{\mathcal{B}} = \begin{bmatrix} 1 \\ 8 \end{bmatrix}, \quad \lambda = \begin{bmatrix} 0 \\ -1/2 \end{bmatrix}, \quad \mu_{\mathcal{V}} = \begin{bmatrix} 1 \\ 1/2 \end{bmatrix}.$$

The vertex is optimal because $\mu_{\mathcal{V}}$ has no negative entries. Our algorithm thus terminates with $\mathcal{B} = \{2, 3\}$, for which the design point is $x^* = [0, 1, 8, 0]$.

```julia
function step_lp!(B, LP)
    A, b, c = LP.A, LP.b, LP.c
    n = size(A, 2)
    b_inds = sort!(B)
    n_inds = sort!(setdiff(1:n, B))
    AB, AV = A[:,b_inds], A[:,n_inds]
    xB = AB\b
    cB = c[b_inds]
    λ = AB' \ cB
    cV = c[n_inds]
    μV = cV - AV'*λ

    q, p, xq´, Δ = 0, 0, Inf, Inf
    for i in 1 : length(μV)
        if μV[i] < 0
            pi, xi´ = edge_transition(LP, B, i)
            if μV[i]*xi´ < Δ
                q, p, xq´, Δ = i, pi, xi´, μV[i]*xi´
            end
        end
    end
    if q == 0
        return (B, true) # optimal point found
    end

    if isinf(xq´)
        error("unbounded")
    end

    j = findfirst(isequal(b_inds[p]), B)
    B[j] = n_inds[q] # swap indices
    return (B, false) # new vertex but not optimal
end
```

Algorithm 11.3. A single iteration of the simplex algorithm in which the set $\mathcal{B}$ is moved from one vertex to a neighbor while maximally decreasing the objective function. Here, step_lp! takes a partition defined by B and a linear program LP.

```julia
function minimize_lp!(B, LP)
    done = false
    while !done
        B, done = step_lp!(B, LP)
    end
    return B
end
```

Algorithm 11.4. Minimizing a linear program given a vertex partition defined by B and a linear program LP.

where $\mathbf{Z}$ is a diagonal matrix whose diagonal entries are

$$Z_{ii} = \begin{cases} +1 & \text{if } b_i \geq 0 \\ -1 & \text{otherwise.} \end{cases} \tag{11.29}$$

The auxiliary linear program is solved with a partition defined by $\mathcal{B}$, which selects only the z values. The corresponding vertex has $\mathbf{x} = \mathbf{0}$, and each z-element is the absolute value of the corresponding b-value: $z_j = |b_j|$. This initial vertex can easily be shown to be feasible.

Example 11.8 demonstrates using an auxiliary linear program to obtain a feasible vertex.

Consider the equality-form linear program:

$$\underset{x_1, x_2, x_3}{\text{minimize}} \quad c_1 x_1 + c_2 x_2 + c_3 x_3$$

$$\text{subject to} \quad 2x_1 - 1x_2 + 2x_3 = 1$$
$$5x_1 + 1x_2 - 3x_3 = -2$$
$$x_1, x_2, x_3 \geq 0$$

We can identify a feasible vertex by solving:

$$\underset{x_1, x_2, x_3, z_1, z_2}{\text{minimize}} \quad z_1 + z_2$$

$$\text{subject to} \quad 2x_1 - 1x_2 + 2x_3 + z_1 = 1$$
$$5x_1 + 1x_2 - 3x_3 - z_2 = -2$$
$$x_1, x_2, x_3, z_1, z_2 \geq 0$$

with an initial vertex defined by $\mathcal{B} = \{4, 5\}$.

The initial vertex has:

$$\mathbf{x}_{\mathcal{B}}^{(1)} = \mathbf{A}_{\mathcal{B}}^{-1} \mathbf{b}_{\mathcal{B}} = \begin{bmatrix} 1 & 0 \\ 0 & -1 \end{bmatrix}^{-1} \begin{bmatrix} 1 \\ -2 \end{bmatrix} = \begin{bmatrix} 1 \\ 2 \end{bmatrix}$$

and is thus $\mathbf{x}^{(1)} = [0, 0, 0, 1, 2]$. Solving the auxiliary problem yields $\mathbf{x}^* \approx [0.045, 1.713, 1.312, 0, 0]$. Thus $[0.045, 1.713, 1.312]$ is a feasible vertex in the original problem.

Example 11.8. Using an auxiliary linear program to obtain a feasible vertex.

The partition obtained by solving the auxiliary linear program will produce a feasible design point, $\mathbf{Ax} = \mathbf{b}$ because $\mathbf{z}$ will have been zeroed out. If $\mathbf{z}$ is nonzero, then the original linear program is infeasible. If $\mathbf{z}$ is zero, then the resulting partition can be used as the initial partition for the optimization phase of the simplex algorithm. The original problem must be slightly modified to incorporate the new $\mathbf{z}$ variables:

$$
\begin{aligned}
\underset{\mathbf{x},\mathbf{z}}{\text{minimize}} \quad & \begin{bmatrix} \mathbf{c}^\top & \mathbf{0}^\top \end{bmatrix} \begin{bmatrix} \mathbf{x} \\ \mathbf{z} \end{bmatrix} \\
\text{subject to} \quad & \begin{bmatrix} \mathbf{A} & \mathbf{I} \\ \mathbf{0} & \mathbf{I} \end{bmatrix} \begin{bmatrix} \mathbf{x} \\ \mathbf{z} \end{bmatrix} = \begin{bmatrix} \mathbf{b} \\ \mathbf{0} \end{bmatrix} \\
& \begin{bmatrix} \mathbf{x} \\ \mathbf{z} \end{bmatrix} \geq \mathbf{0}
\end{aligned}
\tag{11.30}
$$

The $\mathbf{z}$ values must be included. Despite their vector counterparts being zero, it is possible that some indices in the components of $\mathbf{z}$ are included in the initial partition $\mathcal{B}$. One can inspect the initial partition and include only the specific components that are needed.

The solution $(\mathbf{x}^*, \mathbf{z}^*)$ obtained by solving the second LP will have $\mathbf{z}^* = \mathbf{0}$. Thus, $\mathbf{x}^*$ will be a solution to the original linear problem.

An implementation for the complete simplex algorithm is given in algorithm 11.5.

11.3   Dual Certificates

Suppose we have a candidate solution and we want to verify that it is optimal. Verifying optimality using *dual certificates* (algorithm 11.6) is useful in many cases, such as debugging our linear program code.

We know from the FONCs for constrained optimization that the optimal value of the dual problem d^* is a lower bound of the optimal value of the primal problem p^*. Linear programs are linear and convex, and one can show that the optimal value of the dual problem is also the optimal value of the primal problem, $d^* = p^*$.

```
function minimize_lp(LP)
    A, b, c = LP.A, LP.b, LP.c
    m, n = size(A)
    z = ones(m)
    Z = Matrix(Diagonal([j ≥ 0 ? 1 : -1 for j in b]))

    A′ = hcat(A, Z)
    b′ = b
    c′ = vcat(zeros(n), z)
    LP_init = LinearProgram(A′, b′, c′)
    B = collect(1:m).+n
    minimize_lp!(B, LP_init)

    if any(i-> i > n, B)
        error("infeasible")
    end

    A′′ = [A           Matrix(1.0I, m, m);
           zeros(m,n) Matrix(1.0I, m, m)]
    b′′ = vcat(b, zeros(m))
    c′′ = c′
    LP_opt = LinearProgram(A′′, b′′, c′′)
    minimize_lp!(B, LP_opt)
    return get_vertex(B, LP_opt)[1:n]
end
```

Algorithm 11.5. The simplex algorithm for solving linear programs in equality form when an initial partition is not known.

The primal linear program can be converted to its dual form:

$$\max_{\mu \geq 0, \lambda} \min_{x} \mathcal{L}(x, \mu, \lambda) = \max_{\mu \geq 0, \lambda} \min_{x} c^\top x - \mu^\top x - \lambda^\top (Ax - b) \qquad (11.31)$$

$$= \max_{\mu \geq 0, \lambda} \min_{x} (c - \mu - A^\top \lambda)^\top x + \lambda^\top b \qquad (11.32)$$

From the FONC, we know $c - \mu - A^\top \lambda = 0$, which allows us to drop the first term in the objective above. In addition, we know $\mu = c - A^\top \lambda \geq 0$, which implies $A^\top \lambda \leq c$. In summary, we have:

Primal Form (equality)   Dual Form

$$\begin{aligned}\underset{x}{\text{minimize}} \quad & c^\top x \\ \text{subject to} \quad & Ax = b \\ & x \geq 0\end{aligned}$$

$$\begin{aligned}\underset{\lambda}{\text{maximize}} \quad & b^\top \lambda \\ \text{subject to} \quad & A^\top \lambda \leq c\end{aligned}$$

If the primal problem has n variables and m equality constraints, then the dual problem has m variables and n constraints.[16] Furthermore, the dual of the dual is the primal problem.

Optimality can be assessed by verifying three properties. If someone claims (x^*, λ^*) is optimal, we can quickly verify the claim by checking whether all three of the following conditions are satisfied:

1. x^* is feasible in the primal problem.

2. λ^* is feasible in the dual problem.

3. $p^* = c^\top x^* = b^\top \lambda^* = d^*$.

Dual certificates are used in example 11.9 to verify the solution to a linear program.

```
function dual_certificate(LP, x, λ, ϵ=1e-6)
    A, b, c = LP.A, LP.b, LP.c
    primal_feasible = all(x .≥ 0) && A*x ≈ b
    dual_feasible = all(A'*λ .≤ c)
    return primal_feasible && dual_feasible &&
           isapprox(c·x, b·λ, atol=ϵ)
end
```

[16] An alternative to the simplex algorithm, the *self-dual simplex algorithm*, tends to be faster in practice. It does not require that the matrix A_B satisfy $x_B = A_B^{-1} b \geq 0$. The self-dual simplex algorithm is a modification of the simplex algorithm for the dual of the linear programming problem in standard form.

Algorithm 11.6. A method for checking whether a candidate solution given by design point x and dual point λ for the linear program LP in equality form is optimal. The parameter ϵ controls the tolerance for the equality constraint.

Example 11.9. Verifying a solution using dual certificates.

Consider the standard-form linear program with

$$\mathbf{A} = \begin{bmatrix} 1 & 1 & -1 \\ -1 & 2 & 0 \\ 1 & 2 & 3 \end{bmatrix}, \quad \mathbf{b} = \begin{bmatrix} 1 \\ -2 \\ 5 \end{bmatrix}, \quad \mathbf{c} = \begin{bmatrix} 1 \\ 1 \\ -1 \end{bmatrix}$$

We would like to determine whether $\mathbf{x}^* = [2, 0, 1]$ and $\boldsymbol{\lambda}^* = [1, 0, 0]$ are an optimal solution pair. We first verify that $\mathbf{x}^*$ is feasible:

$$\mathbf{A}\mathbf{x}^* = [1, -2, 5] = \mathbf{b}, \qquad \mathbf{x}^* \geq \mathbf{0}$$

We then verify that $\boldsymbol{\lambda}^*$ is dual-feasible:

$$\mathbf{A}^\top \boldsymbol{\lambda}^* = [1, 1, -1] \leq \mathbf{c}$$

Finally, we verify that p^* and d^* are the same:

$$p^* = \mathbf{c}^\top \mathbf{x}^* = 1 = \mathbf{b}^\top \boldsymbol{\lambda}^* = d^*$$

We conclude that $(\mathbf{x}^*, \boldsymbol{\lambda}^*)$ are optimal.

11.4 Summary

- Linear programs are problems consisting of a linear objective function and linear constraints.

- The simplex algorithm can optimize linear programs globally in an efficient manner.

- Dual certificates allow us to verify that a candidate primal-dual solution pair is optimal.

11.5 Exercises

Exercise 11.1. Suppose you do not know any optimization algorithm for solving a linear program. You decide to evaluate all the vertices and determine, by inspection, which one minimizes the objective function. Give a loose upper bound on the number of possible minimizers you will examine. Furthermore, does this method properly handle all linear constrained optimization problems?

Exercise 11.2. If the program in example 11.1 is bounded below, argue that the simplex method must converge.

Exercise 11.3. Suppose we want to minimize $6x_1 + 5x_2$ subject to the constraint $3x_1 - 2x_2 \geq 5$. How would we translate this problem into a linear program in equality form with the same minimizer?

Exercise 11.4. Suppose your optimization algorithm has found a search direction $\mathbf{d}$ and you want to conduct a line search. However, you know that there is a linear constraint $\mathbf{w}^\top \mathbf{x} \geq 0$. How would you modify the line search to take this constraint into account? You can assume that your current design point is feasible.

Exercise 11.5. Reformulate the linear program

$$\begin{aligned} \underset{\mathbf{x}}{\text{minimize}} \quad & \mathbf{c}^\top \mathbf{x} \\ \text{subject to} \quad & \mathbf{A}\mathbf{x} \geq \mathbf{0} \end{aligned} \tag{11.33}$$

into an unconstrained optimization problem with a log barrier penalty.

12   *Multiobjective Optimization*

Previous chapters have developed methods for optimizing single-objective functions, but this chapter is concerned with *multiobjective optimization*, or *vector optimization*, where we must optimize with respect to several objectives simultaneously. Engineering is often a tradeoff between cost, performance, and time-to-market, and it is often unclear how to prioritize different objectives. We will discuss various methods for transforming vector-valued objective functions to scalar-valued objective functions so that we can use the algorithms discussed in previous chapters to arrive at an optimum. In addition, we will discuss algorithms for identifying the set of design points that represent the best tradeoff between objectives, without having to commit to a particular prioritization of objectives. These design points can then be presented to experts who can then identify the most desirable design.[1]

12.1   *Pareto Optimality*

The notion of *Pareto optimality* is useful when discussing problems where there are multiple objectives. A design is Pareto optimal if it is impossible to improve in one objective without worsening at least one other objective. In multiobjective design optimization, we can generally focus our efforts on designs that are Pareto optimal without having to commit to a particular tradeoff between objectives. This section introduces some definitions and concepts that are helpful when discussing approaches to identifying Pareto-optimal designs.

[1] Additional methods are surveyed in R. T. Marler and J. S. Arora, "Survey of Multi-Objective Optimization Methods for Engineering," *Structural and Multidisciplinary Optimization*, vol. 26, no. 6, pp. 369–395, 2004. For a textbook dedicated entirely to multiobjective optimization, see K. Miettinen, *Nonlinear Multiobjective Optimization*. Kluwer Academic Publishers, 1999.

12.1.1   Dominance

In single-objective optimization, two design points $\mathbf{x}$ and $\mathbf{x}'$ can be ranked objectively based on their scalar function values. The point $\mathbf{x}'$ is better whenever $f(\mathbf{x}')$ is less than $f(\mathbf{x})$.

In multiobjective optimization, our objective function $\mathbf{f}$ returns an m-dimensional vector of values $\mathbf{y}$ when evaluated at a design point $\mathbf{x}$. The different dimensions of $\mathbf{y}$ correspond to different objectives, sometimes also referred to as metrics or criteria. We can objectively rank two design points $\mathbf{x}$ and $\mathbf{x}'$ only when one is better in at least one objective and no worse in any other. That is, $\mathbf{x}$ *dominates* $\mathbf{x}'$ if and only if

$$f_i(\mathbf{x}) \le f_i(\mathbf{x}') \ \text{ for all } i \text{ in } \{1, \ldots, m\}$$
$$\text{and } f_i(\mathbf{x}) < f_i(\mathbf{x}') \ \text{ for some } i \tag{12.1}$$

as compactly implemented in algorithm 12.1.

```
dominates(y, y′) = all(y .≤ y′) && any(y .< y′)
```

Algorithm 12.1. A method for checking whether $\mathbf{x}$ dominates $\mathbf{x}'$, where y is the vector of objective values for $\mathbf{f}(\mathbf{x})$ and y′ is the vector of objective values for $\mathbf{f}(\mathbf{x}')$.

Figure 12.1 shows that in multiple dimensions there are regions with dominance ambiguity. This ambiguity arises whenever $\mathbf{x}$ is better in some objectives and $\mathbf{x}'$ is better in others. Several methods exist for resolving these ambiguities.

Single Objective

Multiple Objectives

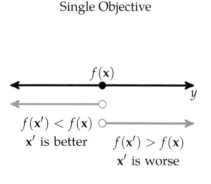
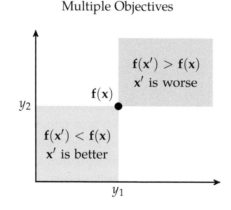

Figure 12.1. Design points can be objectively ranked in single-objective optimization but can be objectively ranked in multiobjective optimization only in some cases.

12.1.2 Pareto Frontier

In mathematics, an *image* of an input set through some function is the set of all possible outputs of that function when evaluated on elements of that input set. We will denote the image of $\mathcal{X}$ through $\mathbf{f}$ as $\mathcal{Y}$, and we will refer to $\mathcal{Y}$ as the *criterion space*. Figure 12.2 shows examples of criterion space for problems with single and multiple objectives. As illustrated, the criterion space in single-objective optimization is one dimensional. All of the global optima share a single objective function value, y^*. In multiobjective optimization, the criterion space is m-dimensional, where m is the number of objectives. There is typically no globally best objective function value because there may be ambiguity when tradeoffs between objectives are not specified.

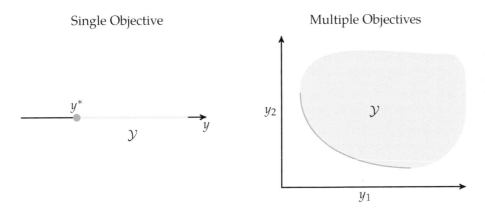

Figure 12.2. The criterion space is the set of all objective values obtained by feasible design points. Well-posed problems have criterion spaces that are bounded from below, but they do not have to be bounded from above. The Pareto frontier is highlighted in dark blue.

In multiobjective optimization, we can define the notion of *Pareto optimality*. A design point $\mathbf{x}$ is Pareto-optimal when no point dominates it. That is, $\mathbf{x} \in \mathcal{X}$ is Pareto-optimal if there does not exist an $\mathbf{x}' \in \mathcal{X}$ such that $\mathbf{x}'$ dominates $\mathbf{x}$. The set of Pareto-optimal points forms the *Pareto frontier*. The Pareto frontier is valuable for helping decision-makers make design trade decisions as discussed in example 12.1. In two dimensions, the Pareto frontier is also referred to as a *Pareto curve*.

All Pareto-optimal points lie on the boundary of the criterion space. Some multiobjective optimization methods also find *weakly Pareto-optimal* points. Whereas Pareto-optimal points are those such that no other point improves at least one objective, weakly Pareto-optimal points are those such that no other point improves all of the objectives (figure 12.3). That is, $\mathbf{x} \in \mathcal{X}$ is weakly Pareto-optimal if there does not exist an $\mathbf{x}' \in \mathcal{X}$ such that $\mathbf{f}(\mathbf{x}') < \mathbf{f}(\mathbf{x})$. Pareto-optimal points are also weakly Pareto optimal. Weakly Pareto-optimal points are not necessarily Pareto optimal.

Several methods discussed below use another special point. We define the *utopia point* to be the point in the criterion space consisting of the component-wise optima:

$$y_i^{\text{utopia}} = \underset{\mathbf{x} \in \mathcal{X}}{\text{minimize}}\, f_i(\mathbf{x}) \tag{12.2}$$

The utopia point is often not attainable; optimizing one component typically requires a tradeoff in another component.

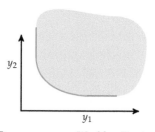

Figure 12.3. Weakly Pareto-optimal points, shown in red, cannot be improved simultaneously in all objectives.

12.1.3   Pareto Frontier Generation

There are several methods for generating Pareto frontiers. A naive approach is to sample design points throughout the design space and then to identify the nondominated points (algorithm 12.2). This approach is typically wasteful, leading to many dominated design points as shown in figure 12.4. In addition, this approach does not guarantee a smooth or correct Pareto frontier. The remainder of this chapter discusses more effective ways to generate Pareto frontiers.

When constructing a collision avoidance system for aircraft, one must minimize both the collision rate and the alert rate. Although more alerts can result in preventing more collisions if the alerts are followed, too many alerts can result in pilots losing trust in the system and lead to decreased compliance with the system. Hence, the designers of the system must carefully trade alerts and collision risk.

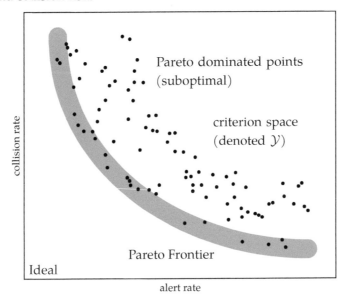

Example 12.1. An approximate Pareto frontier obtained from evaluating many different design points for an aircraft collision avoidance system.

By varying the collision avoidance system's design parameters, we can obtain many different collision avoidance systems, but, as the figure shows, some of these will be better than others. A Pareto frontier can be extracted to help domain experts and regulators understand the effects that objective tradeoffs will have on the optimized system.

```
function naive_pareto(xs, ys)
    pareto_xs, pareto_ys = similar(xs, 0), similar(ys, 0)
    for (x,y) in zip(xs,ys)
        if !any(dominates(y′,y) for y′ in ys)
            push!(pareto_xs, x)
            push!(pareto_ys, y)
        end
    end
    return (pareto_xs, pareto_ys)
end
```

Algorithm 12.2. A method for generating a Pareto frontier using randomly sampled design points xs and their multiobjective values ys. Both the Pareto-optimal design points and their objective values are returned.

12.2 Constraint Methods

Constraints can be used to cut out sections of the Pareto frontier and obtain a single optimal point in the criterion space. Constraints can be supplied either by the problem designer or automatically obtained based on an ordering of the objectives.

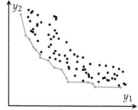

Figure 12.4. Generating Pareto frontiers with naively scattered points is straightforward but inefficient and approximate.

12.2.1 Constraint Method

The *constraint method* constrains all but one of the objectives. Here we choose f_1 without loss of generality:

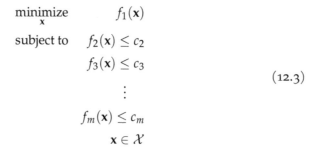

$$
\begin{aligned}
\underset{\mathbf{x}}{\text{minimize}} \quad & f_1(\mathbf{x}) \\
\text{subject to} \quad & f_2(\mathbf{x}) \leq c_2 \\
& f_3(\mathbf{x}) \leq c_3 \\
& \quad \vdots \\
& f_m(\mathbf{x}) \leq c_m \\
& \mathbf{x} \in \mathcal{X}
\end{aligned}
\tag{12.3}
$$

Given the vector $\mathbf{c}$, the constraint method produces a unique optimal point in the criterion space, provided that the constraints are feasible. The constraint method can be used to generate Pareto frontiers by varying $\mathbf{c}$ as shown in figure 12.5.

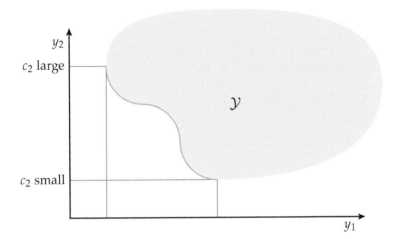

Figure 12.5. The constraint method for generating a Pareto frontier. This method can identify points in the concave region of the Pareto frontier.

12.2.2 Lexicographic Method

The *lexicographic method* ranks the objectives in order of importance. A series of single-objective optimizations are performed on the objectives in order of importance. Each optimization problem includes constraints to preserve the optimality with respect to previously optimized objectives as shown in figure 12.6.

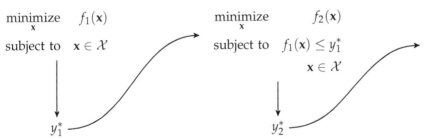

Figure 12.6. The lexicographic method for an optimization problem with three objectives.

Iterations are always feasible because the minimum point from the previous optimization is always feasible. The constraints could also be replaced with equalities, but inequalities are often easier for optimizers to enforce. In addition, if the optimization method used is not optimal, then subsequent optimizations may encounter better solutions that would otherwise be rejected. The lexicographic method is sensitive to the ordering of the objective functions.

12.3   Weight Methods

A designer can sometimes identify preferences between the objectives and encode these preferences as a vector of weights. In cases where the choice of weights is not obvious, we can generate a Pareto frontier by sweeping over the space of weights. This section also discusses a variety of alternative methods for transforming multiobjective functions into single-objective functions.

12.3.1   Weighted Sum Method

The *weighted sum method* (algorithm 12.3) uses a vector of weights $\mathbf{w}$ to convert $\mathbf{f}$ to a single objective f:[2]

$$f(\mathbf{x}) = \mathbf{w}^\top \mathbf{f}(\mathbf{x}) \qquad (12.4)$$

where the weights are nonnegative and sum to 1. The weights can be interpreted as costs associated with each objective. The Pareto frontier can be extracted by varying $\mathbf{w}$ and solving the associated optimization problem with the objective in equation (12.4). In two dimensions, we vary w_1 from 0 to 1, setting $w_2 = 1 - w_1$. This approach is illustrated in figure 12.7.

[2] L. Zadeh, "Optimality and Non-Scalar-Valued Performance Criteria," *IEEE Transactions on Automatic Control*, vol. 8, no. 1, pp. 59–60, 1963.

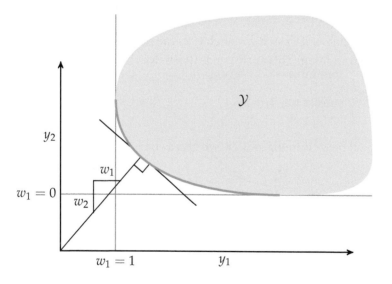

Figure 12.7.   The weighted sum method used to generate a Pareto frontier. Varying the weights allows us to trace the Pareto frontier.

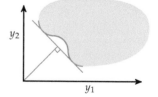

Figure 12.8. The points in red are Pareto optimal but cannot be obtained using the weighted sum method.

In contrast with the constraint method, the weighted sum method cannot obtain points in nonconvex regions of the Pareto frontier as shown in figure 12.8.

A given set of weights forms a linear objective function with parallel contour lines marching away from the origin. If the feasible set bends away from the origin, it will have other Pareto optimal points on the boundary that cannot be recovered by minimizing equation (12.4).

```
function weight_pareto(f1, f2, npts)
    return [
        optimize(x->w1*f1(x) + (1-w1)*f2(x))
        for w1 in range(0,stop=1,length=npts)
    ]
end
```

Algorithm 12.3. The weighted sum method for generating a Pareto frontier, which takes objective functions f1 and f2 and number of Pareto points npts.

12.3.2   Goal Programming

Goal programming[3] is a method for converting a multiobjective function to a single-objective function by minimizing an L_p norm between $\mathbf{f}(\mathbf{x})$ and a goal point:

$$\underset{\mathbf{x}\in\mathcal{X}}{\text{minimize}}\left\|\mathbf{f}(\mathbf{x}) - \mathbf{y}^{\text{goal}}\right\|_p \tag{12.5}$$

where the goal point is typically the utopia point. The equation above does not involve a vector of weights, but the other methods discussed in this chapter can be thought of as generalizations of goal programming. This approach is illustrated in figure 12.9.

[3] Goal programming generally refers to using $p = 1$. An overview is presented in D. Jones and M. Tamiz, *Practical Goal Programming*. Springer, 2010.

12.3.3   Weighted Exponential Sum

The *weighted exponential sum* combines goal programming and the weighted sum method[4]

$$f(\mathbf{x}) = \sum_{i=1}^{m} w_i\left(f_i(\mathbf{x}) - y_i^{\text{goal}}\right)^p \tag{12.6}$$

where $\mathbf{w}$ is a vector of positive weights that sum to 1 and $p \geq 1$ is an exponent similar to that used in L_p norms. As before, zero-valued weights can result in weakly Pareto-optimal points.

[4] P. L. Yu, "Cone Convexity, Cone Extreme Points, and Nondominated Solutions in Decision Problems with Multiobjectives," *Journal of Optimization Theory and Applications*, vol. 14, no. 3, pp. 319–377, 1974.

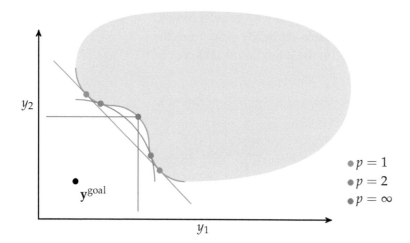

Figure 12.9. Solutions to goal programming as the value for p is changed.

The weighted exponential sum weighs each component of the distance between the solution point and the goal point in the criterion space. Increasing p increases the relative penalty of the largest coordinate deviation between $\mathbf{f}(\mathbf{x})$ and the goal point. While portions of the Pareto-optimal set can be obtained by continuously varying p, we are not guaranteed to obtain the complete Pareto frontier, and it is generally preferable to vary $\mathbf{w}$ using a constant p.

12.3.4   Weighted Min-Max Method

Using higher values of p with the weighted exponential sum objective tends to produce better coverage of the Pareto frontier because the distance contours are able to enter nonconvex regions of the Pareto frontier. The *weighted min-max method*, also called the *weighted Tchebycheff method*, is the limit as p approaches infinity:[5]

$$f(\mathbf{x}) = \max_i \left[w_i \left(f_i(\mathbf{x}) - y_i^{\text{goal}} \right) \right] \tag{12.7}$$

The weighted min-max method can provide the complete Pareto-optimal set by scanning over the weights but will also produce weakly Pareto-optimal points. The method can be augmented to produce only the Pareto frontier

$$f(\mathbf{x}) = \max_i \left[w_i \left(f_i(\mathbf{x}) - y_i^{\text{goal}} \right) \right] + \rho\, \mathbf{f}(\mathbf{x})^\top \mathbf{y}^{\text{goal}} \tag{12.8}$$

[5] The maximization can be removed by including an additional parameter λ:

$$\underset{\mathbf{x}, \lambda}{\text{minimize}} \quad \lambda$$
$$\text{subject to} \quad \mathbf{x} \in \mathcal{X}$$
$$\mathbf{w} \odot \left(\mathbf{f}(\mathbf{x}) - \mathbf{y}^{\text{goal}} \right) - \lambda \mathbf{1} \leq 0$$

where ρ is a small positive scalar with values typically between 0.0001 and 0.01. The added term requires that all terms in $\mathbf{y}_{\text{goal}}$ be positive, which can be accomplished by shifting the objective function. By definition, $\mathbf{f}(\mathbf{x}) \geq \mathbf{y}_{\text{goal}}$ for all $\mathbf{x}$. Any weakly Pareto-optimal point will have $\mathbf{f}(\mathbf{x})^\top \mathbf{y}^{\text{goal}}$ larger than a strongly Pareto-optimal point closer to $\mathbf{y}_{\text{goal}}$.

12.3.5   Exponential Weighted Criterion

The *exponential weighted criterion*[6] was motivated by the inability of the weighted sum method to obtain points on nonconvex portions of the Pareto frontier. It constructs a scalar objective function according to

$$f(\mathbf{x}) = \sum_{i=1}^{m} (e^{pw_i} - 1)e^{pf_i(\mathbf{x})} \tag{12.9}$$

[6] T. W. Athan and P. Y. Papalambros, "A Note on Weighted Criteria Methods for Compromise Solutions in Multi-Objective Optimization," *Engineering Optimization*, vol. 27, no. 2, pp. 155–176, 1996.

Each objective is individually transformed and reweighted. High values of p can lead to numerical overflow.

12.4   Multiobjective Population Methods

Population methods have also been applied to multiobjective optimization.[7] We can adapt the standard algorithms to encourage populations to spread over the Pareto frontier.

[7] Population methods are covered in chapter 9.

12.4.1   Subpopulations

Population methods can divide their attention over several potentially competing objectives. The population can be partitioned into *subpopulations*, where each subpopulation is optimized with respect to different objectives. A traditional genetic algorithm, for example, can be modified to bias the selection of individuals for recombination toward the fittest individuals within each subpopulation. Those selected can form offspring with individuals from different subpopulations.

One of the first adaptations of population methods to multiobjective optimization is the *vector evaluated genetic algorithm*[8] (algorithm 12.4). Figure 12.10 shows how subpopulations are used in a vector evaluated genetic algorithm to maintain diversity over multiple objectives. The progression of a vector evaluated genetic algorithm is shown in figure 12.11.

[8] J. D. Schaffer, "Multiple Objective Optimization with Vector Evaluated Genetic Algorithms," in *International Conference on Genetic Algorithms and Their Applications*, 1985.

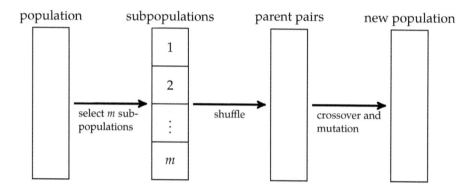

Figure 12.10. Using subpopulations in a vector evaluated genetic algorithm.

```
function vector_evaluated_genetic_algorithm(f, population,
    k_max, S, C, M)
    m = length(f(population[1]))
    m_pop = length(population)
    m_subpop = m_pop ÷ m
    for k in 1 : k_max
        ys = f.(population)
        parents = select(S, [y[1] for y in ys])[1:m_subpop]
        for i in 2 : m
            subpop = select(S,[y[i] for y in ys])[1:m_subpop]
            append!(parents, subpop)
        end

        p = randperm(2m_pop)
        p_ind=i->parents[mod(p[i]-1,m_pop)+1][(p[i]-1)÷m_pop + 1]
        parents = [[p_ind(i), p_ind(i+1)] for i in 1 : 2 : 2m_pop]
        children = [crossover(C,population[p[1]],population[p[2]])
                for p in parents]
        population = [mutate(M, c) for c in children]
    end
    return population
end
```

Algorithm 12.4. The vector evaluated genetic algorithm, which takes a vector-valued objective function f, an initial population, number of iterations k_max, a SelectionMethod S, a CrossoverMethod C, and a MutationMethod M. The resulting population is returned.

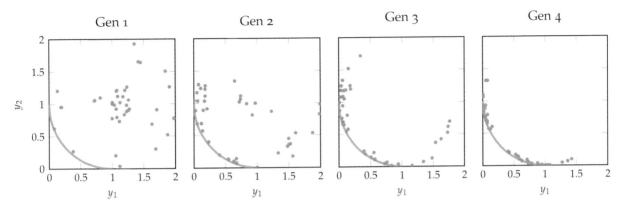

Figure 12.11. A vector evaluated genetic algorithm applied to the circle function defined in appendix B.8. The Pareto frontier is shown in blue.

12.4.2 Nondomination Ranking

One can compute naive Pareto frontiers using the individuals in a population. A design point that lies on the approximate Pareto frontier is considered better than a value deep within the criterion space. We can use *nondomination ranking* (algorithm 12.5) to rank individuals according to the following levels:[9]

Level 1. Nondominated individuals in the population.

Level 2. Nondominated individuals except those in Level 1.

Level 3. Nondominated individuals except those in Levels 1 or 2.

$\vdots$

Level k. Nondominated individuals except those in Levels 1 to $k-1$.

Level 1 is obtained by applying algorithm 12.2 to the population. Subsequent levels are generated by removing all previous levels from the population and then applying algorithm 12.2 again. This process is repeated until all individuals have been ranked. An individual's objective function value is proportional to its rank.

The nondomination levels for an example population are shown in figure 12.12.

[9] K. Deb, A. Pratap, S. Agarwal, and T. Meyarivan, "A Fast and Elitist Multiobjective Genetic Algorithm: NSGA-II," *IEEE Transactions on Evolutionary Computation*, vol. 6, no. 2, pp. 182–197, 2002.

```
function get_non_domination_levels(ys)
    L, m = 0, length(ys)
    levels = zeros(Int, m)
    while minimum(levels) == 0
        L += 1
        for (i,y) in enumerate(ys)
            if levels[i] == 0 &&
                !any((levels[i] == 0 || levels[i] == L) &&
                    dominates(ys[i],y) for i in 1 : m)
                levels[i] = L
            end
        end
    end
    return levels
end
```

Algorithm 12.5. A function for getting the nondomination levels of an array of multiobjective function evaluations, ys.

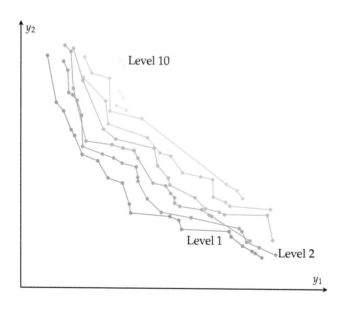

Figure 12.12. The nondomination levels for a population. Darker levels have lower (better) rankings.

12.4.3 Pareto Filters

Population methods can be augmented with a *Pareto filter*, which is a population that approximates the Pareto frontier.[10] The filter is typically updated with every generation (algorithm 12.7). Individuals in the population that are not dominated by any individuals in the filter are added. Any dominated points in the filter are removed. Individuals from the Pareto filter can be injected into the population, thereby reducing the chance that portions of the Pareto frontier are lost between generations.

The filter often has a maximum capacity.[11] Filters that are overcapacity can be reduced by finding the closest pair of design points in the criterion space and removing one individual from that pair. This pruning method is implemented in algorithm 12.6. A Pareto filter obtained using a genetic algorithm is shown in figure 12.13.

[10] H. Ishibuchi and T. Murata, "A Multi-Objective Genetic Local Search Algorithm and Its Application to Flowshop Scheduling," *IEEE Transactions on Systems, Man, and Cybernetics*, vol. 28, no. 3, pp. 392–403, 1998.

[11] Typically the size of the population.

```
function discard_closest_pair!(xs, ys)
    index, min_dist = 0, Inf
    for (i,y) in enumerate(ys)
        for (j, y′) in enumerate(ys[i+1:end])
            dist = norm(y - y′)
            if dist < min_dist
                index, min_dist = rand([i,j]), dist
            end
        end
    end
    deleteat!(xs, index)
    deleteat!(ys, index)
    return (xs, ys)
end
```

Algorithm 12.6. The method `discard_closest_pair!` is used to remove one individual from a filter that is above capacity. The method takes the filter's list of design points xs and associated objective function values ys.

```
function update_pareto_filter!(filter_xs, filter_ys, xs, ys;
    capacity=length(xs),
    )
    for (x,y) in zip(xs, ys)
        if !any(dominates(y′,y) for y′ in filter_ys)
            push!(filter_xs, x)
            push!(filter_ys, y)
        end
    end
    filter_xs, filter_ys = naive_pareto(filter_xs, filter_ys)
    while length(filter_xs) > capacity
        discard_closest_pair!(filter_xs, filter_ys)
    end
    return (filter_xs, filter_ys)
end
```

Algorithm 12.7. A method for updating a Pareto filter with design points filter_xs, corresponding objective function values filter_ys, a population with design points xs and objective function values ys, and filter capacity capacity which defaults to the population size.

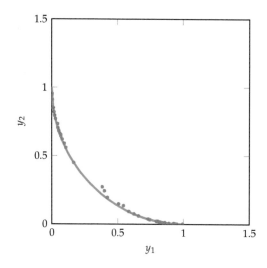

Figure 12.13. A Pareto filter used on the genetic algorithm in figure 12.11 to approximate the Pareto frontier.

12.4.4 Niche Techniques

The term *niche* refers to a focused cluster of points, typically in the criterion space, as shown in figure 12.14. Population methods can converge on a few niches, which limits their spread over the Pareto frontier. *Niche techniques* help encourage an even spread of points.

In *fitness sharing*,[12] shown in figure 12.15, an individual's objective values are penalized by a factor equal to the number of other points within a specified distance in the criterion space.[13] This scheme causes all points in a local region to share the fitness of the other points within the local region. Fitness sharing can be used together with nondomination ranking and subpopulation evaluation.

Equivalence class sharing can be applied to nondomination ranking. When comparing two individuals, the fitter individual is first determined based on the nondomination ranking. If they are equal, the better individual is the one with the fewest number of individuals within a specified distance in the criterion space.

Another niche technique has been proposed for genetic algorithms in which parents selected for crossover cannot be too close together in the criterion space. Selecting only nondominated individuals is also recommended.[14]

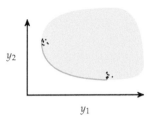

Figure 12.14. Two clear niches for a population in a two-dimensional criterion space.

[12] Fitness is inversely related to the objective being minimized.

[13] D. E. Goldberg and J. Richardson, "Genetic Algorithms with Sharing for Multimodal Function Optimization," in *International Conference on Genetic Algorithms*, 1987.

[14] S. Narayanan and S. Azarm, "On Improving Multiobjective Genetic Algorithms for Design Optimization," *Structural Optimization*, vol. 18, no. 2-3, pp. 146–155, 1999.

Figure 12.15. The results of applying fitness sharing to the Pareto filter in figure 12.13, thereby significantly improving its coverage.

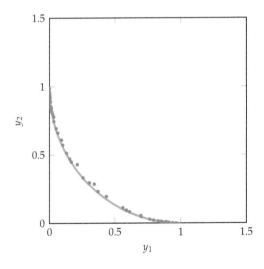

12.5   Preference Elicitation

Preference elicitation involves inferring a scalar-valued objective function from preferences of experts about the tradeoffs between objectives.[15] There are many different ways to represent the scalar-valued objective function, but this section will focus on the weighted sum model where $f(\mathbf{x}) = \mathbf{w}^\top \mathbf{f}(\mathbf{x})$. Once we identify a suitable $\mathbf{w}$, we can use this scalar-valued objective function to find an optimal design.

12.5.1   Model Identification

A common approach for identifying the weight vector $\mathbf{w}$ in our preference model involves asking experts to state their preference between two points $\mathbf{a}$ and $\mathbf{b}$ in the criterion space $\mathcal{Y}$ (figure 12.16). Each of these points is the result of optimizing for a point on the Pareto frontier using an associated weight vector $\mathbf{w_a}$ and $\mathbf{w_b}$. The expert's response is either a preference for $\mathbf{a}$ or a preference for $\mathbf{b}$. There are other schemes for eliciting preference information, such as ranking points in the criterion space, but this binary preference query has been shown to pose minimal cognitive burden on the expert.[16]

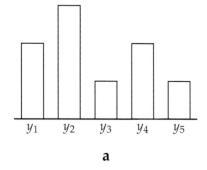

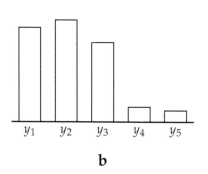

a   b

[15] This chapter overviews non-Bayesian approaches to preference elicitation. For Bayesian approaches, see: S. Guo and S. Sanner, "Real-Time Multiattribute Bayesian Preference Elicitation with Pairwise Comparison Queries," in *International Conference on Artificial Intelligence and Statistics (AISTATS)*, 2010. J. R. Lepird, M. P. Owen, and M. J. Kochenderfer, "Bayesian Preference Elicitation for Multiobjective Engineering Design Optimization," *Journal of Aerospace Information Systems*, vol. 12, no. 10, pp. 634–645, 2015.

[16] V. Conitzer, "Eliciting Single-Peaked Preferences Using Comparison Queries," *Journal of Artificial Intelligence Research*, vol. 35, pp. 161–191, 2009.

Figure 12.16.   Preference elicitation schemes often involve asking experts their preferences between two points in the criterion space.

Suppose the outcomes of the expert queries have resulted in a set of criterion pairs

$$\left\{ (\mathbf{a}^{(1)}, \mathbf{b}^{(1)}), \dots, (\mathbf{a}^{(n)}, \mathbf{b}^{(n)}) \right\} \tag{12.10}$$

where $\mathbf{a}^{(i)}$ is preferable to $\mathbf{b}^{(i)}$ in each pair. For each of these preferences, the weight vector must satisfy

$$\mathbf{w}^\top \mathbf{a}^{(i)} < \mathbf{w}^\top \mathbf{b}^{(i)} \implies (\mathbf{a}^{(i)} - \mathbf{b}^{(i)})^\top \mathbf{w} < 0 \tag{12.11}$$

In order to be consistent with the data, the weight vector must satisfy

$$
\begin{cases}
(\mathbf{a}^{(i)} - \mathbf{b}^{(i)})^\top \mathbf{w} < 0 \text{ for all } i \text{ in } \{1, \dots, n\} \\
\mathbf{1}^\top \mathbf{w} = 1 \\
\mathbf{w} \geq \mathbf{0}
\end{cases}
\tag{12.12}
$$

Many different weight vectors could potentially satisfy the above equation. One approach is to choose a $\mathbf{w}$ that best separates $\mathbf{w}^\top \mathbf{a}^{(i)}$ from $\mathbf{w}^\top \mathbf{b}^{(i)}$

$$
\begin{aligned}
\underset{\mathbf{w}}{\text{minimize}} \quad & \sum_{i=1}^{n} (\mathbf{a}^{(i)} - \mathbf{b}^{(i)})^\top \mathbf{w} \\
\text{subject to} \quad & (\mathbf{a}^{(i)} - \mathbf{b}^{(i)})^\top \mathbf{w} < 0 \text{ for } i \in \{1, \dots, n\} \\
& \mathbf{1}^\top \mathbf{w} = 1 \qquad \mathbf{w} \geq \mathbf{0}
\end{aligned}
\tag{12.13}
$$

It is often desirable to choose the next weight vector such that it minimizes the distance from the previous weight vector. We can replace the objective function in equation (12.13) with $\|\mathbf{w} - \mathbf{w}^{(n)}\|_1$, thereby ensuring that our new weight vector $\mathbf{w}^{(n+1)}$ is as close as possible to our current one.[17]

12.5.2 *Paired Query Selection*

We generally want to choose the two points in the criterion space so that the outcome of the query is as informative as possible. There are many different approaches for such *paired query selection*, but we will focus on methods that attempt to reduce the space of weights consistent with *expert responses*, preference information supplied by a domain expert.

We will denote the set of weights consistent with expert responses as $\mathcal{W}$, which is defined by the linear constraints in equation (12.12). Because weights are bounded between 0 and 1, the feasible set is an enclosed region forming a convex polytope with finite volume. We generally want to reduce the volume of $\mathcal{W}$ in as few queries as possible.

Q-Eval[18], shown in figure 12.17, is a greedy elicitation strategy that heuristically seeks to reduce the volume of $\mathcal{W}$ as quickly as possible with each iteration. It chooses the query that comes closest to bisecting $\mathcal{W}$ into two equal parts. The method operates on a finite sampling of Pareto-optimal design points. The procedure for choosing a query pair is:

[17] The previous weight vector may or may not be consistent with the added constraint $(\mathbf{a}^{(n)} - \mathbf{b}^{(n)})^\top \mathbf{w} < 0$.

[18] V. S. Iyengar, J. Lee, and M. Campbell, "Q-EVAL: Evaluating Multiple Attribute Items Using Queries," in *ACM Conference on Electronic Commerce*, 2001.

1. Compute the *prime analytic center* **c** of $\mathcal{W}$, which is the point that maximizes the sum of the logarithms of the distances between itself and the closest point on each nonredundant constraint in $\mathcal{W}$:

$$\mathbf{c} = \arg\max_{\mathbf{w} \in \mathcal{W}} \sum_{i=1}^{n} \ln\left((\mathbf{b}^{(i)} - \mathbf{a}^{(i)})^{\top} \mathbf{w} \right) \tag{12.14}$$

2. Compute the normal distance from the bisecting hyperplane between each pair of points and the center.

3. Sort the design-point pairs in order of increasing distance.

4. For each of the k hyperplanes closest to **c**, compute the volume ratio of the two polytopes formed by splitting $\mathcal{W}$ along the hyperplane.

5. Choose the design-point pair with split ratio closest to 1.

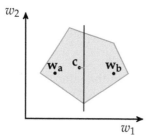

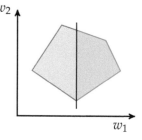

The *polyhedral method*[19] works by approximating $\mathcal{W}$ with a bounding ellipsoid centered at the analytic center of $\mathcal{W}$ as shown in figure 12.18. Queries are designed to partition the bounding ellipsoid into approximately equal parts and to favor cuts that are perpendicular to the longest axis of the ellipsoid to reduce both uncertainty and to balance the breadth in each dimension.

12.5.3   Design Selection

The previous section discussed query methods that select query pairs for efficiently reducing the search space. After query selection is complete, one must still select a final design. This process is known as *design selection*.

Figure 12.17. Visualizations of the Q-Eval greedy elicitation strategy. The figure progression shows the initial set of weights $\mathcal{W}$ consistent with previous preferences, a pair of weight vectors and their corresponding bisecting hyperplane, and the two polytopes formed by splitting along the bisecting hyperplane. The algorithm considers all possible pairs from a finite sampling of Pareto-optimal design points and chooses the query that most evenly splits $\mathcal{W}$.

[19] D. Braziunas and C. Boutilier, "Elicitation of Factored Utilities," *AI Magazine*, vol. 29, no. 4, pp. 79–92, 2009.

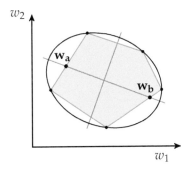

Figure 12.18. The polyhedral method uses a bounding ellipsoid for $\mathcal{W}$.

One such method, *decision quality improvement*,[20] is based on the idea that if we have to commit to a particular weight, we should commit to the one for which the worst-case objective value is lowest:

$$\mathbf{x}^* = \arg\min_{\mathbf{x} \in \mathcal{X}} \max_{\mathbf{w} \in \mathcal{W}} \mathbf{w}^\top \mathbf{f}(\mathbf{x}) \qquad (12.15)$$

This *minimax decision* is robust because it provides an upper bound on the objective value.

The *minimax regret*[21] instead minimizes the maximum amount of regret the user can have when selecting a particular design:

$$\mathbf{x}^* = \arg\min_{\mathbf{x} \in \mathcal{X}} \underbrace{\max_{\mathbf{w} \in \mathcal{W}} \max_{\mathbf{x}' \in \mathcal{X}} \mathbf{w}^\top \mathbf{f}(\mathbf{x}) - \mathbf{w}^\top \mathbf{f}(\mathbf{x}')}_{\text{maximum regret}} \qquad (12.16)$$

where $\mathbf{w}^\top \mathbf{f}(\mathbf{x}) - \mathbf{w}^\top \mathbf{f}(\mathbf{x}')$ is the *regret* associated with choosing design $\mathbf{x}$ instead of design $\mathbf{x}'$ under the preference weight vector $\mathbf{w}$. Minimax regret can be viewed as accounting for the decision system's uncertainty with respect to the designer's true utility function.

The minimax regret can be used as stopping criteria for preference elicitation strategies. We can terminate the preference elicitation procedure once the minimax regret drops below a certain threshold.

[20] D. Braziunas and C. Boutilier, "Minimax Regret-Based Elicitation of Generalized Additive Utilities," in *Conference on Uncertainty in Artificial Intelligence (UAI)*, 2007.

[21] C. Boutilier, R. Patrascu, P. Poupart, and D. Schuurmans, "Constraint-Based Optimization and Utility Elicitation Using the Minimax Decision Criterion," *Artificial Intelligence*, vol. 170, no. 8-9, pp. 686–713, 2006.

12.6 Summary

- Design problems with multiple objectives often involve trading performance between different objectives.

- The Pareto frontier represents the set of potentially optimal solutions.

- Vector-valued objective functions can be converted to scalar-valued objective functions using constraint-based or weight-based methods.

- Population methods can be extended to produce individuals that span the Pareto frontier.

- Knowing the preferences of experts between pairs of points in the criterion space can help guide the inference of a scalar-valued objective function.

12.7 Exercises

Exercise 12.1. The weighted sum method is a very simple approach, and it is indeed used by engineers for multiobjective optimization. What is one disadvantage of the procedure when it is used to compute the Pareto frontier?

Exercise 12.2. Why are population methods well-suited for multiobjective optimization?

Exercise 12.3. Suppose you have the points $\{[1, 2], [2, 1], [2, 2], [1, 1]\}$ in the criterion space and you wish to approximate a Pareto frontier. Which points are Pareto optimal with respect to the rest of the points? Are any weakly Pareto-optimal points?

Exercise 12.4. Multiobjective optimization is not easily done with second-order methods. Why is this the case?

Exercise 12.5. Consider a square criterion space $\mathcal{Y}$ with $y_1 \in [0, 1]$ and $y_2 \in [0, 1]$. Plot the criterion space, indicate the Pareto-optimal points, and indicate the weakly Pareto optimal points.

Exercise 12.6. Enforcing $\mathbf{w} \geq 0$ and $\|\mathbf{w}\|_1 = 1$ in the weighted sum method is not sufficient for Pareto optimality. Give an example where coordinates with zero-valued weights find weakly Pareto-optimal points.

Exercise 12.7. Provide an example where goal programming does not produce a Pareto-optimal point.

Exercise 12.8. Use the constraint method to obtain the Pareto curve for the optimization problem:

$$\underset{x}{\text{minimize}} \ \ [x^2, (x-2)^2] \tag{12.17}$$

Exercise 12.9. Suppose we have a multiobjective optimization problem where the two objectives are as follows:

$$f_1(x) = -(x-2)\sin(x) \tag{12.18}$$
$$f_2(x) = -(x+3)^2 \sin(x) \tag{12.19}$$

With $x \in \{-5, -3, -1, 1, 3, 5\}$, plot the points in the criterion space. How many points are on the Pareto frontier?

13 Sampling Plans

For many optimization problems, function evaluations can be quite expensive. For example, evaluating a hardware design may require a lengthy fabrication process, an aircraft design may require a wind tunnel test, and new deep learning hyperparameters may require a week of GPU training. A common approach for optimizing in contexts where evaluating design points is expensive is to build a *surrogate model*, which is a model of the optimization problem that can be efficiently optimized in lieu of the true objective function. Further evaluations of the true objective function can be used to improve the model. Fitting such models requires an initial set of points, ideally points that are *space-filling*; that is, points that cover the region as well as possible. This chapter covers different *sampling plans* for covering the search space when we have limited resources.[1]

13.1 Full Factorial

The *full factorial* sampling plan (algorithm 13.1) places a grid of evenly spaced points over the search space. This approach is easy to implement, does not rely on randomness, and covers the space, but it uses a large number of points. A grid of evenly spaced points is spread over the search space as shown in figure 13.1. Optimization over the points in a full factorial sampling plan is referred to as *grid search*.

The sampling grid is defined by a lower-bound vector $\mathbf{a}$ and an upper-bound vector $\mathbf{b}$ such that $a_i \leq x_i \leq b_i$ for each component i. For a grid with m_i samples in the ith dimension, the nearest points are separated by a distance $(b_i - a_i)/(m_i - 1)$.

The full factorial method requires a sample count exponential in the number of dimensions.[2] For n dimensions with m samples per dimension, we have m^n

[1] There are other references that discuss the topics in this chapter in greater detail. See, for example: G. E. P. Box, W. G. Hunter, and J. S. Hunter, *Statistics for Experimenters: An Introduction to Design, Data Analysis, and Model Building*, 2nd ed. Wiley, 2005. A. Dean, D. Voss, and D. Draguljić, *Design and Analysis of Experiments*, 2nd ed. Springer, 2017. D. C. Montgomery, *Design and Analysis of Experiments*. Wiley, 2017.

[2] The full factorial method gets its name not from a factorial sample count (it is exponential) but from designing with two or more *discrete factors*. Here the factors are the m discretized levels associated with each variable.

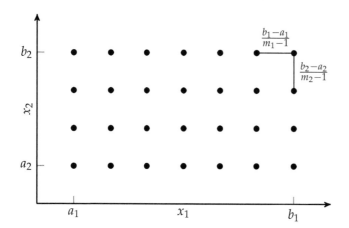

Figure 13.1. Full factorial search covers the search space in a grid of points.

total samples. This exponential growth is far too high to be of practical use when there are more than a few variables. Even when full factorial sampling is able to be used, the grid points are generally forced to be quite coarse and therefore can easily miss small, local features of the optimization landscape.

```
function samples_full_factorial(a, b, m)
    ranges = [range(a[i], stop=b[i], length=m[i])
              for i in 1 : length(a)]
    collect.(collect(Iterators.product(ranges...)))
end
```

Algorithm 13.1. A function for obtaining all sample locations for the full factorial grid. Here, a is a vector of variable lower bounds, b is a vector of variable upper bounds, and m is a vector of sample counts for each dimension.

13.2   Random Sampling

A straightforward alternative to full factorial sampling is *random sampling*, which simply draws m random samples over the design space using a pseudorandom number generator. To generate a random sample $\mathbf{x}$, we can sample each variable independently from a distribution. If we have bounds on the variables, such as $a_i \leq x_i \leq b_i$, a common approach is to use a uniform distribution over $[a_i, b_i]$, although other distributions may be used. For some variables, it may make sense to use a log-uniform distribution.[3] The samples of design points are uncorrelated with each other. The hope is that the randomness, on its own, will result in an adequate cover of the design space.

[3] Some parameters, such as the learning rate for deep neural networks, are best searched in log-space.

13.3   Uniform Projection Plans

Suppose we have a two-dimensional optimization problem discretized into an $m \times m$ sampling grid as with the full factorial method, but, instead of taking all m^2 samples, we want to sample only m positions. We could choose the samples at random, but not all arrangements are equally useful. We want the samples to be spread across the space, and we want the samples to be spread across each individual component.

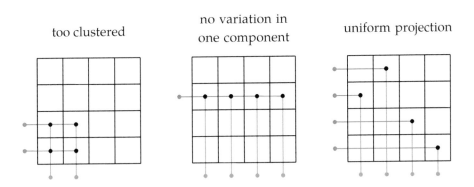

too clustered   no variation in one component   uniform projection

Figure 13.2. Several ways to choose m samples from a two-dimensional grid. We generally prefer sampling plans that cover the space and vary across each component.

A *uniform projection plan* is a sampling plan over a discrete grid where the distribution over each dimension is uniform. For example, in the rightmost sampling plan in figure 13.2, each row has exactly one entry and each column has exactly one entry.

A uniform projection plan with m samples on an $m \times m$ grid can be constructed using an m-element permutation as shown in figure 13.3. There are therefore $m!$ possible uniform projection plans.[4]

[4] For $m = 5$ this is already $5! = 120$ possible plans. For $m = 10$ there are 3,628,800 plans.

$p = 4\ 2\ 1\ 3\ 5$

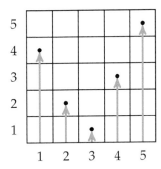

Figure 13.3. Constructing a uniform projection plan using a permutation.

Sampling with uniform projection plans is sometimes called *Latin-hypercube sampling* because of the connection to *Latin squares* (figure 13.4). A Latin square is an $m \times m$ grid where each row contains each integer 1 through m and each column contains each integer 1 through m. Latin-hypercubes are a generalization to any number of dimensions.

Uniform projection plans for n dimensions can be constructed using a permutation for each dimension (algorithm 13.2).

```
function uniform_projection_plan(m, n)
    perms = [randperm(m) for i in 1 : n]
    [[perms[i][j] for i in 1 : n] for j in 1 : m]
end
```

4	1	3	2
1	4	2	3
3	2	1	4
2	3	4	1

Figure 13.4. A 4×4 Latin square. A uniform projection plan can be constructed by choosing a value $i \in \{1, 2, 3, 4\}$ and sampling all cells with that value.

Algorithm 13.2. A function for constructing a uniform projection plan for an n-dimensional hypercube with m samples per dimension. It returns a vector of index vectors.

13.4   Stratified Sampling

Many sampling plans, including uniform projection and full factorial plans, are based on an $m \times m$ grid. Such a grid, even if fully sampled, could miss important information due to systematic regularities as shown in figure 13.5. One method for providing an opportunity to hit every point is to use *stratified sampling*.

Stratified sampling modifies any grid-based sampling plan, including full factorial and uniform projection plans. Cells are sampled at a point chosen uniformly at random from within the cell rather than at the cell's center as shown in figure 13.6.

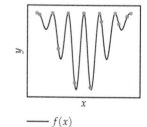

—— $f(x)$
• sampling on grid
• stratified sampling

Figure 13.5. Using an evenly-spaced grid on a function with systematic regularities can miss important information.

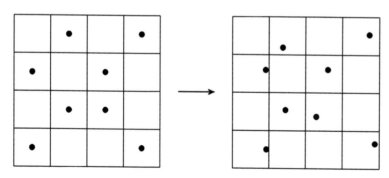

Figure 13.6. Stratified sampling applied to a uniform projection plan.

13.5 Space-Filling Metrics

A good sampling plan fills the design space since the ability for a surrogate model to generalize from samples decays with the distance from those samples. Not all plans, even uniform projection plans, are equally good at covering the search space. For example, a grid diagonal (figure 13.7) is a uniform projection plan but only covers a narrow strip. This section discusses different *space-filling metrics* for measuring the degree to which a sampling plan $X \subseteq \mathcal{X}$ fills the design space.

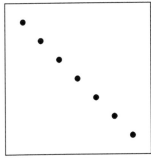

Figure 13.7. A uniform projection plan that is not space-filling.

13.5.1 Discrepancy

The ability of the sampling plan to fill a hyper-rectangular design space can be measured by its *discrepancy*.[5] If X has low discrepancy, then a randomly chosen subset of the design space should contain a fraction of samples proportional to the subset's volume.[6] The discrepancy associated with X is the maximum difference between the fraction of samples in a hyper-rectangular subset $\mathcal{H}$ and that subset's volume:

$$d(X) = \underset{\mathcal{H}}{\text{supremum}} \left| \frac{\#(X \cap \mathcal{H})}{\#X} - \lambda(\mathcal{H}) \right| \tag{13.1}$$

where $\#X$ and $\#(X \cap \mathcal{H})$ are the number of points in X and the number of points in X that lie in $\mathcal{H}$, respectively. The value $\lambda(\mathcal{H})$ is the *n*-dimensional volume of $\mathcal{H}$, the product of the side lengths of $\mathcal{H}$. The term *supremum* is very similar to maximization but allows a solution to exist for problems where $\mathcal{H}$ merely approaches a particular rectangular subset, as seen in example 13.1.[7]

Computing the discrepancy of a sampling plan over the unit hyper-rectangle is often difficult, and it is not always clear how we can compute the discrepancy for nonrectangular feasible sets.

13.5.2 Pairwise Distances

An alternative method determining which of two *m*-point sampling plans is more space-filling is to compare the *pairwise distances* between all points within each sampling plan. Sampling plans that are more spread out will tend to have larger pairwise distances.

The comparison is typically done by sorting each set's pairwise distances in ascending order. The plan with the first pairwise distance that exceeds the other is considered more space-filling.

[5] L. Kuipers and H. Niederreiter, *Uniform Distribution of Sequences*. Dover, 2012.

[6] In arbitrary dimensions, we can use the *Lebesgue measure*, which is a generalization of volume to any subset of *n*-dimensional Euclidean space. It is length in one-dimensional space, area in two-dimensional space, and volume in three-dimensional space.

[7] The definition of discrepancy requires hyper-rectangles and typically assumes that X is a finite subset of a unit hypercube. The notion of discrepancy can be extended to allow $\mathcal{H}$ to include other sets, such as convex polytopes.

Consider the set:

$$X = \left\{ \left[\frac{1}{5}, \frac{1}{5}\right], \left[\frac{2}{5}, \frac{1}{5}\right], \left[\frac{1}{10}, \frac{3}{5}\right], \left[\frac{9}{10}, \frac{3}{10}\right], \left[\frac{1}{50}, \frac{1}{50}\right], \left[\frac{3}{5}, \frac{4}{5}\right] \right\}$$

The discrepancy of X with respect to the unit square is determined by a rectangular subset $\mathcal{H}$ that either has very small area but contains very many points or has very large area and contains very few points.

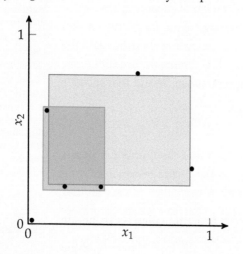

Example 13.1. Computing the discrepancy for a sampling plan over the unit square. The sizes of the rectangles are slightly exaggerated to clearly show which points they contain.

The blue rectangle, $x_1 \in \left[\frac{1}{10}, \frac{2}{5}\right]$, $x_2 \in \left[\frac{1}{5}, \frac{3}{5}\right]$, has a volume of 0.12 and contains 3 points. Its corresponding discrepancy measure is thus 0.38.

The purple rectangle, $x_1 \in \left[\frac{1}{10} + \epsilon, \frac{9}{10} - \epsilon\right]$, $x_2 \in \left[\frac{1}{5} + \epsilon, \frac{4}{5} - \epsilon\right]$, produces an even higher discrepancy. As ϵ approaches zero, the volume and the discrepancy approach 0.48 because the rectangle contains no points. Note that the limit was required, reflecting the need to use a supremum in the definition of discrepancy.

Algorithm 13.3 computes all pairwise distances between points in a sampling plan. Algorithm 13.4 compares how well two sampling plans fill space using their respective pairwise distances.

```
import LinearAlgebra: norm
function pairwise_distances(X, p=2)
    m = length(X)
    [norm(X[i]-X[j], p) for i in 1:(m-1) for j in (i+1):m]
end
```

Algorithm 13.3. A function for obtaining the list of pairwise distances between points in sampling plan X using the L_p norm specified by p.

```
function compare_sampling_plans(A, B, p=2)
    pA = sort(pairwise_distances(A, p))
    pB = sort(pairwise_distances(B, p))
    for (dA, dB) in zip(pA, pB)
        if dA < dB
            return 1
        elseif dA > dB
            return -1
        end
    end
    return 0
end
```

Algorithm 13.4. A function for comparing the degree to which two sampling plans A and B are space-filling using the L_p norm specified by p. The function returns -1 if A is more space-filling than B. It returns 1 if B is more space-filling than A. It returns 0 if they are equivalent.

One method for generating a space-filling uniform projection plan is to generate several candidates at random and then use the one that is most space-filling.

We can search for a space-filling uniform projection plan by repeatedly mutating a uniform projection plan in a way that preserves the uniform projection property (algorithm 13.5). Simulated annealing, for example, could be used to search the space for a sampling plan with good coverage.

```
function mutate!(X)
    m, n = length(X), length(X[1])
    j = rand(1:n)
    i = randperm(m)[1:2]
    X[i[1]][j], X[i[2]][j] = X[i[2]][j], X[i[1]][j]
    return X
end
```

Algorithm 13.5. A function for mutating uniform projection plan X, while maintaining its uniform projection property.

13.5.3 *Morris-Mitchell Criterion*

The comparison scheme in section 13.5.2 typically results in a challenging optimization problem with many local minima. An alternative is to optimize with respect to the *Morris-Mitchell criterion* (algorithm 13.6):[8]

$$\Phi_q(X) = \left(\sum_i d_i^{-q} \right)^{1/q}$$

(13.2)

[8] M. D. Morris and T. J. Mitchell, "Exploratory Designs for Computational Experiments," *Journal of Statistical Planning and Inference*, vol. 43, no. 3, pp. 381–402, 1995.

where d_i is the ith pairwise distance between points in X and $q > 0$ is a tunable parameter.[9] Morris and Mitchell recommend optimizing:

[9] Larger values of q will give higher penalties to large distances.

$$\underset{X}{\text{minimize}} \quad \underset{q \in \{1,2,5,10,20,50,100\}}{\text{maximize}} \quad \Phi_q(X)$$

(13.3)

```
function phiq(X, q=1, p=2)
    dists = pairwise_distances(X, p)
    return sum(dists.^(-q))^(1/q)
end
```

Algorithm 13.6. An implementation of the Morris-Mitchell criterion which takes a list of design points X, the criterion parameter q > 0, and a norm parameter p ≥ 1.

Figure 13.8 shows the Morris-Mitchell criterion evaluated for several randomly generated uniform projection plans.

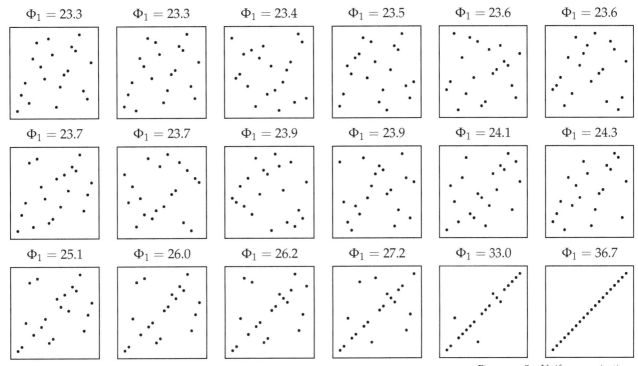

Figure 13.8. Uniform projection plans sorted from best to worst according to Φ_1.

13.6   Space-Filling Subsets

In some cases, we have a set of points X and want to find a subset of points $S \subset X$ that still maximally fills X. The need for identifying *space-filling subsets* of X arises in the context of multifidelity models.[10] For example, suppose we used a sampling plan X to identify a variety of aircraft wing designs to evaluate using computational fluid dynamic models in simulation. We can choose only a subset of these design points S to build and test in a wind tunnel. We still want S to be space filling.

The degree to which S fills the design space can be quantified using the maximum distance between a point in X and the closest point in S. This metric generalizes to any two finite sets A and B (algorithm 13.7). We can use any L_p norm, but we typically use L_2, the Euclidean distance:

$$d_{\max}(X, S) = \underset{\mathbf{x} \in X}{\text{maximize}} \, \underset{\mathbf{s} \in S}{\text{minimize}} \|\mathbf{s} - \mathbf{x}\|_p \tag{13.4}$$

[10] A. I. J. Forrester, A. Sóbester, and A. J. Keane, "Multi-Fidelity Optimization via Surrogate Modelling," *Proceedings of the Royal Society of London A: Mathematical, Physical and Engineering Sciences*, vol. 463, no. 2088, pp. 3251–3269, 2007.

```
min_dist(a, B, p) = minimum(norm(a-b, p) for b in B)
d_max(A, B, p=2) = maximum(min_dist(a, B, p) for a in A)
```

Algorithm 13.7. The set L_p distance metrics between two discrete sets, where A and B are lists of design points and p is the L_p norm parameter.

A space-filling sampling plan is one that minimizes this metric.[11] Finding a space-filling sampling plan with m elements is an optimization problem

[11] We can also minimize the Morris-Mitchell criterion for S.

$$\begin{aligned} \underset{S}{\text{minimize}} \quad & d_{\max}(X, S) \\ \text{subject to} \quad & S \subseteq X \\ & \#S = m \end{aligned} \tag{13.5}$$

Optimizing equation (13.5) is typically computationally intractable. A brute force approach would try all $d! / m! (d - m)!$ size-m subsets for a dataset of d design points. Both *greedy local search* (algorithm 13.8) and the *exchange algorithm* (algorithm 13.9) are heuristic strategies for overcoming this difficulty. They typically find acceptable space-filling subsets of X.

Greedy local search starts with a point selected randomly from X and incrementally adds the next best point that minimizes the distance metric. Points are added until the desired number of points is reached. Because the points are initialized randomly, the best results are obtained by running greedy local search several times and keeping the best sampling plan (algorithm 13.10).

```
function greedy_local_search(X, m, d=d_max)
    S = [X[rand(1:m)]]
    for i in 2 : m
        j = argmin([x ∈ S ? Inf : d(X, push!(copy(S), x))
                    for x in X])
        push!(S, X[j])
    end
    return S
end
```

Algorithm 13.8.   Greedy local search, for finding m-element sampling plans that minimize a distance metric d for discrete set X.

The exchange algorithm initializes S to a random subset of X and repeatedly replaces points that are in S with a different point in X that is not already in S to improve on the distance metric. The exchange algorithm is also typically run multiple times.

Figure 13.9 compares space-filling subsets obtained using greedy local search and the exchange algorithm.

13.7   Quasi-Random Sequences

Quasi-random sequences,[12] also called *low-discrepancy sequences*, are often used in the context of trying to approximate an integral over a multidimensional space:

[12] C. Lemieux, *Monte Carlo and Quasi-Monte Carlo Sampling*. Springer, 2009.

$$\int_{\mathcal{X}} f(\mathbf{x})\, d\mathbf{x} \approx \frac{v}{m} \sum_{i=1}^{m} f(\mathbf{x}^{(i)}) \tag{13.6}$$

where each $\mathbf{x}^{(i)}$ is sampled uniformly at random over the domain $\mathcal{X}$ and v is the volume of $\mathcal{X}$. This approximation is known as *Monte Carlo integration*.

Rather than relying on random or pseudorandom numbers to generate integration points, quasi-random sequences are deterministic sequences that fill the space in a systematic manner so that the integral converges as fast as possible in the number of points m.[13] These *quasi-Monte Carlo methods* have an error convergence of $O(1/m)$ as opposed to $O(1/\sqrt{m})$ for typical Monte Carlo integration, as shown in figure 13.10.

[13] Pseudorandom number sequences, such as those produced by a sequence of calls to rand, are deterministic given a particular seed, but they appear random. Quasi-random numbers are also deterministic but do not appear random.

```
function exchange_algorithm(X, m, d=d_max)
    S = X[randperm(m)]
    δ, done = d(X, S), false
    while !done
        best_pair = (0,0)
        for i in 1 : m
            s = S[i]
            for (j,x) in enumerate(X)
                if !in(x, S)
                    S[i] = x
                    δ' = d(X, S)
                    if δ' < δ
                        δ = δ'
                        best_pair = (i,j)
                    end
                end
            end
            S[i] = s
        end
        done = best_pair == (0,0)
        if !done
            i,j = best_pair
            S[i] = X[j]
        end
    end
    return S
end
```

Algorithm 13.9. The exchange algorithm for finding m-element sampling plans that minimize a distance metric d for discrete set X.

Algorithm 13.10. Multistart local search runs a particular search algorithm multiple times and returns the best result. Here, X is the list of points, m is the size of the desired sampling plan, alg is either exchange_algorithm or greedy_local_search, k_max is the number of iterations to run, and d is the distance metric.

```
function multistart_local_search(X, m, alg, k_max, d=d_max)
    sets = [alg(X, m, d) for i in 1 : k_max]
    return sets[argmin([d(X, S) for S in sets])]
end
```

greedy local search   exchange algorithm

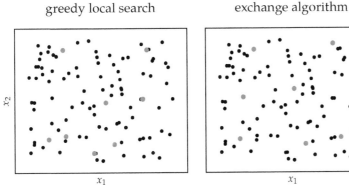

x_2

x_1 x_1

Figure 13.9.   Space-filling subsets obtained with both greedy local search and the exchange algorithm.

Quasi-random sequences are typically constructed for the unit n-dimensional hypercube, $[0,1]^n$. Any multidimensional function with bounds on each variable can be transformed into such a hypercube. This transformation is implemented in algorithm 7.9.

Various methods exist for generating quasi-random sequences. Several such methods are compared to random sampling in figure 13.12.

13.7.1   Additive Recurrence

Simple recurrence relations of the form:

$$x^{(k+1)} = x^{(k)} + c \pmod 1 \tag{13.7}$$

produce space-filling sets provided that c is irrational. The value of c leading to the smallest discrepancy is

$$c = 1 - \varphi = \frac{\sqrt{5}-1}{2} \approx 0.618034 \tag{13.8}$$

where φ is the golden ratio.[14]

We can construct a space-filling set over n dimensions using an *additive recurrence* sequence for each coordinate, each with its own value of c. The square roots of the primes are known to be irrational, and can thus be used to obtain different sequences for each coordinate:

$$c_1 = \sqrt{2}, \quad c_2 = \sqrt{3}, \quad c_3 = \sqrt{5}, \quad c_4 = \sqrt{7}, \quad c_5 = \sqrt{11}, \quad \ldots \tag{13.9}$$

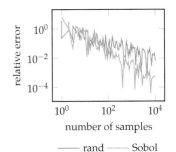

relative error

10^0

10^{-2}

10^{-4}

10^0 10^2 10^4

number of samples

—— rand —— Sobol

Figure 13.10.   The error from estimating $\int_0^1 \sin(10x)\,dx$ using Monte Carlo integration with random numbers from $U(0,1)$ and a Sobol sequence. The Sobol sequence, covered in section 13.7.3, converges faster.

[14] C. Schretter, L. Kobbelt, and P.-O. Dehaye, "Golden Ratio Sequences for Low-Discrepancy Sampling," *Journal of Graphics Tools*, vol. 16, no. 2, pp. 95–104, 2016.

Methods for additive recurrence are implemented in algorithm 13.11.

```julia
using Primes
function get_filling_set_additive_recurrence(m; c=φ-1)
    X = [rand()]
    for i in 2 : m
        push!(X, mod(X[end] + c, 1))
    end
    return X
end
function get_filling_set_additive_recurrence(m, n)
    ps = primes(max(ceil(Int, n*(log(n) + log(log(n)))), 13))
    seqs = [get_filling_set_additive_recurrence(m, c=sqrt(p))
            for p in ps[1:n]]
    return [collect(x) for x in zip(seqs...)]
end
```

Algorithm 13.11. Additive recurrence for constructing m-element filling sequences over n-dimensional unit hypercubes. The Primes package is used to generate the first n prime numbers, where the kth prime number is bounded by

$$k(\log k + \log \log k)$$

for $k > 6$, and primes(a) returns all primes up to a. Note that 13 is the sixth prime number.

13.7.2 Halton Sequence

The *Halton sequence* is a multidimensional quasi-random space-filling set.[15] The single-dimensional version, called *van der Corput sequences*, generates sequences where the unit interval is divided into powers of base b. For example, $b = 2$ produces:

$$X = \left\{ \frac{1}{2}, \frac{1}{4}, \frac{3}{4}, \frac{1}{8}, \frac{5}{8}, \frac{3}{8}, \frac{7}{8}, \frac{1}{16}, \cdots \right\} \tag{13.10}$$

whereas $b = 5$ produces:

$$X = \left\{ \frac{1}{5}, \frac{2}{5}, \frac{3}{5}, \frac{4}{5}, \frac{1}{25}, \frac{6}{25}, \frac{11}{25}, \cdots \right\} \tag{13.11}$$

Multi-dimensional space-filling sequences use one van der Corput sequence for each dimension, each with its own base b. The bases, however, must be *coprime*[16] in order to be uncorrelated. Methods for constructing Halton sequences are implemented in algorithm 13.12.

[15] J. H. Halton, "Algorithm 247: Radical-Inverse Quasi-Random Point Sequence," *Communications of the ACM*, vol. 7, no. 12, pp. 701–702, 1964.

[16] Two integers are coprime if the only positive integer that divides them both is 1.

```
using Primes
function halton(i, b)
    result, f = 0.0, 1.0
    while i > 0
        f = f / b;
        result = result + f * mod(i, b)
        i = floor(Int, i / b)
    end
    return result
end
get_filling_set_halton(m; b=2) = [halton(i,b) for i in 1: m]
function get_filling_set_halton(m, n)
    bs = primes(max(ceil(Int, n*(log(n) + log(log(n)))), 6))
    seqs = [get_filling_set_halton(m, b=b) for b in bs[1:n]]
    return [collect(x) for x in zip(seqs...)]
end
```

Algorithm 13.12. Halton quasi-random m-element filling sequences over n-dimensional unit hypercubes, where b is the base. The bases bs must be coprime.

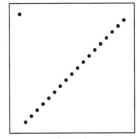

Figure 13.11. The Halton sequence with $\mathbf{b} = [19, 23]$ for which the first 18 samples are perfectly linearly correlated.

For large primes, we can get correlation in the first few numbers. Such a correlation is shown in figure 13.11. Correlation can be avoided by the *leaped Halton method*,[17] which takes every *p*th point, where *p* is a prime different from all coordinate bases.

13.7.3 Sobol Sequences

Sobol sequences are quasi-random space-filling sequences for *n*-dimensional hypercubes.[18] They are generated by xor-ing the previous Sobol number with a set of direction numbers:[19]

$$X_j^{(i)} = X_j^{(i-1)} \veebar v_j^{(k)} \tag{13.12}$$

where $v_j^{(k)}$ is the *j*th bit of the *k*th direction number. Tables of good direction numbers have been provided by various authors.[20]

A comparison of these and previous approaches is shown in figure 13.12. For high values several methods exhibit a clear underlying structure.

[17] L. Kocis and W.J. Whiten, "Computational Investigations of Low-Discrepancy Sequences," *ACM Transactions on Mathematical Software*, vol. 23, no. 2, pp. 266–294, 1997.

[18] I. M. Sobol, "On the Distribution of Points in a Cube and the Approximate Evaluation of Integrals," *USSR Computational Mathematics and Mathematical Physics*, vol. 7, no. 4, pp. 86–112, 1967.

[19] The symbol $\veebar$ denotes the xor operation, which returns true if and only if both inputs are different.

[20] The Sobol.jl package provides an implementation for up to 1,111 dimensions.

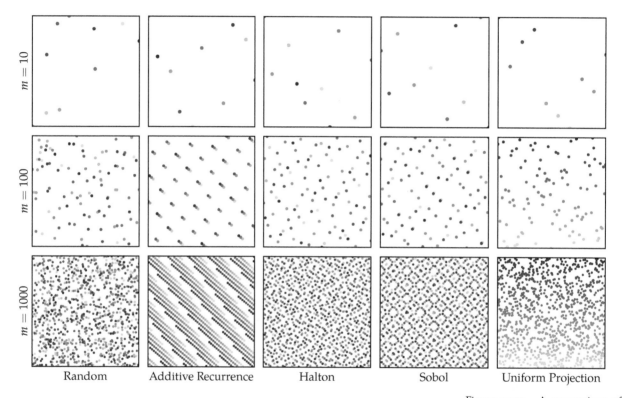

Figure 13.12. A comparison of space-filling sampling plans in two dimensions. Samples are colored according to the order in which they are sampled. The uniform projection plan was generated randomly and is not optimized.

13.8 Summary

- Sampling plans are used to cover search spaces with a limited number of points.

- Full factorial sampling, which involves sampling at the vertices of a uniformly discretized grid, requires a number of points exponential in the number of dimensions.

- Uniform projection plans, which project uniformly over each dimension, can be efficiently generated and can be optimized to be space filling.

- Greedy local search and the exchange algorithm can be used to find a subset of points that maximally fill a space.

- Quasi-random sequences are deterministic procedures by which space-filling sampling plans can be generated.

13.9 Exercises

Exercise 13.1. Filling a multidimensional space requires exponentially more points as the number of dimensions increases. To help build this intuition, determine the side lengths of an n-dimensional hypercube such that it fills half of the volume of the n-dimensional unit hypercube.

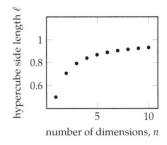

Exercise 13.2. Suppose that you sample randomly inside a unit sphere in n dimensions. Compute the probability that a randomly sampled point is within ϵ distance from the surface of the sphere as $n \to \infty$. Hint: The volume of a sphere is $C(n)r^n$, where r is the radius and $C(n)$ is a function of the dimension n only.

Exercise 13.3. Suppose we have a sampling plan $X = \{x^{(1)}, \ldots, x^{(10)}\}$, where

$$x^{(i)} = [\cos(2\pi i/10), \sin(2\pi i/10)] \tag{13.13}$$

Compute the Morris-Mitchell criterion for X using an L_2 norm when the parameter q is set to 2. In other words, evaluate $\Phi_2(X)$. If we add $[2, 3]$ to each $x^{(i)}$, will $\Phi_2(X)$ change? Why or why not?

Exercise 13.4. Additive recurrence requires that the multiplicative factor c in equation (13.7) be irrational. Why can c not be rational?

14 Surrogate Models

The previous chapter discussed methods for producing a sampling plan. This chapter shows how to use these samples to construct models of the objective function that can be used in place of the real objective function. Such *surrogate models* are designed to be smooth and inexpensive to evaluate so that they can be efficiently optimized. The surrogate model can then be used to help direct the search for the optimum of the real objective function.

14.1 Fitting Surrogate Models

A surrogate model $\hat{f}$ parameterized by θ is designed to mimic the true objective function f. The parameters θ can be adjusted to fit the model based on samples collected from f. An example surrogate model is shown in figure 14.1.

Suppose we have m design points

$$X = \left\{ \mathbf{x}^{(1)}, \mathbf{x}^{(2)}, \ldots, \mathbf{x}^{(m)} \right\} \tag{14.1}$$

and associated function evaluations

$$\mathbf{y} = \left\{ y^{(1)}, y^{(2)}, \ldots, y^{(m)} \right\} \tag{14.2}$$

For a particular set of parameters, the model will predict

$$\hat{\mathbf{y}} = \left\{ \hat{f}_\theta(\mathbf{x}^{(1)}), \hat{f}_\theta(\mathbf{x}^{(2)}), \ldots, \hat{f}_\theta(\mathbf{x}^{(m)}) \right\} \tag{14.3}$$

Fitting a model to a set of points requires tuning the parameters to minimize the difference between the true evaluations and those predicted by the model, typically according to an L_p norm:[1]

$$\operatorname*{minimize}_{\theta} \quad \|\mathbf{y} - \hat{\mathbf{y}}\|_p \tag{14.4}$$

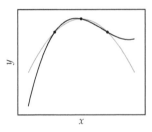

- design points
- —— surrogate model
- —— true objective function

Figure 14.1. Surrogate models approximate the true objective function. The model is fitted to the evaluated design points but deviates farther away from them.

[1] It is common to use the L_2 norm. Minimizing this equation with an L_2 norm is equivalent to minimizing the mean squared error at those data points.

Equation (14.4) penalizes the deviation of the model only at the data points. There is no guarantee that the model will continue to fit well away from observed data, and model accuracy typically decreases the farther we go from the sampled points.

This form of model fitting is called *regression*. A large body of work exists for solving regression problems, and it is extensively studied in machine learning.[2] The rest of this chapter covers several popular surrogate models and algorithms for fitting surrogate models to data, and concludes with methods for choosing between types of models.

[2] K. P. Murphy, *Machine Learning: A Probabilistic Perspective*. MIT Press, 2012.

14.2   Linear Models

A simple surrogate model is the *linear model*, which has the form[3]

[3] This equation may seem familiar. It is the equation for a hyperplane.

$$\hat{f} = w_0 + \mathbf{w}^\top \mathbf{x} \qquad \theta = \{w_0, \mathbf{w}\} \tag{14.5}$$

For an n-dimensional design space, the linear model has $n + 1$ parameters, and thus requires at least $n + 1$ samples to fit unambiguously.

Instead of having both $\mathbf{w}$ and w_0 as parameters, it is common to construct a single vector of parameters $\theta = [w_0, \mathbf{w}]$ and prepend 1 to the vector $\mathbf{x}$ to get

$$\hat{f} = \theta^\top \mathbf{x} \tag{14.6}$$

Finding an optimal θ requires solving a *linear regression* problem:

$$\underset{\theta}{\text{minimize}} \quad \|\mathbf{y} - \hat{\mathbf{y}}\|_2^2 \tag{14.7}$$

which is equivalent to solving

$$\underset{\theta}{\text{minimize}} \quad \|\mathbf{y} - \mathbf{X}\theta\|_2^2 \tag{14.8}$$

where $\mathbf{X}$ is a *design matrix* formed from m data points

$$\mathbf{X} = \begin{bmatrix} (\mathbf{x}^{(1)})^\top \\ (\mathbf{x}^{(2)})^\top \\ \vdots \\ (\mathbf{x}^{(m)})^\top \end{bmatrix} \tag{14.9}$$

```
function design_matrix(X)
    n, m = length(X[1]), length(X)
    return [j==0 ? 1.0 : X[i][j] for i in 1:m, j in 0:n]
end
function linear_regression(X, y)
    θ = pinv(design_matrix(X))*y
    return x -> θ·[1; x]
end
```

Algorithm 14.1. A method for constructing a design matrix from a list of design points X and a method for fitting a surrogate model using linear regression to a list of design points X and a vector of objective function values y.

Algorithm 14.1 implements methods for computing a design matrix and for solving a linear regression problem. Several cases for linear regression are shown in figure 14.2.

Linear regression has an analytic solution

$$\theta = \mathbf{X}^+ \mathbf{y} \tag{14.10}$$

where $\mathbf{X}^+$ is the Moore-Penrose *pseudoinverse* of $\mathbf{X}$.

If $\mathbf{X}^\top \mathbf{X}$ is invertible, the pseudoinverse can be computed as

$$\mathbf{X}^+ = \left(\mathbf{X}^\top \mathbf{X}\right)^{-1} \mathbf{X}^\top \tag{14.11}$$

If $\mathbf{X}\mathbf{X}^\top$ is invertible, the pseudoinverse can be computed as

$$\mathbf{X}^+ = \mathbf{X}^\top \left(\mathbf{X}\mathbf{X}^\top\right)^{-1} \tag{14.12}$$

The function pinv computes the pseudoinverse of a given matrix.[4]

[4] The function pinv uses the singular value decomposition, $\mathbf{X} = \mathbf{U}\mathbf{\Sigma}\mathbf{V}^*$, to compute the pseudoinverse:

$$\mathbf{X}^+ = \mathbf{V}\mathbf{\Sigma}^+\mathbf{U}^*$$

where the pseudoinverse of the diagonal matrix $\mathbf{\Sigma}$ is obtained by taking the reciprocal of each nonzero element of the diagonal and then transposing the result.

14.3   Basis Functions

The linear model is a linear combination of the components of $\mathbf{x}$:

$$\hat{f}(\mathbf{x}) = \theta_1 x_1 + \cdots + \theta_n x_n = \sum_{i=1}^n \theta_i x_i = \theta^\top \mathbf{x} \tag{14.13}$$

which is a specific example of a more general linear combination of *basis functions*

$$\hat{f}(\mathbf{x}) = \theta_1 b_1(\mathbf{x}) + \cdots + \theta_q b_q(\mathbf{x}) = \sum_{i=1}^q \theta_i b_i(\mathbf{x}) = \theta^\top \mathbf{b}(\mathbf{x}) \tag{14.14}$$

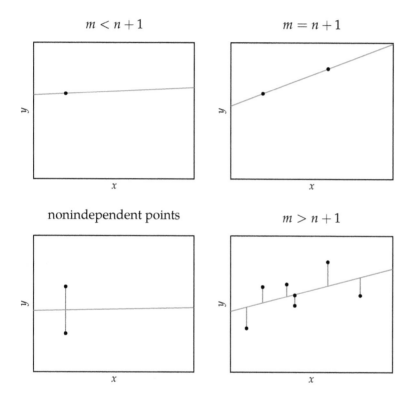

Figure 14.2. Models resulting from linear regression, which minimizes the square vertical distance of the model from each point. The pseudoinverse produces a unique solution for any nonempty point configuration.

The bottom-left subfigure shows the model obtained for two repeated points, in this case, $m = n + 1$. Because the two entries are repeated, the matrix $\mathbf{X}$ is nonsingular. Although $\mathbf{X}$ does not have an inverse in this case, the pseudoinverse produces a unique solution that passes between the two points.

In the case of linear regression, the basis functions simply extract each component, $b_i(\mathbf{x}) = x_i$.

Any surrogate model represented as a linear combination of basis functions can be fit using regression:

$$\underset{\theta}{\text{minimize}} \quad \|\mathbf{y} - \mathbf{B}\theta\|_2^2 \qquad (14.15)$$

where $\mathbf{B}$ is the basis matrix formed from m data points:

$$\mathbf{B} = \begin{bmatrix} \mathbf{b}(\mathbf{x}^{(1)})^\top \\ \mathbf{b}(\mathbf{x}^{(2)})^\top \\ \vdots \\ \mathbf{b}(\mathbf{x}^{(m)})^\top \end{bmatrix} \qquad (14.16)$$

The weighting parameters can be obtained using the pseudoinverse

$$\theta = \mathbf{B}^+\mathbf{y} \qquad (14.17)$$

Algorithm 14.2 implements this more general regression procedure.

```
using LinearAlgebra
function regression(X, y, bases)
    B = [b(x) for x in X, b in bases]
    θ = pinv(B)*y
    return x -> sum(θ[i] * bases[i](x) for i in 1 : length(θ))
end
```

Algorithm 14.2. A method for fitting a surrogate model to a list of design points X and corresponding objective function values y using regression with basis functions contained in the bases array.

Linear models cannot capture nonlinear relations. There are a variety of other families of basis functions that can represent more expressive surrogate models. The remainder of this section discusses a few common families.

14.3.1 Polynomial Basis Functions

Polynomial basis functions consist of a product of design vector components, each raised to a power. Linear basis functions are a special case of polynomial basis functions.

From the Taylor series expansion[5] we know that any infinitely differentiable function can be closely approximated by a polynomial of sufficient degree. We can construct these bases using algorithm 14.3.

[5] Covered in appendix C.2.

In one dimension, a polynomial model of degree k has the form

$$\hat{f}(x) = \theta_0 + \theta_1 x + \theta_2 x^2 + \theta_3 x^3 + \cdots + \theta_k x^k = \sum_{i=0}^{k} \theta_i x^i \qquad (14.18)$$

Hence, we have a set of basis functions $b_i(x) = x^i$ for i ranging from 0 to k.

In two dimensions, a polynomial model of degree k has basis functions of the form

$$b_{ij}(\mathbf{x}) = x_1^i x_2^j \ \text{ for } i, j \in \{0, \ldots, k\}, \ \ i + j \leq k \qquad (14.19)$$

Fitting a polynomial surrogate model is a regression problem, so a polynomial model is linear in higher dimensional space (figure 14.3). Any linear combination of basis functions can be viewed as linear regression in a higher dimensional space.

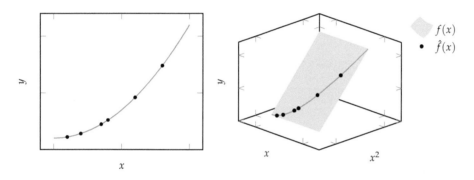

Figure 14.3. A polynomial model is linear in higher dimensions. The function exists in the plane formed from its bases, but it does not occupy the entire plane because the terms are not independent.

```
polynomial_bases_1d(i, k) = [x->x[i]^p for p in 0:k]
function polynomial_bases(n, k)
    bases = [polynomial_bases_1d(i, k) for i in 1 : n]
    terms = Function[]
    for ks in Iterators.product([0:k for i in 1:n]...)
        if sum(ks) ≤ k
            push!(terms,
                x->prod(b[j+1](x) for (j,b) in zip(ks,bases)))
        end
    end
    return terms
end
```

Algorithm 14.3. A method for constructing an array of polynomial basis functions up to a degree k for the ith component of a design point, and a method for constructing a list of n-dimensional polynomial bases for terms up to degree k.

14.3.2   Sinusoidal Basis Functions

Any continuous function over a finite domain can be represented using an infinite set of *sinusoidal basis functions*.[6] A *Fourier series* can be constructed for any integrable univariate function f on an interval $[a, b]$

[6] The Fourier series is also exact for functions defined over the entire real line if the function is periodic.

$$f(x) = \frac{\theta_0}{2} + \sum_{i=1}^{\infty} \theta_i^{(\sin)} \sin\left(\frac{2\pi i x}{b-a}\right) + \theta_i^{(\cos)} \cos\left(\frac{2\pi i x}{b-a}\right) \tag{14.20}$$

where

$$\theta_0 = \frac{2}{b-a} \int_a^b f(x)\, dx \tag{14.21}$$

$$\theta_i^{(\sin)} = \frac{2}{b-a} \int_a^b f(x) \sin\left(\frac{2\pi i x}{b-a}\right) dx \tag{14.22}$$

$$\theta_i^{(\cos)} = \frac{2}{b-a} \int_a^b f(x) \cos\left(\frac{2\pi i x}{b-a}\right) dx \tag{14.23}$$

Just as the first few terms of a Taylor series are used in polynomial models, so too are the first few terms of the Fourier series used in sinusoidal models. The bases for a single component over the domain $x \in [a, b]$ are:

$$\begin{cases} b_0(x) & = 1/2 \\ b_i^{(\sin)}(x) & = \sin\left(\frac{2\pi i x}{b-a}\right) \\ b_i^{(\cos)}(x) & = \cos\left(\frac{2\pi i x}{b-a}\right) \end{cases} \tag{14.24}$$

We can combine the terms for multidimensional sinusoidal models in the same way we combine terms in polynomial models. Algorithm 14.4 can be used to construct sinusoidal basis functions. Several cases for sinusoidal regression are shown in figure 14.4.

14.3.3   Radial Basis Functions

A *radial function* ψ is one which depends only on the distance of a point from some center point $\mathbf{c}$, such that it can be written $\psi(\mathbf{x}, \mathbf{c}) = \psi(\|\mathbf{x} - \mathbf{c}\|) = \psi(r)$. Radial functions are convenient basis functions because placing a radial function contributes a hill or valley to the function landscape. Some common radial basis functions are shown in figure 14.5.

```
function sinusoidal_bases_1d(j, k, a, b)
    T = b[j] - a[j]
    bases = Function[x->1/2]
    for i in 1 : k
        push!(bases, x->sin(2π*i*x[j]/T))
        push!(bases, x->cos(2π*i*x[j]/T))
    end
    return bases
end
function sinusoidal_bases(k, a, b)
    n = length(a)
    bases = [sinusoidal_bases_1d(i, k, a, b) for i in 1 : n]
    terms = Function[]
    for ks in Iterators.product([0:2k for i in 1:n]...)
        powers = [div(k+1,2) for k in ks]
        if sum(powers) ≤ k
            push!(terms,
                x->prod(b[j+1](x) for (j,b) in zip(ks,bases)))
        end
    end
    return terms
end
```

Algorithm 14.4.   The method sinusoidal_bases_1d produces a list of basis functions up to degree k for the ith component of the design vector given lower bound a and upper bound b. The method sinusoidal_bases produces all base function combinations up to degree k for lower-bound vector a and upper-bound vector b.

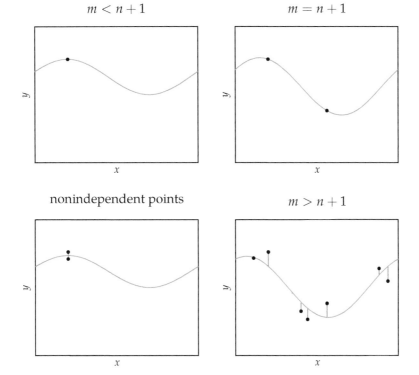

Figure 14.4.   Fitting sinusoidal models to noisy points.

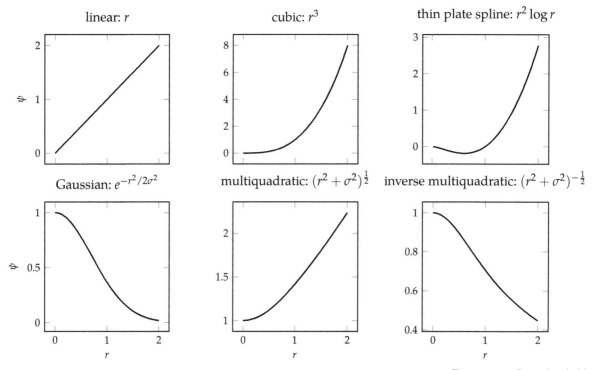

Figure 14.5. Several radial basis functions.

Radial basis functions require specifying the center points. One approach when fitting radial basis functions to a set of data points is to use the data points as the centers. For a set of m points, one thus constructs m radial basis functions

$$b_i(\mathbf{x}) = \psi(\|\mathbf{x} - \mathbf{x}^{(i)}\|) \quad \text{for } i \in \{1, \ldots, m\} \tag{14.25}$$

The corresponding $m \times m$ basis matrix is always semidefinite. Algorithm 14.5 can be used to construct radial basis functions with known center points. Surrogate models with different radial basis functions are shown in figure 14.6.

```
radial_bases(ψ, C, p=2) = [x->ψ(norm(x - c, p)) for c in C]
```

Algorithm 14.5. A method for obtaining a list of basis functions given a radial basis function ψ, a list of centers C, and an L_p norm parameter p.

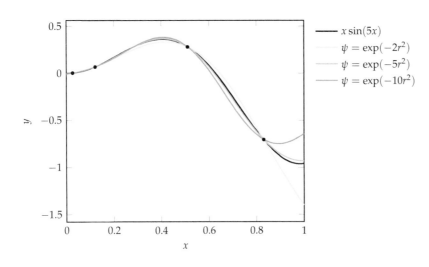

Figure 14.6. Several different Gaussian radial basis functions used to fit $x\sin(5x)$ based on 4 noise-free samples.

14.4   Fitting Noisy Objective Functions

Models fit using regression will pass as close as possible to every design point. When the objective function evaluations are noisy, complex models are likely to excessively contort themselves to pass through every point. However, smoother fits are often better predictors of the true underlying objective function.

The basis regression problem specified in equation (14.15) can be augmented to prefer smoother solutions. A *regularization term* is added in addition to the prediction error in order to give preference to solutions with lower weights. The resulting basis regression problem with L_2 *regularization*[7] is:

$$\underset{\theta}{\text{minimize}} \quad \|\mathbf{y} - \mathbf{B}\theta\|_2^2 + \lambda\|\theta\|_2^2 \qquad (14.26)$$

where $\lambda \geq 0$ is a smoothing parameter, with $\lambda = 0$ resulting in no smoothing.

The optimal parameter vector is given by:[8]

$$\theta = \left(\mathbf{B}^\top\mathbf{B} + \lambda\mathbf{I}\right)^{-1}\mathbf{B}^\top\mathbf{y} \qquad (14.27)$$

where $\mathbf{I}$ is the identity matrix.

Algorithm 14.6 implements regression with L_2 regularization. Surrogate models with different radial basis functions fit to noisy samples are shown in figure 14.7.

```
function regression(X, y, bases, λ)
    B = [b(x) for x in X, b in bases]
    θ = (B'B + λ*I)\B'y
    return x -> sum(θ[i] * bases[i](x) for i in 1 : length(θ))
end
```

[7] Other L_p-norms, covered in appendix C.4, can be used as well. Using the L_1 norm will encourage sparse solutions with less influential component weights set to zero, which can be useful in identifying important basis functions.

[8] The matrix $\left(\mathbf{B}^\top\mathbf{B} + \lambda\mathbf{I}\right)$ is not always invertible if $\lambda = 0$. However, we can always produce an invertible matrix with a positive λ.

Algorithm 14.6. A method for regression in the presence of noise, where λ is a smoothing term. It returns a surrogate model fitted to a list of design points X and corresponding objective function values y using regression with basis functions bases.

14.5 Model Selection

So far, we have discussed how to fit a particular model to data. This section explains how to select which model to use. We generally want to minimize *generalization error*, which is a measure of the error of the model on the full design space, including points that may not be included in the data used to train the model. One way to measure generalization error is to use the expected squared error of its predictions:

$$\epsilon_{\text{gen}} = \mathbb{E}_{\mathbf{x}\sim\mathcal{X}}\left[\left(f(\mathbf{x}) - \hat{f}(\mathbf{x})\right)^2\right] \qquad (14.28)$$

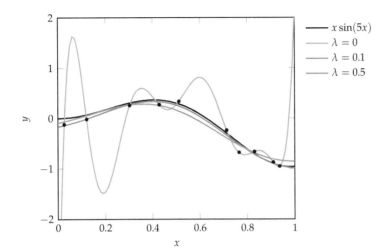

Figure 14.7. Several different Gaussian radial basis functions used to fit $x \sin(5x)$ with zero mean, standard deviation 0.1 error based on 10 noisy samples and radial basis function $\psi = \exp(-5r^2)$.

Of course, we cannot calculate this generalization error exactly because it requires knowing the function we are trying to approximate. It may be tempting to estimate the generalization error of a model from the *training error*. One way to measure training error is to use the *mean squared error* (MSE) of the model evaluated on the m samples used for training:

$$\epsilon_{\text{train}} = \frac{1}{m} \sum_{i=1}^{m} \left(f(\mathbf{x}^{(i)}) - \hat{f}(\mathbf{x}^{(i)}) \right)^2 \tag{14.29}$$

However, performing well on the training data does not necessarily correspond to low generalization error. Complex models may reduce the error on the training set, but they may not provide good predictions in other points in the design space as illustrated in example 14.1.[9]

This section discusses several methods for estimating generalization error. These methods train and test on subsets of the data. We introduce the `TrainTest` type (algorithm 14.7), which contains a list of training indices and a list of test indices. The method `fit` takes in a training set and produces a model. The method `metric` takes a model and a test set and produces a metric, such as the mean squared error. The method `train_and_validate` (algorithm 14.7) is a utility function for training and then evaluating a model. Although we train on subsets of the data when estimating the generalization error, once we have decided which model to use, we can train on the full dataset.

[9] A major theme in machine learning is balancing model complexity to avoid *overfitting* the training data. K. P. Murphy, *Machine Learning: A Probabilistic Perspective*. MIT Press, 2012.

Consider fitting polynomials of varying degrees to evaluations of the objective function

$$f(x) = x/10 + \sin(x)/4 + \exp\left(-x^2\right)$$

Example 14.1. A comparison of training and generalization error as the degree of a polynomial surrogate model is varied.

Below we plot polynomial surrogate models of varying degrees using the same nine evaluations evenly spaced over $[-4, 4]$. The training and generalization error are shown as well, where generalization is calculated over $[-5, 5]$.

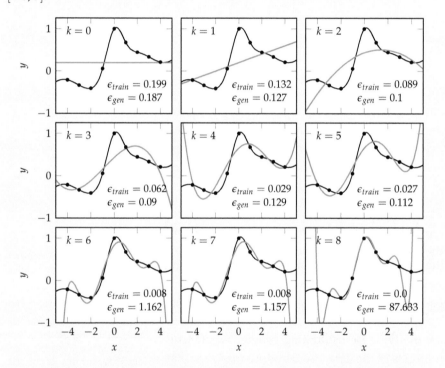

The plot shows that the generalization error is high for both very low and high values of k, and that training error decreases as we increase the polynomial degree. The high-degree polynomials are particularly poor predictors for designs outside $[-4, 4]$.

```
struct TrainTest
    train
    test
end
function train_and_validate(X, y, tt, fit, metric)
    model = fit(X[tt.train], y[tt.train])
    return metric(model, X[tt.test], y[tt.test])
end
```

Algorithm 14.7. A utility type and method for training a model and then validating it on a metric. Here, `train` and `test` are lists of indices into the training data, X is a list of design points, y is the vector of corresponding function evaluations, `tt` is a train-test partition, `fit` is a model fitting function, and `metric` evaluates a model on a test set to produce an estimate of generalization error.

14.5.1   Holdout

train($\bullet$) $\longrightarrow$ test($\hat{f}, \bullet$) $\longrightarrow$ generalization error estimate

Figure 14.8. The holdout method (left) partitions the data into train and test sets.

A simple approach to estimating the generalization error is the *holdout method*, which partitions the available data into a *test set* $\mathcal{D}_h$ with h samples and a *training set* $\mathcal{D}_t$ consisting of all remaining $m - h$ samples as shown in figure 14.8. The training set is used to fit model parameters. The held out test set is not used during model fitting, and can thus be used to estimate the generalization error. Different split ratios are used, typically ranging from 50% train, 50% test to 90% train, 10% test, depending on the size and nature of the dataset. Using too few samples for training can result in poor fits (figure 14.9), whereas using too many will result in poor generalization estimates.

The holdout error for a model $\hat{f}$ fit to the training set is

$$\epsilon_{\text{holdout}} = \frac{1}{h} \sum_{(\mathbf{x}, y) \in \mathcal{D}_h} \left(y - \hat{f}(\mathbf{x}) \right)^2 \qquad (14.30)$$

Even if the partition ratio is fixed, the holdout error will depend on the particular train-test partition chosen. Choosing a partition at random (algorithm 14.8) will only give a point estimate. In *random subsampling* (algorithm 14.9), we apply the holdout method multiple times with randomly selected train-test partitions. The estimated generalization error is the mean over all runs.[10] Because the validation sets are chosen randomly, this method does not guarantee that we validate on all of the data points.

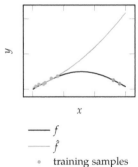

Figure 14.9. Poor train-test splits can result in poor model performance.

[10] The standard deviation over all runs can be used to estimate the standard deviation of the estimated generalization error.

```
function holdout_partition(m, h=div(m,2))
    p = randperm(m)
    train = p[(h+1):m]
    holdout = p[1:h]
    return TrainTest(train, holdout)
end
```

Algorithm 14.8. A method for randomly partitioning m data samples into training and holdout sets, where h samples are assigned to the holdout set.

```
function random_subsampling(X, y, fit, metric;
    h=div(length(X),2), k_max=10)
    m = length(X)
    mean(train_and_validate(X, y, holdout_partition(m, h),
        fit, metric) for k in 1 : k_max)
end
```

Algorithm 14.9. The random subsampling method used to obtain mean and standard deviation estimates for model generalization error using k_max runs of the holdout method.

14.5.2   Cross Validation

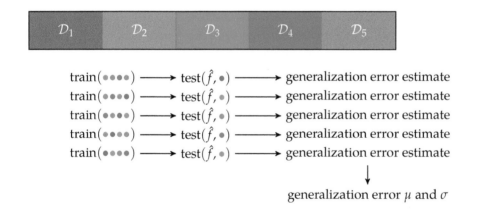

Figure 14.10. Cross-validation partitions the data into equally sized sets. Each set is the holdout set once. Here we show 5-fold cross-validation.

Using a train-test partition can be wasteful because our model tuning can take advantage only of a segment of our data. Better results are often obtained by using *k-fold cross validation.*[11] Here, the original dataset $\mathcal{D}$ is randomly partitioned into k sets $\mathcal{D}_1, \mathcal{D}_2, \ldots, \mathcal{D}_k$ of equal, or approximately equal, size, as shown in figure 14.10 and implemented in algorithm 14.10. We then train k models, one on each subset of $k-1$ sets, and we use the withheld set to estimate the generalization error. The cross-validation estimate of generalization error is the mean generalization error over all folds:[12]

[11] Also known as *rotation estimation.*

[12] As with random subsampling, an estimate of variance can be obtained from the standard deviation over folds.

$$\epsilon_{\text{cross-validation}} = \frac{1}{k} \sum_{i=1}^{k} \epsilon_{\text{cross-validation}}^{(i)}$$

$$\epsilon_{\text{cross-validation}}^{(i)} = \frac{1}{|\mathcal{D}_{\text{test}}^{(i)}|} \sum_{(\mathbf{x},y) \in \mathcal{D}_{\text{test}}^{(i)}} \left(y - \hat{f}^{(i)}(\mathbf{x}) \right)^2 \qquad (14.31)$$

where $\epsilon_{\text{cross-validation}}^{(i)}$ and $\mathcal{D}_{\text{test}}^{(i)}$ are the cross-validation estimate and the withheld test set, respectively, for the ith fold.

```
function k_fold_cross_validation_sets(m, k)
    perm = randperm(m)
    sets = TrainTest[]
    for i = 1:k
        validate = perm[i:k:m];
        train = perm[setdiff(1:m, i:k:m)]
        push!(sets, TrainTest(train, validate))
    end
    return sets
end
function cross_validation_estimate(X, y, sets, fit, metric)
    mean(train_and_validate(X, y, tt, fit, metric)
            for tt in sets)
end
```

Algorithm 14.10. The method `k_fold_cross_validation_sets` constructs the sets needed for k-fold cross validation on `m` samples, with k ≤ m. The method `cross_validation_estimate` computes the mean of the generalization error estimate by training and validating on the list of train-validate sets contained in `sets`. The other variables are the list of design points `X`, the corresponding objective function values `y`, a function `fit` that trains a surrogate model, and a function `metric` that evaluates a model on a data set.

Cross-validation also depends on the particular data partition. An exception is *leave-one-out cross-validation* with $k = m$, which has a deterministic partition. It trains on as much data as possible, but it requires training m models.[13] Averaging over all $\binom{m}{m/k}$ possible partitions, known as *complete cross-validation*, is often too expensive. While one can average multiple cross-validation runs, it is more common to average the models from a single cross-validation partition.

Cross-validation is demonstrated in example 14.2.

[13] M. Stone, "Cross-Validatory Choice and Assessment of Statistical Predictions," *Journal of the Royal Statistical Society*, vol. 36, no. 2, pp. 111–147, 1974.

14.5.3   The Bootstrap

The *bootstrap method*[14] uses multiple *bootstrap samples*, which consist of m indices into a dataset of size m independently chosen uniformly at random. The indices are chosen with replacement, so some indices may be chosen multiple times and some indices may not be chosen at all as shown in figure 14.11. The bootstrap sample is used to fit a model that is then evaluated on the original training set. A method for obtaining bootstrap samples is given in algorithm 14.11.

If b bootstrap samples are made, then the bootstrap estimate of the generalization error is the mean of the corresponding generalization error estimates $\epsilon_{\text{test}}^{(1)}, \ldots, \epsilon_{\text{test}}^{(b)}$:

[14] B. Efron, "Bootstrap Methods: Another Look at the Jackknife," *The Annals of Statistics*, vol. 7, pp. 1–26, 1979.

$$\epsilon_{\text{boot}} = \frac{1}{b} \sum_{i=1}^{b} \epsilon_{\text{test}}^{(i)} \tag{14.32}$$

$$= \frac{1}{m} \sum_{j=1}^{m} \frac{1}{b} \sum_{i=1}^{b} \left(y^{(j)} - \hat{f}^{(i)}(\mathbf{x}^{(j)}) \right)^2 \tag{14.33}$$

where $\hat{f}^{(i)}$ is the model fit to the ith bootstrap sample. The bootstrap method is implemented in algorithm 14.12.

The bootstrap error in equation (14.32) tests models on data points to which they were fit. The *leave-one-out bootstrap estimate* removes this source of bias by only evaluating fitted models to withheld data:

$$\epsilon_{\text{leave-one-out-boot}} = \frac{1}{m} \sum_{j=1}^{m} \frac{1}{c_{-j}} \sum_{i=1}^{b} \begin{cases} \left(y^{(j)} - \hat{f}^{(i)}(\mathbf{x}^{(j)}) \right)^2 & \text{if } j\text{th index was not in the } i\text{th bootstrap sample} \\ 0 & \text{otherwise} \end{cases}$$

$$\tag{14.34}$$

where c_{-j} is the number of bootstrap samples that do not contain index j. The leave-one-out bootstrap method is implemented in algorithm 14.13.

The probability of a particular index not being in a bootstrap sample is:

$$\left(1 - \frac{1}{m} \right)^m \approx e^{-1} \approx 0.368 \tag{14.35}$$

so a bootstrap sample is expected to have on average $0.632m$ distinct indices from the original dataset.

Suppose we want to fit a noisy objective function using radial basis functions with the noise hyperparameter λ (section 14.4). We can use cross validation to determine λ. We are given ten samples from our noisy objective function. In practice, the objective function will be unknown, but this example uses

$$f(x) = \sin(2x)\cos(10x) + \epsilon/10$$

where $x \in [0, 1]$ and ϵ is random noise with zero mean and unit variance, $\epsilon \sim \mathcal{N}(0, 1)$.

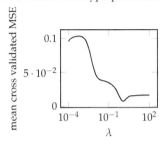

Example 14.2. Cross validation used to fit a hyperparameter.

```
Random.seed!(0)
f = x->sin(2x)*cos(10x)
X = rand(10)
y = f.(X) + randn(length(X))/10
```

We will use three folds assigned randomly:

```
sets = k_fold_cross_validation_sets(length(X), 3)
```

Next, we implement our metric. We use the mean squared error:

```
metric = (f, X, y)->begin
    m = length(X)
    return sum((f(X[i]) - y[i])^2 for i in m)/m
end
```

We now loop through different values of λ and fit different radial basis functions. We will use the Gaussian radial basis. Cross validation is used to obtain the MSE for each value:

```
λs = 10 .^ range(-4, stop=2, length=101)
es = []
basis = r->exp(-5r^2)
for λ in λs
    fit = (X, y)->regression(X, y, radial_bases(basis, X), λ)
    push!(es,
        cross_validation_estimate(X, y, sets, fit, metric)[1])
end
```

The resulting curve has a minimum at $\lambda \approx 0.2$.

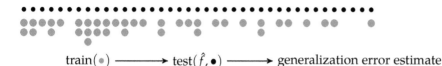

Figure 14.11. A single bootstrap sample consists of m indices into the dataset sampled with replacement. A bootstrap sample is used to train a model, which is evaluated on the full dataset to obtain an estimate of the generalization error.

```
bootstrap_sets(m, b) = [TrainTest(rand(1:m, m), 1:m) for i in 1:b]
```

Algorithm 14.11. A method for obtaining b bootstrap samples, each for a data set of size m.

```
function bootstrap_estimate(X, y, sets, fit, metric)
    mean(train_and_validate(X, y, tt, fit, metric) for tt in sets)
end
```

Algorithm 14.12. A method for computing the bootstrap generalization error estimate by training and validating on the list of train-validate sets contained in sets. The other variables are the list of design points X, the corresponding objective function values y, a function fit that trains a surrogate model, and a function metric that evaluates a model on a data set.

```
function leave_one_out_bootstrap_estimate(X, y, sets, fit, metric)
    m, b = length(X), length(sets)
    ε = 0.0
    models = [fit(X[tt.train], y[tt.train]) for tt in sets]
    for j in 1 : m
        c = 0
        δ = 0.0
        for i in 1 : b
            if j ∉ sets[i].train
                c += 1
                δ += metric(models[i], [X[j]], [y[j]])
            end
        end
        ε += δ/c
    end
    return ε/m
end
```

Algorithm 14.13. A method for computing the leave-one-out bootstrap generalization error estimate using the train-validate sets sets. The other variables are the list of design points X, the corresponding objective function values y, a function fit that trains a surrogate model, and a function metric that evaluates a model on a data set.

Unfortunately, the leave-one-out bootstrap estimate introduces a new bias due to the varying test set sizes. The *0.632 bootstrap estimate*[15] (algorithm 14.14) alleviates this bias:

$$\epsilon_{0.632\text{-boot}} = 0.632\epsilon_{\text{leave-one-out-boot}} + 0.368\epsilon_{\text{boot}} \tag{14.36}$$

```
function bootstrap_632_estimate(X, y, sets, fit, metric)
    models = [fit(X[tt.train], y[tt.train]) for tt in sets]
    ε_loob = leave_one_out_bootstrap_estimate(X,y,sets,fit,metric)
    ε_boot = bootstrap_estimate(X,y,sets,fit,metric)
    return 0.632ε_loob + 0.368ε_boot
end
```

Several generalization estimation methods are compared in example 14.3.

Consider ten evenly spread samples of $f(x) = x^2 + \epsilon/2$ over $x \in [-3,3]$, where ϵ is zero-mean, unit-variance Gaussian noise. We would like to test several different generalization error estimation methods when fitting a linear model to this data. Our metric is the *root mean squared error*, which is the square root of the mean squared error.

The methods used are the holdout method with eight training samples, five-fold cross validation, and the bootstrap methods each with ten bootstrap samples. Each method was fitted 100 times and the resulting statistics are shown below.

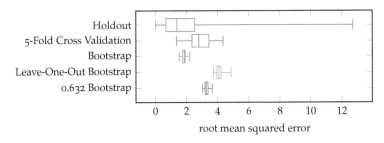

root mean squared error

[15] The 0.632 bootstrap estimate was introduced in B. Efron, "Estimating the Error Rate of a Prediction Rule: Improvement on Cross-Validation," *Journal of the American Statistical Association*, vol. 78, no. 382, pp. 316–331, 1983. A variant, the 0.632+ bootstrap estimate, was introduced in B. Efron and R. Tibshirani, "Improvements on Cross-Validation: The .632+ Bootstrap Method," *Journal of the American Statistical Association*, vol. 92, no. 438, pp. 548–560, 1997.

Algorithm 14.14. A method for obtaining the 0.632 bootstrap estimate for data points X, objective function values y, number of bootstrap samples b, fitting function `fit`, and metric function `metric`.

Example 14.3. A comparison of generalization error estimation methods. The vertical lines in the *box and whisker* plots indicate the minimum, maximum, first and third quartiles, and median of every generalization error estimation method among 50 trials.

14.6   Summary

- Surrogate models are function approximations that can be optimized instead of the true, potentially expensive objective function.

- Many surrogate models can be represented using a linear combination of basis functions.

- Model selection involves a bias-variance tradeoff between models with low complexity that cannot capture important trends and models with high complexity that overfit to noise.

- Generalization error can be estimated using techniques such as holdout, k-fold cross validation, and the bootstrap.

14.7   Exercises

Exercise 14.1. Derive an expression satisfied by the optimum of the regression problem equation (14.8) by setting the gradient to zero. Do not invert any matrices. The resulting relation is called the *normal equation*.

Exercise 14.2. When would we use a more descriptive model, for example, with polynomial features, versus a simpler model like linear regression?

Exercise 14.3. A linear regression problem of the form in equation (14.8) is not always solved analytically, and optimization techniques are used instead. Why is this the case?

Exercise 14.4. Suppose we evaluate our objective function at four points: 1, 2, 3, and 4, and we get back 0, 5, 4, and 6. We want to fit a polynomial model $f(x) = \sum_{i=0}^{k} \theta_i x^i$. Compute the leave-one-out cross validation estimate of the mean squared error as k varies between 0 and 4. According to this metric, what is the best value for k, and what are the best values for the elements of θ?

15 *Probabilistic Surrogate Models*

The previous chapter discussed how to construct surrogate models from evaluated design points. When using surrogate models for the purpose of optimization, it is often useful to quantify our confidence in the predictions of these models. One way to quantify our confidence is by taking a probabilistic approach to surrogate modeling. A common probabilistic surrogate model is the *Gaussian process*, which represents a probability distribution over functions. This chapter will explain how to use Gaussian processes to infer a distribution over the values of different design points given the values of previously evaluated design points. We will discuss how to incorporate gradient information as well as noisy measurements of the objective function. Since the predictions made by a Gaussian process are governed by a set of parameters, we will discuss how to infer these parameters directly from data.

15.1 *Gaussian Distribution*

Before introducing Gaussian processes, we will first review some relevant properties of the multivariate *Gaussian distribution*, often also referred to as the multivariate *normal distribution*.[1] An n-dimensional Gaussian distribution is parameterized by its mean μ and its covariance matrix Σ. The probability density at $\mathbf{x}$ is

[1] The univariate Gaussian distribution is discussed in appendix C.7.

$$\mathcal{N}(\mathbf{x} \mid \mu, \Sigma) = (2\pi)^{-n/2} |\Sigma|^{-1/2} \exp\left(-\frac{1}{2}(\mathbf{x} - \mu)^\top \Sigma^{-1}(\mathbf{x} - \mu) \right) \qquad (15.1)$$

Figure 15.1 shows contour plots of the density functions with different covariance matrices. Covariance matrices are always positive semidefinite.

A value sampled from a Gaussian is written

$$\mathbf{x} \sim \mathcal{N}(\mu, \Sigma) \qquad (15.2)$$

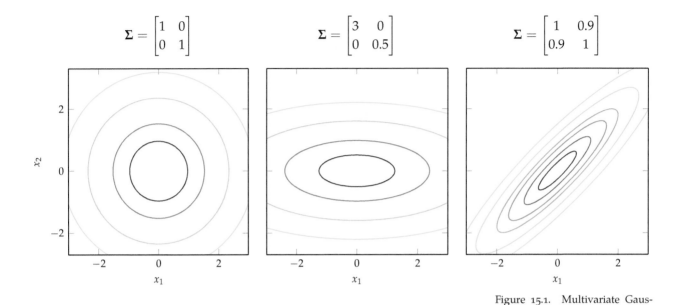

Figure 15.1. Multivariate Gaussians with different covariance matrices.

Two jointly Gaussian random variables $\mathbf{a}$ and $\mathbf{b}$ can be written

$$\begin{bmatrix} \mathbf{a} \\ \mathbf{b} \end{bmatrix} \sim \mathcal{N}\left(\begin{bmatrix} \mathbf{\mu_a} \\ \mathbf{\mu_b} \end{bmatrix}, \begin{bmatrix} \mathbf{A} & \mathbf{C} \\ \mathbf{C}^\top & \mathbf{B} \end{bmatrix} \right) \tag{15.3}$$

The *marginal distribution*[2] for a vector of random variables is given by its corresponding mean and covariance

$$\mathbf{a} \sim \mathcal{N}(\mathbf{\mu_a}, \mathbf{A}) \qquad \mathbf{b} \sim \mathcal{N}(\mathbf{\mu_b}, \mathbf{B}) \tag{15.4}$$

The *conditional distribution* for a multivariate Gaussian also has a convenient closed-form solution:

$$\mathbf{a} \mid \mathbf{b} \sim \mathcal{N}\left(\mathbf{\mu_{a|b}}, \mathbf{\Sigma_{a|b}} \right) \tag{15.5}$$

$$\mathbf{\mu_{a|b}} = \mathbf{\mu_a} + \mathbf{C}\mathbf{B}^{-1}(\mathbf{b} - \mathbf{\mu_b}) \tag{15.6}$$

$$\mathbf{\Sigma_{a|b}} = \mathbf{A} - \mathbf{C}\mathbf{B}^{-1}\mathbf{C}^\top \tag{15.7}$$

Example 15.1 illustrates how to extract the marginal and conditional distributions from a multivariate Gaussian.

[2] The marginal distribution is the distribution of a subset of the variables when the rest are integrated, or marginalized, out. For a distribution over two variables a and b the marginal distribution over a is:

$$p(a) = \int p(a,b)\, db$$

For example, consider

$$\begin{bmatrix} x_1 \\ x_2 \end{bmatrix} \sim \mathcal{N}\left(\begin{bmatrix} 0 \\ 1 \end{bmatrix}, \begin{bmatrix} 3 & 1 \\ 1 & 2 \end{bmatrix} \right)$$

The marginal distribution for x_1 is $\mathcal{N}(0, 3)$, and the marginal distribution for x_2 is $\mathcal{N}(1, 2)$.

The conditional distribution for x_1 given $x_2 = 2$ is

$$\mu_{x_1 | x_2 = 2} = 0 + 1 \cdot 2^{-1} \cdot (2 - 1) = 0.5$$

$$\Sigma_{x_1 | x_2 = 2} = 3 - 1 \cdot 2^{-1} \cdot 1 = 2.5$$

$$x_1 \mid (x_2 = 2) \sim \mathcal{N}(0.5, 2.5)$$

Example 15.1. Marginal and conditional distributions for a multivariate Gaussian.

15.2   Gaussian Processes

In the previous chapter, we approximated the objective function f using a surrogate model function $\hat{f}$ fitted to previously evaluated design points. A special type of surrogate model known as a *Gaussian process* allows us not only to predict f but also to quantify our uncertainty in that prediction using a probability distribution.[3]

A Gaussian process is a distribution over functions. For any finite set of points $\{\mathbf{x}^{(1)}, \ldots, \mathbf{x}^{(m)}\}$, the associated function evaluations $\{y_1, \ldots, y_m\}$ are distributed according to:

[3] A more extensive introduction to Gaussian processes is provided by C. E. Rasmussen and C. K. I. Williams, *Gaussian Processes for Machine Learning*. MIT Press, 2006.

$$\begin{bmatrix} y_1 \\ \vdots \\ y_m \end{bmatrix} \sim \mathcal{N}\left(\begin{bmatrix} m(\mathbf{x}^{(1)}) \\ \vdots \\ m(\mathbf{x}^{(m)}) \end{bmatrix}, \begin{bmatrix} k(\mathbf{x}^{(1)}, \mathbf{x}^{(1)}) & \cdots & k(\mathbf{x}^{(1)}, \mathbf{x}^{(m)}) \\ \vdots & \ddots & \vdots \\ k(\mathbf{x}^{(m)}, \mathbf{x}^{(1)}) & \cdots & k(\mathbf{x}^{(m)}, \mathbf{x}^{(m)}) \end{bmatrix} \right) \qquad (15.8)$$

where $m(\mathbf{x})$ is a *mean function* and $k(\mathbf{x}, \mathbf{x}')$ is the *covariance function*, or *kernel*.[4] The mean function can represent prior knowledge about the function. The kernel controls the smoothness of the functions. Methods for constructing the mean vector and covariance matrix using mean and covariance functions are given in algorithm 15.1.

[4] The mean function produces the expectation:

$$m(\mathbf{x}) = \mathbb{E}[f(\mathbf{x})]$$

and the covariance function produces the covariance:

$$k(\mathbf{x}, \mathbf{x}') = \mathbb{E}\left[(f(\mathbf{x}) - m(\mathbf{x}))(f(\mathbf{x}') - m(\mathbf{x}')) \right]$$

```
μ(X, m) = [m(x) for x in X]
Σ(X, k) = [k(x,x') for x in X, x' in X]
K(X, X', k) = [k(x,x') for x in X, x' in X']
```

Algorithm 15.1. The function μ for constructing a mean vector given a list of design points and a mean function m, and the function Σ for constructing a covariance matrix given one or two lists of design points and a covariance function k.

A common kernel function is the *squared exponential kernel*, where

$$k(x, x') = \exp\left(-\frac{(x - x')^2}{2\ell^2}\right) \qquad (15.9)$$

The parameter ℓ corresponds to what is called the *characteristic length-scale*, which can be thought of as the distance we have to travel in design space until the objective function value changes significantly.[5] Hence, larger values of ℓ result in smoother functions. Figure 15.2 shows functions sampled from a Gaussian process with a zero-mean function and a squared exponential kernel with different characteristic length-scales.

[5] A mathematical definition of characteristic length-scale is provided by C.E. Rasmussen and C.K.I. Williams, *Gaussian Processes for Machine Learning*. MIT Press, 2006.

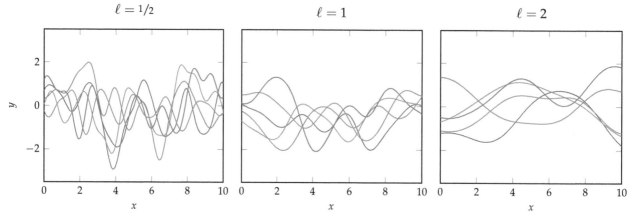

Figure 15.2. Functions sampled from Gaussian processes with squared exponential kernels.

There are many other kernel functions besides the squared exponential. Several are shown in figure 15.3. Many kernel functions use r, which is the distance between $\mathbf{x}$ and $\mathbf{x}'$. Usually the Euclidean distance is used. The *Matérn kernel* uses the *gamma function* Γ, implemented by gamma from the SpecialFunctions.jl package, and $K_\nu(x)$ is the *modified Bessel function of the second kind*, implemented by besselk(ν,x). The *neural network kernel* augments each design vector with a 1 for ease of notation: $\bar{\mathbf{x}} = [1, x_1, x_2, \ldots]$ and $\bar{\mathbf{x}}' = [1, x_1', x_2', \ldots]$.

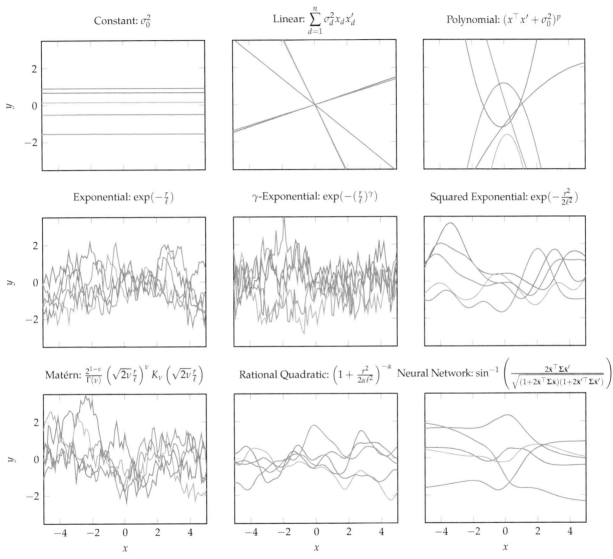

Figure 15.3. Functions sampled from Gaussian processes with different kernel functions. Shown functions are for $\sigma_0^2 = \sigma_d^2 = \ell = 1$, $p = 2$, $\gamma = \nu = \alpha = 0.5$, and $\mathbf{\Sigma} = \mathbf{I}$.

This chapter will focus on examples of Gaussian processes with single-dimensional design spaces for ease of plotting. However, Gaussian processes can be defined over multidimensional design spaces, as illustrated in figure 15.4.

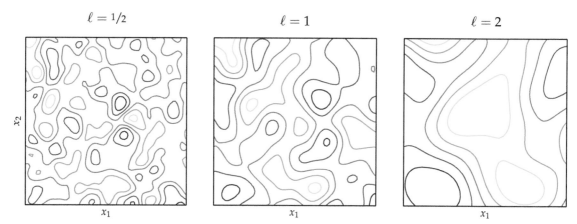

Figure 15.4. Functions sampled from a Gaussian process with zero-mean and squared-exponential kernels over a two-dimensional design space.

As we will see in section 15.5, Gaussian processes can also incorporate prior independent noise variance, denoted v. A Gaussian process is thus defined by mean and covariance functions, prior design points and their function evaluations, and a noise variance. The associated type is given in algorithm 15.2.

```
mutable struct GaussianProcess
    m # mean
    k # covariance function
    X # design points
    y # objective values
    v # noise variance
end
```

Algorithm 15.2. A Gaussian process is defined by a mean function m, a covariance function k, sampled design vectors X and their corresponding values y, and a noise variance v.

15.3 Prediction

Gaussian processes are able to represent distributions over functions using conditional probabilities. Suppose we already have a set of points X and the corresponding $\mathbf{y}$, but we wish to predict the values $\hat{\mathbf{y}}$ at points X^*. The joint distribution is

$$\begin{bmatrix} \hat{\mathbf{y}} \\ \mathbf{y} \end{bmatrix} \sim \mathcal{N}\left(\begin{bmatrix} \mathbf{m}(X^*) \\ \mathbf{m}(X\ ) \end{bmatrix}, \begin{bmatrix} \mathbf{K}(X^*, X^*) & \mathbf{K}(X^*, X) \\ \mathbf{K}(X\ , X^*) & \mathbf{K}(X\ , X) \end{bmatrix} \right) \tag{15.10}$$

In the equation above, we use the functions $\mathbf{m}$ and $\mathbf{K}$, which are defined as follows:

$$\mathbf{m}(X) = [m(\mathbf{x}^{(1)}), \ldots, m(\mathbf{x}^{(n)})] \tag{15.11}$$

$$\mathbf{K}(X, X') = \begin{bmatrix} k(\mathbf{x}^{(1)}, \mathbf{x}'^{(1)}) & \cdots & k(\mathbf{x}^{(1)}, \mathbf{x}'^{(m)}) \\ \vdots & \ddots & \vdots \\ k(\mathbf{x}^{(n)}, \mathbf{x}'^{(1)}) & \cdots & k(\mathbf{x}^{(n)}, \mathbf{x}'^{(m)}) \end{bmatrix} \tag{15.12}$$

[6] In the language of Bayesian statistics, the posterior distribution is the distribution of possible unobserved values conditioned on observed values.

The conditional distribution is given by

$$\hat{\mathbf{y}} \mid \mathbf{y} \sim \mathcal{N}\left(\underbrace{\mathbf{m}(X^*) + \mathbf{K}(X^*, X)\mathbf{K}(X, X)^{-1}(\mathbf{y} - \mathbf{m}(X))}_{\text{mean}}, \underbrace{\mathbf{K}(X^*, X^*) - \mathbf{K}(X^*, X)\mathbf{K}(X, X)^{-1}\mathbf{K}(X, X^*)}_{\text{covariance}} \right) \tag{15.13}$$

Note that the covariance does not depend on $\mathbf{y}$. This distribution is often referred to as the *posterior distribution*.[6] A method for computing and sampling from the posterior distribution defined by a Gaussian process is given in algorithm 15.3.

```
function mvnrand(μ, Σ, inflation=1e-6)
    N = MvNormal(μ, Σ + inflation*I)
    return rand(N)
end
Base.rand(GP, X) = mvnrand(μ(X, GP.m), Σ(X, GP.k))
```

Algorithm 15.3.   The function mvnrand samples from a multivariate Gaussian with an added inflation factor to prevent numerical issues. The method rand samples a Gaussian process GP at the given design points in matrix X.

The predicted mean can be written as a function of $\mathbf{x}$:

$$\hat{\mu}(\mathbf{x}) = m(\mathbf{x}) + \mathbf{K}(\mathbf{x}, X)\mathbf{K}(X, X)^{-1}(\mathbf{y} - \mathbf{m}(X)) \tag{15.14}$$

$$= m(\mathbf{x}) + \boldsymbol{\theta}^\top \mathbf{K}(X, \mathbf{x}) \tag{15.15}$$

where $\boldsymbol{\theta} = \mathbf{K}(X, X)^{-1}(\mathbf{y} - \mathbf{m}(X))$ can be computed once and reused for different values of $\mathbf{x}$. Notice the similarity to the surrogate models in the previous chapter. The value of the Gaussian process beyond the surrogate models discussed previously is that it also quantifies our uncertainty in our predictions.

The variance of the predicted mean can also be obtained as a function of $\mathbf{x}$:

$$\hat{v}(\mathbf{x}) = \mathbf{K}(\mathbf{x}, \mathbf{x}) - \mathbf{K}(\mathbf{x}, X)\mathbf{K}(X, X)^{-1}\mathbf{K}(X, \mathbf{x}) \tag{15.16}$$

In some cases, it is more convenient to formulate equations in terms of the *standard deviation*, which is the square root of the variance:

$$\hat{\sigma}(\mathbf{x}) = \sqrt{\hat{v}(\mathbf{x})} \tag{15.17}$$

The standard deviation has the same units as the mean. From the standard deviation, we can compute the 95% *confidence region*, which is an interval containing 95% of the probability mass associated with the distribution over y given $\mathbf{x}$. For a particular $\mathbf{x}$, the 95% confidence region is given by $\hat{\mu}(\mathbf{x}) \pm 1.96\hat{\sigma}(\mathbf{x})$. One may want to use a confidence level different from 95%, but we will use 95% for the plots in this chapter. Figure 15.5 shows a plot of a confidence region associated with a Gaussian process fit to four function evaluations.

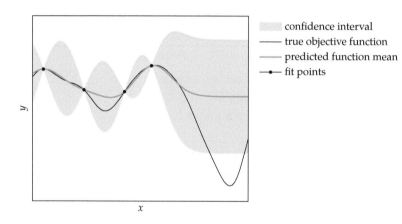

Figure 15.5. A Gaussian process using the squared exponential kernel and its 95% confidence interval. Uncertainty increases the farther we are from a data point. The expected function value approaches zero as we move far away from the data point.

15.4 Gradient Measurements

Gradient observations can be incorporated into Gaussian processes in a manner consistent with the existing Gaussian process machinery.[7] The Gaussian process is extended to include both the function value and its gradient:

$$\begin{bmatrix} \mathbf{y} \\ \nabla \mathbf{y} \end{bmatrix} \sim \mathcal{N}\left(\begin{bmatrix} \mathbf{m}_f \\ \mathbf{m}_\nabla \end{bmatrix}, \begin{bmatrix} \mathbf{K}_{ff} & \mathbf{K}_{f\nabla} \\ \mathbf{K}_{\nabla f} & \mathbf{K}_{\nabla\nabla} \end{bmatrix} \right) \tag{15.18}$$

[7] For an overview, see for example A. O'Hagan, "Some Bayesian Numerical Analysis," *Bayesian Statistics*, vol. 4, J. M. Bernardo, J. O. Berger, A. P. Dawid, and A. F. M. Smith, eds., pp. 345–363, 1992.

where $\mathbf{y} \sim \mathcal{N}\left(\mathbf{m}_f, \mathbf{K}_{ff}\right)$ is a traditional Gaussian process, $\mathbf{m}_\nabla$ is a mean function for the gradient,[8] $\mathbf{K}_{f\nabla}$ is the covariance matrix between function values and gradients, $\mathbf{K}_{\nabla f}$ is the covariance matrix between function gradients and values, and $\mathbf{K}_{\nabla\nabla}$ is the covariance matrix between function gradients.

These covariance matrices are constructed using covariance functions. The linearity of Gaussians causes these covariance functions to be related:

$$k_{ff}(\mathbf{x}, \mathbf{x}') = k(\mathbf{x}, \mathbf{x}') \tag{15.19}$$
$$k_{\nabla f}(\mathbf{x}, \mathbf{x}') = \nabla_\mathbf{x} k(\mathbf{x}, \mathbf{x}') \tag{15.20}$$
$$k_{f\nabla}(\mathbf{x}, \mathbf{x}') = \nabla_{\mathbf{x}'} k(\mathbf{x}, \mathbf{x}') \tag{15.21}$$
$$k_{\nabla\nabla}(\mathbf{x}, \mathbf{x}') = \nabla_\mathbf{x} \nabla_{\mathbf{x}'} k(\mathbf{x}, \mathbf{x}') \tag{15.22}$$

Example 15.2 uses these relations to derive the higher-order covariance functions for a particular kernel.

Consider the squared exponential covariance function

$$k_{ff}(\mathbf{x}, \mathbf{x}') = \exp\left(-\frac{1}{2}\|\mathbf{x} - \mathbf{x}'\|^2\right)$$

We can use equations (15.19) to (15.22) to obtain the other covariance functions necessary for using Gaussian processes with gradient information:

$$k_{\nabla f}(\mathbf{x}, \mathbf{x}')_i = -(\mathbf{x}_i - \mathbf{x}'_i)\exp\left(-\frac{1}{2}\|\mathbf{x} - \mathbf{x}'\|^2\right)$$

$$k_{\nabla\nabla}(\mathbf{x}, \mathbf{x}')_{ij} = -\left((i = j) - (\mathbf{x}_i - \mathbf{x}'_i)(\mathbf{x}_j - \mathbf{x}'_j)\right)\exp\left(-\frac{1}{2}\|\mathbf{x} - \mathbf{x}'\|^2\right)$$

As a reminder, Boolean expressions, such as $(i = j)$, return 1 if true and 0 if false.

Example 15.2. Deriving covariance functions for a Gaussian process with gradient observations.

Prediction can be accomplished in the same manner as with a traditional Gaussian process. We first construct the joint distribution

$$\begin{bmatrix} \hat{\mathbf{y}} \\ \mathbf{y} \\ \nabla\mathbf{y} \end{bmatrix} \sim \mathcal{N}\left(\begin{bmatrix} \mathbf{m}_f(X^*) \\ \mathbf{m}_f(X) \\ \mathbf{m}_\nabla(X) \end{bmatrix}, \begin{bmatrix} \mathbf{K}_{ff}(X^*, X^*) & \mathbf{K}_{ff}(X^*, X) & \mathbf{K}_{f\nabla}(X^*, X) \\ \mathbf{K}_{ff}(X, X^*) & \mathbf{K}_{ff}(X, X) & \mathbf{K}_{f\nabla}(X, X) \\ \mathbf{K}_{\nabla f}(X, X^*) & \mathbf{K}_{\nabla f}(X, X) & \mathbf{K}_{\nabla\nabla}(X, X) \end{bmatrix}\right) \tag{15.23}$$

For a Gaussian process over n-dimensional design vectors given m pairs of function and gradient evaluations and ℓ query points, the covariance blocks have the following dimensions:

$$
\begin{matrix}
\ell \times \ell & \ell \times m & \ell \times nm \\
m \times \ell & m \times m & m \times nm \\
nm \times \ell & nm \times m & nm \times nm
\end{matrix}
\tag{15.24}
$$

Example 15.3 constructs such a covariance matrix.

Suppose we have evaluated a function and its gradient at two locations, $\mathbf{x}^{(1)}$ and $\mathbf{x}^{(2)}$, and we wish to predict the function value at $\hat{\mathbf{x}}$. We can infer the joint distribution over $\hat{\mathbf{y}}$, $\mathbf{y}$, and $\nabla \mathbf{y}$ using a Gaussian process. The covariance matrix is:

$$
\begin{bmatrix}
k_{ff}(\hat{\mathbf{x}},\hat{\mathbf{x}}) & k_{ff}(\hat{\mathbf{x}},\mathbf{x}^{(1)}) & k_{ff}(\hat{\mathbf{x}},\mathbf{x}^{(2)}) & k_{f\nabla}(\hat{\mathbf{x}},\mathbf{x}^{(1)})_1 & k_{f\nabla}(\hat{\mathbf{x}},\mathbf{x}^{(1)})_2 & k_{f\nabla}(\hat{\mathbf{x}},\mathbf{x}^{(2)})_1 & k_{f\nabla}(\hat{\mathbf{x}},\mathbf{x}^{(2)})_2 \\
k_{ff}(\mathbf{x}^{(1)},\hat{\mathbf{x}}) & k_{ff}(\mathbf{x}^{(1)},\mathbf{x}^{(1)}) & k_{ff}(\mathbf{x}^{(1)},\mathbf{x}^{(2)}) & k_{f\nabla}(\mathbf{x}^{(1)},\mathbf{x}^{(1)})_1 & k_{f\nabla}(\mathbf{x}^{(1)},\mathbf{x}^{(1)})_2 & k_{f\nabla}(\mathbf{x}^{(1)},\mathbf{x}^{(2)})_1 & k_{f\nabla}(\mathbf{x}^{(1)},\mathbf{x}^{(2)})_2 \\
k_{ff}(\mathbf{x}^{(2)},\hat{\mathbf{x}}) & k_{ff}(\mathbf{x}^{(2)},\mathbf{x}^{(1)}) & k_{ff}(\mathbf{x}^{(2)},\mathbf{x}^{(2)}) & k_{f\nabla}(\mathbf{x}^{(2)},\mathbf{x}^{(1)})_1 & k_{f\nabla}(\mathbf{x}^{(2)},\mathbf{x}^{(1)})_2 & k_{f\nabla}(\mathbf{x}^{(2)},\mathbf{x}^{(2)})_1 & k_{f\nabla}(\mathbf{x}^{(2)},\mathbf{x}^{(2)})_2 \\
k_{\nabla f}(\mathbf{x}^{(1)},\hat{\mathbf{x}})_1 & k_{\nabla f}(\mathbf{x}^{(1)},\mathbf{x}^{(1)})_1 & k_{\nabla f}(\mathbf{x}^{(1)},\mathbf{x}^{(2)})_1 & k_{\nabla\nabla}(\mathbf{x}^{(1)},\mathbf{x}^{(1)})_{11} & k_{\nabla\nabla}(\mathbf{x}^{(1)},\mathbf{x}^{(1)})_{12} & k_{\nabla\nabla}(\mathbf{x}^{(1)},\mathbf{x}^{(2)})_{11} & k_{\nabla\nabla}(\mathbf{x}^{(1)},\mathbf{x}^{(2)})_{12} \\
k_{\nabla f}(\mathbf{x}^{(1)},\hat{\mathbf{x}})_2 & k_{\nabla f}(\mathbf{x}^{(1)},\mathbf{x}^{(1)})_2 & k_{\nabla f}(\mathbf{x}^{(1)},\mathbf{x}^{(2)})_2 & k_{\nabla\nabla}(\mathbf{x}^{(1)},\mathbf{x}^{(1)})_{21} & k_{\nabla\nabla}(\mathbf{x}^{(1)},\mathbf{x}^{(1)})_{22} & k_{\nabla\nabla}(\mathbf{x}^{(1)},\mathbf{x}^{(2)})_{21} & k_{\nabla\nabla}(\mathbf{x}^{(1)},\mathbf{x}^{(2)})_{22} \\
k_{\nabla f}(\mathbf{x}^{(2)},\hat{\mathbf{x}})_1 & k_{\nabla f}(\mathbf{x}^{(2)},\mathbf{x}^{(1)})_1 & k_{\nabla f}(\mathbf{x}^{(2)},\mathbf{x}^{(2)})_1 & k_{\nabla\nabla}(\mathbf{x}^{(2)},\mathbf{x}^{(1)})_{11} & k_{\nabla\nabla}(\mathbf{x}^{(2)},\mathbf{x}^{(1)})_{12} & k_{\nabla\nabla}(\mathbf{x}^{(2)},\mathbf{x}^{(2)})_{11} & k_{\nabla\nabla}(\mathbf{x}^{(2)},\mathbf{x}^{(2)})_{12} \\
k_{\nabla f}(\mathbf{x}^{(2)},\hat{\mathbf{x}})_2 & k_{\nabla f}(\mathbf{x}^{(2)},\mathbf{x}^{(1)})_2 & k_{\nabla f}(\mathbf{x}^{(2)},\mathbf{x}^{(2)})_2 & k_{\nabla\nabla}(\mathbf{x}^{(2)},\mathbf{x}^{(1)})_{21} & k_{\nabla\nabla}(\mathbf{x}^{(2)},\mathbf{x}^{(1)})_{22} & k_{\nabla\nabla}(\mathbf{x}^{(2)},\mathbf{x}^{(2)})_{21} & k_{\nabla\nabla}(\mathbf{x}^{(2)},\mathbf{x}^{(2)})_{22}
\end{bmatrix}
$$

Example 15.3. Constructing the covariance matrix for a Gaussian process with gradient observations.

The conditional distribution follows the same Gaussian relations as in equation (15.13):

$$
\hat{\mathbf{y}} \mid \mathbf{y}, \nabla \mathbf{y} \sim \mathcal{N}(\mu_\nabla, \Sigma_\nabla)
\tag{15.25}
$$

where:

$$
\mu_\nabla = \mathbf{m}_f(X^*) + \begin{bmatrix} \mathbf{K}_{ff}(X,X^*) \\ \mathbf{K}_{\nabla f}(X,X^*) \end{bmatrix}^\top \begin{bmatrix} \mathbf{K}_{ff}(X,X) & \mathbf{K}_{f\nabla}(X,X) \\ \mathbf{K}_{\nabla f}(X,X) & \mathbf{K}_{\nabla\nabla}(X,X) \end{bmatrix}^{-1} \begin{bmatrix} \mathbf{y} - \mathbf{m}_f(X) \\ \nabla\mathbf{y} - \mathbf{m}_\nabla(X) \end{bmatrix}
\tag{15.26}
$$

$$
\Sigma_\nabla = \mathbf{K}_{ff}(X^*,X^*) - \begin{bmatrix} \mathbf{K}_{ff}(X,X^*) \\ \mathbf{K}_{\nabla f}(X,X^*) \end{bmatrix}^\top \begin{bmatrix} \mathbf{K}_{ff}(X,X) & \mathbf{K}_{f\nabla}(X,X) \\ \mathbf{K}_{\nabla f}(X,X) & \mathbf{K}_{\nabla\nabla}(X,X) \end{bmatrix}^{-1} \begin{bmatrix} \mathbf{K}_{ff}(X,X^*) \\ \mathbf{K}_{\nabla f}(X,X^*) \end{bmatrix}
\tag{15.27}
$$

The regions obtained when including gradient observations are compared to those without gradient observations in figure 15.6.

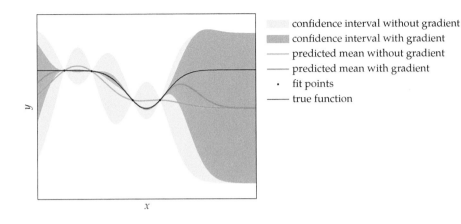

confidence interval without gradient
confidence interval with gradient
predicted mean without gradient
predicted mean with gradient
· fit points
true function

Figure 15.6. Gaussian processes with and without gradient information using squared exponential kernels. Incorporating gradient information can significantly reduce the confidence intervals.

15.5   Noisy Measurements

So far we have assumed that the objective function f is deterministic. In practice, however, evaluations of f may include measurement noise, experimental error, or numerical roundoff.

We can model noisy evaluations as $y = f(\mathbf{x}) + z$, where f is deterministic but z is zero-mean Gaussian noise, $z \sim \mathcal{N}(0, \nu)$. The variance of the noise ν can be adjusted to control the uncertainty.[9]

The new joint distribution is:

$$\begin{bmatrix} \hat{\mathbf{y}} \\ \mathbf{y} \end{bmatrix} \sim \mathcal{N}\left(\begin{bmatrix} \mathbf{m}(X^*) \\ \mathbf{m}(X) \end{bmatrix}, \begin{bmatrix} \mathbf{K}(X^*, X^*) & \mathbf{K}(X^*, X) \\ \mathbf{K}(X, X^*) & \mathbf{K}(X, X) + \nu\mathbf{I} \end{bmatrix} \right) \tag{15.28}$$

with conditional distribution:

$$\hat{\mathbf{y}} \mid \mathbf{y}, \nu \sim \mathcal{N}(\boldsymbol{\mu}^*, \boldsymbol{\Sigma}^*) \tag{15.29}$$

$$\boldsymbol{\mu}^* = \mathbf{m}(X^*) + \mathbf{K}(X^*, X)(\mathbf{K}(X, X) + \nu\mathbf{I})^{-1}(\mathbf{y} - \mathbf{m}(X)) \tag{15.30}$$

$$\boldsymbol{\Sigma}^* = \mathbf{K}(X^*, X^*) - \mathbf{K}(X^*, X)(\mathbf{K}(X, X) + \nu\mathbf{I})^{-1}\mathbf{K}(X, X^*) \tag{15.31}$$

As the equations above show, accounting for Gaussian noise is straightforward and the posterior distribution can be computed analytically. Figure 15.7 shows a noisy Gaussian process. Algorithm 15.4 implements prediction for Gaussian processes with noisy measurements.

[9] The techniques covered in section 14.5 can be used to tune the variance of the noise.

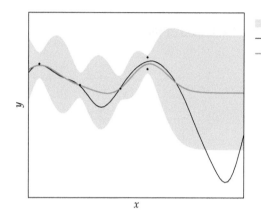

Figure 15.7. A noisy Gaussian process using a squared exponential kernel.

```
function predict(GP, X_pred)
    m, k, ν = GP.m, GP.k, GP.ν
    tmp = K(X_pred, GP.X, k) / (K(GP.X, GP.X, k) + ν*I)
    μₚ = μ(X_pred, m) + tmp*(GP.y - μ(GP.X, m))
    S = K(X_pred, X_pred, k) - tmp*K(GP.X, X_pred, k)
    νₚ = diag(S) .+ eps() # eps prevents numerical issues
    return (μₚ, νₚ)
end
```

Algorithm 15.4. A method for obtaining the predicted means and standard deviations in f under a Gaussian process. The method takes a Gaussian process GP and a list of points X_pred at which to evaluate the prediction. It returns the mean and variance at each evaluation point.

15.6   Fitting Gaussian Processes

The choice of kernel and parameters has a large effect on the form of the Gaussian process between evaluated design points. Kernels and their parameters can be chosen using cross validation introduced in the previous chapter. Instead of minimizing the squared error on the test data, we maximize the likelihood of the data.[10] That is, we seek the parameters θ that maximize the probability of the function values, $p(\mathbf{y} \mid X, \theta)$. The likelihood of the data is the probability that the observed points were drawn from the model. Equivalently, we can maximize the *log likelihood*, which is generally preferable because multiplying small probabilities in the likelihood calculation can produce extremely small values. Given a dataset $\mathcal{D}$ with n entries, the log likelihood is given by

[10] Alternatively, we could maximize the pseudolikelihood as discussed by C. E. Rasmussen and C. K. I. Williams, *Gaussian Processes for Machine Learning*. MIT Press, 2006.

$$\log p(\mathbf{y} \mid X, \nu, \theta) = -\frac{n}{2}\log 2\pi - \frac{1}{2}\log|\mathbf{K}_\theta(X,X) + \nu\mathbf{I}| - \frac{1}{2}(\mathbf{y} - \mathbf{m}_\theta(X))^\top (\mathbf{K}_\theta(X,X) + \nu\mathbf{I})^{-1}(\mathbf{y} - \mathbf{m}_\theta(X)) \tag{15.32}$$

where the mean and covariance functions are parameterized by θ.

Let us assume a zero mean such that $\mathbf{m}_\theta(X) = \mathbf{0}$ and θ refers only to the parameters for the Gaussian process covariance function. We can arrive at a *maximum likelihood estimate* by gradient ascent. The gradient is then given by

$$\frac{\partial}{\partial\theta_j}\log p(\mathbf{y} \mid X, \theta) = \frac{1}{2}\mathbf{y}^\top \mathbf{K}^{-1}\frac{\partial\mathbf{K}}{\partial\theta_j}\mathbf{K}^{-1}\mathbf{y} - \frac{1}{2}\mathrm{tr}\left(\Sigma_\theta^{-1}\frac{\partial\mathbf{K}}{\partial\theta_j}\right) \tag{15.33}$$

where $\Sigma_\theta = \mathbf{K}_\theta(X,X) + \nu\mathbf{I}$. Above, we use the matrix derivative relations

$$\frac{\partial\mathbf{K}^{-1}}{\partial\theta_j} = -\mathbf{K}^{-1}\frac{\partial\mathbf{K}}{\partial\theta_j}\mathbf{K}^{-1} \tag{15.34}$$

$$\frac{\partial\log|\mathbf{K}|}{\partial\theta_j} = \mathrm{tr}\left(\mathbf{K}^{-1}\frac{\partial\mathbf{K}}{\partial\theta_j}\right) \tag{15.35}$$

where $\mathrm{tr}(\mathbf{A})$ denotes the *trace* of a matrix $\mathbf{A}$, defined to be the sum of the elements on the main diagonal.

15.7   Summary

- Gaussian processes are probability distributions over functions.

- The choice of kernel affects the smoothness of the functions sampled from a Gaussian process.

- The multivariate normal distribution has analytic conditional and marginal distributions.

- We can compute the mean and standard deviation of our prediction of an objective function at a particular design point given a set of past evaluations.

- We can incorporate gradient observations to improve our predictions of the objective value and its gradient.

- We can incorporate measurement noise into a Gaussian process.

- We can fit the parameters of a Gaussian process using maximum likelihood.

15.8   Exercises

Exercise 15.1. Gaussian processes will grow in complexity during the optimization process as more samples accumulate. How can this be an advantage over models based on regression?

Exercise 15.2. How does the computational complexity of prediction with a Gaussian process increase with the number of data points m?

Exercise 15.3. Consider the function $f(x) = \sin(x)/(x^2 + 1)$ over $[-5, 5]$. Plot the 95% confidence bounds for a Gaussian process with derivative information fitted to the evaluations at $\{-5, -2.5, 0, 2.5, 5\}$. What is the maximum standard deviation in the predicted distribution within $[-5, 5]$? How many function evaluations, evenly-spaced over the domain, are needed such that a Gaussian process without derivative information achieves the same maximum predictive standard deviation?

Assume zero-mean functions and noise-free observations, and use the covariance functions:

$$k_{ff}(x, x') = \exp\left(-\frac{1}{2}\|x - x'\|_2^2\right)$$

$$k_{\nabla f}(x, x') = (x' - x)\exp\left(-\frac{1}{2}\|x - x'\|_2^2\right)$$

$$k_{\nabla\nabla}(x, x') = ((x - x')^2 - 1)\exp(-\frac{1}{2}\|x - x'\|_2^2)$$

Exercise 15.4. Derive the relation $k_{f\nabla}(\mathbf{x}, \mathbf{x}')_i = \mathrm{cov}\left(f(\mathbf{x}), \frac{\partial}{\partial x_i'}f(\mathbf{x}')\right) = \frac{\partial}{\partial x_i'}k_{ff}(\mathbf{x}, \mathbf{x}')$.

Exercise 15.5. Suppose we have a multivariate Gaussian distribution over two variables a and b. Show that the variance of the conditional distribution over a given b is no greater than the variance of the marginal distribution over a. Does this make intuitive sense?

Exercise 15.6. Suppose we observe many outliers while sampling, that is, we observe samples that do not fall within the confidence interval given by the Gaussian process. This means the probabilistic model we chose is not appropriate. What can we do?

Exercise 15.7. Consider model selection for the function evaluation pairs (x, y):

$$\{(1, 0), (2, -1), (3, -2), (4, 1), (5, 0)\}$$

Use leave-one-out cross-validation to select the kernel that maximizes the likelihood of predicting the withheld pair given a Gaussian process over the other pairs in the fold. Assume zero mean with no noise. Select from the kernels:

$$\exp(-\|x - x'\|) \qquad \exp(-\|x - x'\|^2) \qquad (1 + \|x - x'\|)^{-1} \qquad (1 + \|x - x'\|^2)^{-1} \qquad (1 + \|x - x'\|)^{-2}$$

16 *Surrogate Optimization*

The previous chapter explained how to use a probabilistic surrogate model, in particular a Gaussian process, to infer probability distributions over the true objective function. These distributions can be used to guide an optimization process toward better design points.[1] This chapter outlines several common techniques for choosing which design point to evaluate next. The techniques we discuss here greedily optimize various metrics.[2] We will also discuss how surrogate models can be used to optimize an objective measure in a safe manner.

[1] A. Forrester, A. Sobester, and A. Keane, *Engineering Design via Surrogate Modelling: A Practical Guide.* Wiley, 2008.

[2] An alternative to greedy optimization is to frame the problem as a *partially observable Markov decision process* and plan ahead some number of steps as outlined by M. Toussaint, "The Bayesian Search Game," in *Theory and Principled Methods for the Design of Metaheuristics,* Y. Borenstein and A. Moraglio, eds. Springer, 2014, pp. 129–144. See also R. Lam, K. Willcox, and D. H. Wolpert, "Bayesian Optimization with a Finite Budget: An Approximate Dynamic Programming Approach," in *Advances in Neural Information Processing Systems (NIPS)*, 2016.

16.1 *Prediction-Based Exploration*

In *prediction-based exploration*, we select the minimizer of the surrogate function. An example of this approach is the quadratic fit search that we discussed earlier in section 3.5. With quadratic fit search, we use a quadratic surrogate model to fit the last three bracketing points and then select the point at the minimum of the quadratic function.

If we use a Gaussian process surrogate model, prediction-based optimization has us select the minimizer of the mean function

$$\mathbf{x}^{(m+1)} = \arg\min_{\mathbf{x} \in \mathcal{X}} \hat{\mu}(\mathbf{x}) \qquad (16.1)$$

where $\hat{\mu}(\mathbf{x})$ is the predicted mean of a Gaussian process at a design point $\mathbf{x}$ based on the previous m design points. The process is illustrated in figure 16.1.

Prediction-based optimization does not take uncertainty into account, and new samples can be generated very close to existing samples. Sampling at locations where we are already confident in the objective value is a waste of function evaluations.

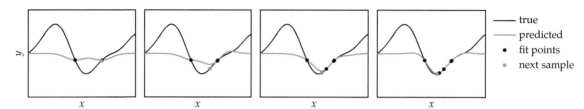

Figure 16.1. Prediction-based optimization selects the point that minimizes the mean of the objective function.

16.2   Error-Based Exploration

Error-based exploration seeks to increase confidence in the true function. A Gaussian process can tell us both the mean and standard deviation at every point. A large standard deviation indicates low confidence, so error-based exploration samples at design points with maximum uncertainty.

The next sample point is:

$$x^{(m+1)} = \arg\max_{\mathbf{x} \in \mathcal{X}} \hat{\sigma}(\mathbf{x}) \qquad (16.2)$$

where $\hat{\sigma}(\mathbf{x})$ is the standard deviation of a Gaussian process at a design point $\mathbf{x}$ based on the previous m design points. The process is illustrated in figure 16.2.

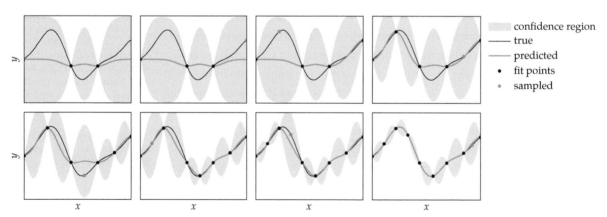

Figure 16.2. Error-based exploration selects a point with maximal uncertainty.

Gaussian processes are often defined over all of $\mathbb{R}^n$. Optimization problems with unbounded feasible sets will always have high uncertainty far away from sampled points, making it impossible to become confident in the true underlying function over the entire domain. Error-based exploration must thus be constrained to a closed region.

16.3   Lower Confidence Bound Exploration

While error-based exploration reduces the uncertainty in the objective function overall, its samples are often in regions that are unlikely to contain a global minimum. *Lower confidence bound exploration* trades off between greedy minimization employed by prediction-based optimization and uncertainty reduction employed by error-based exploration. The next sample minimizes the *lower confidence bound* of the objective function

$$LB(\mathbf{x}) = \hat{\mu}(\mathbf{x}) - \alpha\hat{\sigma}(\mathbf{x}) \tag{16.3}$$

where $\alpha \geq 0$ is a constant that controls the trade-off between *exploration* and *exploitation*. Exploration involves minimizing uncertainty, and exploitation involves minimizing the predicted mean. We have prediction-based optimization with $\alpha = 0$, and we have error-based exploration as α approaches ∞. The process is illustrated in figure 16.3.

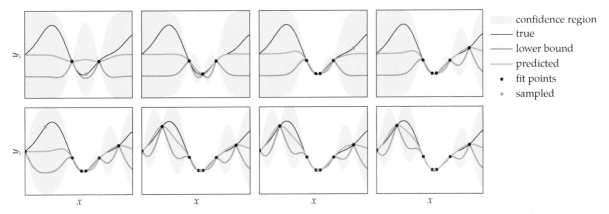

Figure 16.3.   Lower confidence bound exploration trades off between minimizing uncertainty and minimizing the predicted function.

16.4   Probability of Improvement Exploration

We can sometimes obtain faster convergence by selecting the design point that maximizes the chance that the new point will be better than the samples we have seen so far. The *improvement* for a function sampled at $\mathbf{x}$ producing $y = f(\mathbf{x})$ is

$$I(y) = \begin{cases} y_{\min} - y & \text{if } y < y_{\min} \\ 0 & \text{otherwise} \end{cases} \tag{16.4}$$

where y_{min} is the minimum value sampled so far.

The *probability of improvement* at points where $\hat{\sigma} > 0$ is

$$P(y < y_{min}) = \int_{-\infty}^{y_{min}} \mathcal{N}(y \mid \hat{\mu}, \hat{\sigma}^2)dy \qquad (16.5)$$

$$= \Phi\left(\frac{y_{min} - \hat{\mu}}{\hat{\sigma}}\right) \qquad (16.6)$$

where Φ is the *standard normal cumulative distribution function* (see appendix C.7). This calculation (algorithm 16.1) is shown in figure 16.4. Figure 16.5 illustrates this process. When $\hat{\sigma} = 0$, which occurs at points where we have noiseless measurements, the probability of improvement is zero.

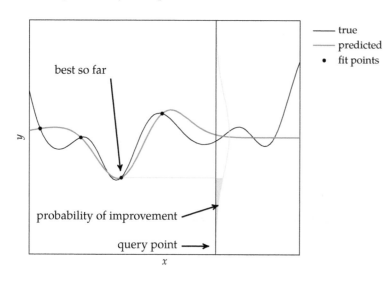

Figure 16.4. The probability of improvement is the probability that evaluating a particular point will yield a better result than the best so far. This figure shows the probability density function predicted at a query point, with the shaded region below y_{min} corresponding to the probability of improvement.

```
prob_of_improvement(y_min, μ, σ) = cdf(Normal(μ, σ), y_min)
```

Algorithm 16.1. Computing the probability of improvement for a given best y value y_min, mean μ, and variance ν.

16.5   Expected Improvement Exploration

Optimization is concerned with finding the minimum of the objective function. While maximizing the probability of improvement will tend to decrease the objective function over time, it does not improve very much with each iteration.

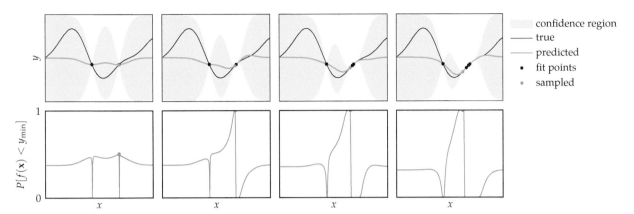

Figure 16.5. Maximizing the probability of improvement selects samples most likely to produce lower objective point values.

We can focus our exploration of points that maximize our *expected improvement* over the current best function value.

Through a substitution

$$z = \frac{y - \hat{\mu}}{\hat{\sigma}} \qquad y'_{\min} = \frac{y_{\min} - \hat{\mu}}{\hat{\sigma}} \tag{16.7}$$

we can write the improvement in equation (16.4) as

$$I(y) = \begin{cases} \hat{\sigma}(y'_{\min} - z) & \text{if } z < y'_{\min} \text{ and } \hat{\sigma} > 0 \\ 0 & \text{otherwise} \end{cases} \tag{16.8}$$

where $\hat{\mu}$ and $\hat{\sigma}$ are the predicted mean and standard deviation at the sample point **x**.

We can calculate the expected improvement using the distribution predicted by the Gaussian process:

$$\mathbb{E}[I(y)] = \hat{\sigma} \int_{-\infty}^{y'_{\min}} (y'_{\min} - z) \mathcal{N}(z \mid 0, 1) \, dz \tag{16.9}$$

$$= \hat{\sigma} \left[y'_{\min} \int_{-\infty}^{y'_{\min}} \mathcal{N}(z \mid 0, 1) \, dz - \int_{-\infty}^{y'_{\min}} z \, \mathcal{N}(z \mid 0, 1) \, dz \right] \tag{16.10}$$

$$= \hat{\sigma} \left[y'_{\min} P(z \le y'_{\min}) + \mathcal{N}(y'_{\min} \mid 0, 1) - \underbrace{\mathcal{N}(-\infty \mid 0, 1)}_{= 0} \right] \tag{16.11}$$

$$= (y_{\min} - \hat{\mu}) P(y \le y_{\min}) + \hat{\sigma}^2 \mathcal{N}(y_{\min} \mid \hat{\mu}, \hat{\sigma}^2) \tag{16.12}$$

Figure 16.6 illustrates this process using algorithm 16.2.

```
function expected_improvement(y_min, μ, σ)
    p_imp = prob_of_improvement(y_min, μ, σ)
    p_ymin = pdf(Normal(μ, σ), y_min)
    return (y_min - μ)*p_imp + σ^2*p_ymin
end
```

Algorithm 16.2. Computing the expected improvement for a given best y value y_min, mean μ, and standard deviation σ.

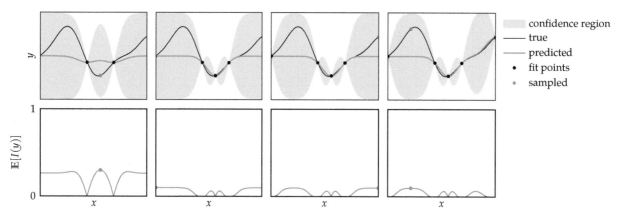

confidence region
true
predicted
• fit points
• sampled

Figure 16.6. Maximizing the expected improvement selects samples which are likely to improve the lower bound by as much as possible.

16.6   Safe Optimization

In some contexts, it may be costly to evaluate points that are deemed unsafe, which may correspond to low performing or infeasible points. Problems such as the in-flight tuning of the controller of a drone or safe movie recommendations require *safe exploration*—searching for an optimal design point while carefully avoiding sampling an unsafe design.

This section outlines the *SafeOpt* algorithm,[3] which addresses a class of safe exploration problems. We sample a series of design points $\mathbf{x}^{(1)}, \dots, \mathbf{x}^{(m)}$ in pursuit of a minimum but without $f(\mathbf{x}^{(i)})$ exceeding a critical safety threshold y_{max}. In addition, we receive only noisy measurements of the objective function, where the noise is zero-mean with variance v. Such an objective function and its associated safe regions are shown in figure 16.7.

The SafeOpt algorithm uses Gaussian process surrogate models for prediction. At each iteration, we fit a Gaussian process to the noisy samples from f. After the

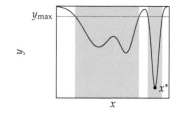

Figure 16.7. SafeOpt solves safe exploration problems that minimize f while remaining within safe regions defined by maximum objective function values.

[3] Y. Sui, A. Gotovos, J. Burdick, and A. Krause, "Safe Exploration for Optimization with Gaussian Processes," in *International Conference on Machine Learning* (ICML), vol. 37, 2015.

ith sample, SafeOpt calculates the upper and lower confidence bounds:

$$u_i(\mathbf{x}) = \hat{\mu}_{i-1}(\mathbf{x}) + \sqrt{\beta \hat{v}_{i-1}(\mathbf{x})} \tag{16.13}$$

$$\ell_i(\mathbf{x}) = \hat{\mu}_{i-1}(\mathbf{x}) - \sqrt{\beta \hat{v}_{i-1}(\mathbf{x})} \tag{16.14}$$

where larger values of β yield wider confidence regions. Such bounds are shown in figure 16.8.

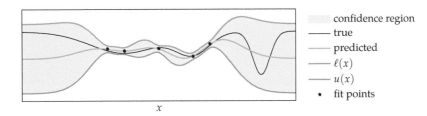

confidence region
—— true
—— predicted
—— $\ell(x)$
—— $u(x)$
• fit points

Figure 16.8. An illustration of functions based on the predictions of a Gaussian process used by SafeOpt.

The Gaussian process predicts a distribution over $f(\mathbf{x})$ for any design point. Being Gaussian, these predictions can provide only a probabilistic guarantee of safety up to an arbitrary factor:[4]

$$P(f(\mathbf{x}) \leq y_{\max}) = \Phi\left(\frac{y_{\max} - \hat{\mu}(\mathbf{x})}{\sqrt{\hat{v}(\mathbf{x})}}\right) \geq P_{\text{safe}} \tag{16.15}$$

[4] Note the similarity to the probability of improvement.

The predicted safe region $\mathcal{S}$ consists of the design points that provide a probability of safety greater than the required level P_{safe}, as illustrated in figure 16.9. The safe region can also be defined in terms of Lipschitz upper bounds constructed from upper bounds evaluated at previously sampled points.

SafeOpt chooses a safe sample point that balances the desire to localize a reachable minimizer of f and to expand the safe region. The set of potential minimizers of f is denoted $\mathcal{M}$ (figure 16.10), and the set of points that will potentially lead to the expansion of the safe regions is denoted $\mathcal{E}$ (figure 16.11). To trade off exploration and exploitation, we choose the design point $\mathbf{x}$ with the largest predictive variance among both sets $\mathcal{M}$ and $\mathcal{E}$.[5]

The set of potential minimizers consists of the safe points whose lower confidence bound is lower than the lowest upper bound:

$$\mathcal{M}_i = \left\{ \mathbf{x} \in \mathcal{S}_i \mid \ell_i(\mathbf{x}) \leq \min_{\mathbf{x}' \in \mathcal{S}_i} u_i(\mathbf{x}') \right\} \tag{16.16}$$

[5] For a variation of this algorithm, see F. Berkenkamp, A. P. Schoellig, and A. Krause, "Safe Controller Optimization for Quadrotors with Gaussian Processes," in *IEEE International Conference on Robotics and Automation (ICRA)*, 2016.

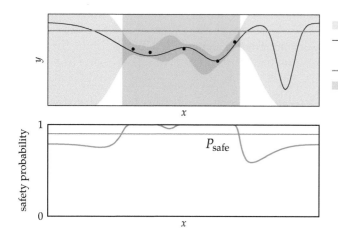

confidence interval

—— objective function f

• fit points

—— safety threshold

estimated safe region $\mathcal{S}$

Figure 16.9. The safety regions (green) predicted by a Gaussian process.

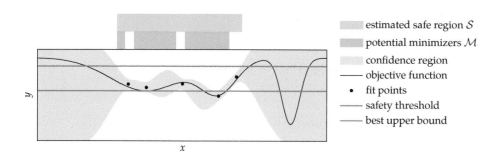

estimated safe region $\mathcal{S}$

potential minimizers $\mathcal{M}$

confidence region

—— objective function

• fit points

—— safety threshold

—— best upper bound

Figure 16.10. The potential minimizers are the safe points whose lower bounds are lower than the best, safe upper bound.

At step i, the set of potential expanders $\mathcal{E}_i$ consists of the safe points that, if added to the Gaussian process, optimistically assuming the lower bound, produce a posterior distribution with a larger safe set. The potential expanders naturally lie near the boundary of the safe region.

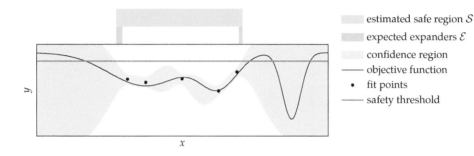

estimated safe region $\mathcal{S}$

expected expanders $\mathcal{E}$

confidence region

—— objective function

• fit points

—— safety threshold

Figure 16.11. The set of potential expanders.

Given an initial safe point[6] $\mathbf{x}^{(1)}$, SafeOpt chooses the design point among sets $\mathcal{M}$ and $\mathcal{E}$ with the greatest uncertainty, as quantified by the width $w_i(x) = u(x) - \ell(x)$:

$$x^{(i)} = \underset{\mathbf{x} \in \mathcal{M}_i \cup \mathcal{E}_i}{\arg\max}\, w_i(\mathbf{x}) \qquad (16.17)$$

SafeOpt proceeds until a termination condition is met. It is common to run the algorithm for a fixed number of iterations or until the maximum width is less than a set threshold.

Maintaining sets in multidimensional spaces can be computationally challenging. SafeOpt assumes a finite design space $\mathcal{X}$ that can be obtained with a sampling method applied over the continuous search domain. Increasing the density of the finite design space leads to more accurate results with respect to the continuous space, but it takes longer per iteration.

SafeOpt is implemented in algorithm 16.3, and calls algorithm 16.4 to update the predicted confidence intervals; algorithm 16.5 to compute the safe, minimizer, and expander regions; and algorithm 16.6 to select a query point. The progression of SafeOpt is shown for one dimension in figure 16.12, and for two dimensions in figure 16.13.

[6] SafeOpt cannot guarantee safety if it is not initialized with at least one point that it knows is safe.

```
function safe_opt(GP, X, i, f, y_max; β=3.0, k_max=10)
    push!(GP, X[i], f(X[i])) # make first observation

    m = length(X)
    u, l = fill(Inf, m), fill(-Inf, m)
    S, M, E = falses(m), falses(m), falses(m)

    for k in 1 : k_max
        update_confidence_intervals!(GP, X, u, l, β)
        compute_sets!(GP, S, M, E, X, u, l, y_max, β)
        i = get_new_query_point(M, E, u, l)
        i != 0 || break
        push!(GP, X[i], f(X[i]))
    end

    # return the best point
    update_confidence_intervals!(GP, X, u, l, β)
    S[:] = u .≤ y_max
    if any(S)
        u_best, i_best = findmin(u[S])
        i_best = findfirst(isequal(i_best), cumsum(S))
        return (u_best, i_best)
    else
        return (NaN,0)
    end
end
```

Algorithm 16.3. The SafeOpt algorithm applied to an empty Gaussian process GP, a finite design space X, index of initial safe point i, objective function f, and safety threshold y_max. The optional parameters are the confidence scalar β and the number of iterations k_max. A tuple containing the best safe upper bound and its index in X is returned.

```
function update_confidence_intervals!(GP, X, u, l, β)
    μₚ, νₚ = predict(GP, X)
    u[:] = μₚ + sqrt.(β*νₚ)
    l[:] = μₚ - sqrt.(β*νₚ)
    return (u, l)
end
```

Algorithm 16.4. A method for updating the lower and upper bounds used in SafeOpt, which takes the Gaussian process GP, the finite search space X, the upper and lower-bound vectors u and l, and the confidence scalar β.

```
function compute_sets!(GP, S, M, E, X, u, l, y_max, β)
    fill!(M, false)
    fill!(E, false)

    # safe set
    S[:] = u .≤ y_max

    if any(S)

        # potential minimizers
        M[S] = l[S] .< minimum(u[S])

        # maximum width (in M)
        w_max = maximum(u[M] - l[M])

        # expanders - skip values in M or those with w ≤ w_max
        E[:] = S .& .~M # skip points in M
        if any(E)
            E[E] .= maximum(u[E] - l[E]) .> w_max
            for (i,e) in enumerate(E)
                if e && u[i] - l[i] > w_max
                    push!(GP, X[i], l[i])
                    μₚ, νₚ = predict(GP, X[.~S])
                    pop!(GP)
                    E[i] = any(μₚ + sqrt.(β*νₚ) .≥ y_max)
                    if E[i]; w_max = u[i] - l[i]; end
                end
            end
        end
    end

    return (S,M,E)
end
```

Algorithm 16.5. A method for updating the safe S, minimizer M, and expander E sets used in SafeOpt. The sets are all Boolean vectors indicating whether the corresponding design point in X is in the set. The method also takes the Gaussian process GP, the upper and lower bounds u and l, respectively, the safety threshold y_max, and the confidence scalar β.

```
function get_new_query_point(M, E, u, l)
    ME = M .| E
    if any(ME)
            v = argmax(u[ME] - l[ME])
        return findfirst(isequal(v), cumsum(ME))
    else
        return 0
    end
end
```

Algorithm 16.6. A method for obtaining the next query point in SafeOpt. The index of the point in X with the greatest width is returned.

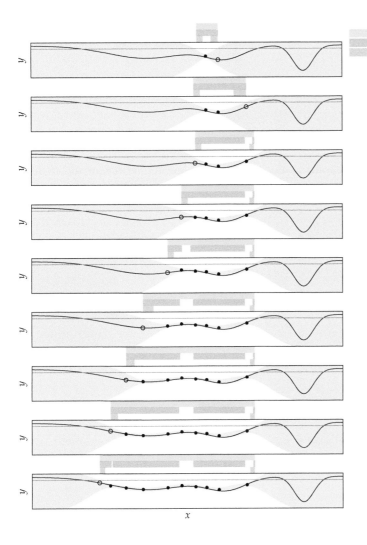

estimated safe region
potential minimizers
potential expanders

Figure 16.12. The first eight iterations of SafeOpt on a univariate function. SafeOpt can never reach the global optimum on the right-hand side because it requires crossing an unsafe region. We can only hope to find the global minima in our locally reachable safe region.

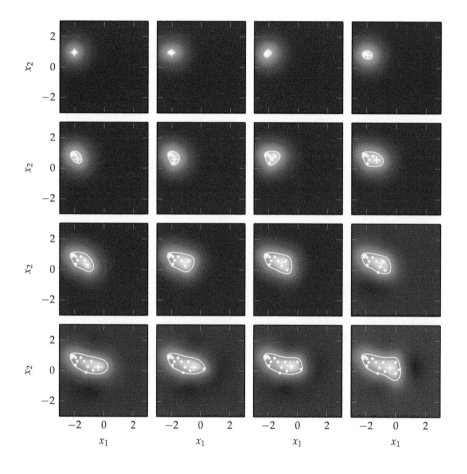

Figure 16.13. SafeOpt applied to the flower function (appendix B.4) with $y_{\max} = 2$, a Gaussian process mean of $\mu(\mathbf{x}) = 2.5$, variance $\nu = 0.01, \beta = 10$, a 51×51 uniform grid over the search space, and an initial point $x^{(1)} = [-2.04, 0.96]$. The color indicates the value of the upper bound, the cross indicates the safe point with the lowest upper bound, and the white contour line is the estimated safe region.

The objective function with the true safe region outlined in white:

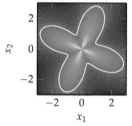

16.7   Summary

- Gaussian processes can be used to guide the optimization process using a variety of strategies that use estimates of quantities such as the lower confidence bound, probability of improvement, and expected improvement.

- Some problems do not allow for the evaluation of unsafe designs, in which case we can use safe exploration strategies that rely on Gaussian processes.

16.8   Exercises

Exercise 16.1. Give an example in which prediction-based optimization fails.

Exercise 16.2. What is the main difference between lower confidence bound exploration and error-based exploration in the context of optimization?

Exercise 16.3. We have a function $f(x) = (x - 2)^2/40 - 0.5$ with $x \in [-5, 5]$, and we have evaluation points at -1 and 1. Assume we use a Gaussian process surrogate model with a zero-mean function, and a squared exponential kernel $\exp(-r^2/2)$, where r is the Euclidean distance between two points. Which value for x would we evaluate next if we were maximizing probability of improvement? Which value for x would we evaluate next if we were maximizing expected improvement?

17   Optimization under Uncertainty

Previous chapters assumed that the optimization objective is to minimize a deterministic function of our design points. In many engineering tasks, however, there may be uncertainty in the objective function or the constraints. Uncertainty may arise due to a number of factors, such as model approximations, imprecision, and fluctuations of parameters over time. This chapter covers a variety of methods for accounting for uncertainty in our optimization to enhance robustness.[1]

17.1   Uncertainty

Uncertainty in the optimization process can arise for a variety of reasons. There may be *irreducible uncertainty*,[2] which is inherent to the system, such as background noise, varying material properties, and quantum effects. These uncertainties cannot be avoided and our design should accommodate them. There may also be *epistemic uncertainty*,[3] which is uncertainty caused by a subjective lack of knowledge by the designer. This uncertainty can arise from approximations in the model[4] used when formulating the design problem and errors introduced by numerical solution methods.

Accounting for these various forms of uncertainty is critical to ensuring robust designs. In this chapter, we will use $\mathbf{z} \in \mathcal{Z}$ to represent a vector of random values. We want to minimize $f(\mathbf{x}, \mathbf{z})$, but we do not have control over $\mathbf{z}$. Feasibility depends on both the design vector $\mathbf{x}$ and the uncertain vector $\mathbf{z}$. This chapter introduces the feasible set over $\mathbf{x}$ and $\mathbf{z}$ pairs as $\mathcal{F}$. We have feasibility if and only if $(\mathbf{x}, \mathbf{z}) \in \mathcal{F}$. We will use $\mathcal{X}$ as the design space, which may include potentially infeasible designs depending on the value of $\mathbf{z}$.

Optimization with uncertainty was briefly introduced in section 15.5 in the context of using a Gaussian process to represent an objective function inferred

[1] Additional references include: H.-G. Beyer and B. Sendhoff, "Robust Optimization—A Comprehensive Survey," *Computer Methods in Applied Mechanics and Engineering*, vol. 196, no. 33, pp. 3190–3218, 2007. G.-J. Park, T.-H. Lee, K. H. Lee, and K.-H. Hwang, "Robust Design: An Overview," *AIAA Journal*, vol. 44, no. 1, pp. 181–191, 2006.

[2] This form of uncertainty is sometimes called *aleatory uncertainty* or *random uncertainty*.

[3] Epistemic uncertainty is also called *reducible uncertainty*.

[4] The statistician George Box famously wrote: *All models are wrong; some models are useful*. G. E. P. Box, W. G. Hunter, and J. S. Hunter, *Statistics for Experimenters: An Introduction to Design, Data Analysis, and Model Building*, 2nd ed. Wiley, 2005. p. 440.

from noisy measurements. We had $f(\mathbf{x}, z) = f(\mathbf{x}) + z$ with the additional assumption that z comes from a zero-mean Gaussian distribution.[5] Uncertainty may be incorporated into the evaluation of a design point in other ways. For example, if we had noise in the input to the objective function,[6] we might have $f(\mathbf{x}, \mathbf{z}) = f(\mathbf{x} + \mathbf{z})$. In general, $f(\mathbf{x}, \mathbf{z})$ can be a complex, nonlinear function of $\mathbf{x}$ and $\mathbf{z}$. In addition, $\mathbf{z}$ may not come from a Gaussian distribution; in fact, it may come from a distribution that is not known.

Figure 17.1 demonstrates how the degree of uncertainty can affect our choice of design. For simplicity, x is a scalar and z is selected from a zero-mean Gaussian distribution. We assume that z corresponds to noise in the input to f, and so $f(x, z) = f(x + z)$. The figure shows the expected value of the objective function for different levels of noise. The global minimum without noise is a. However, aiming for a design near a can be risky since it lies within a steep valley, making it rather sensitive to noise. Even with low noise, it may be better to choose a design near b. Designs near c can provide even greater robustness to larger amounts of noise. If the noise is very high, the best design might even fall between b and c, which corresponds to a local maximum in the absence of noise.

There are a variety of different ways to account for uncertainty in optimization. We will discuss both set-based uncertainty and probabilistic uncertainty.[7]

[5] Here, the two-argument version of f takes as input the design point and random vector, but the single-argument version of f represents a deterministic function of the design point without noise.

[6] For example, there may be variability in the manufacturing of our design.

[7] Other approaches for representing uncertainty include *Dempster-Shafer theory*, *fuzzy-set theory*, and *possibility theory*, which are beyond the scope of this book.

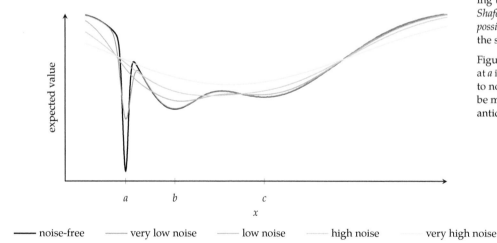

Figure 17.1. The global minimum at a in the noiseless case is sensitive to noise. Other design points may be more robust depending on the anticipated level of noise.

noise-free   very low noise   low noise   high noise   very high noise

17.2   Set-Based Uncertainty

Set-based uncertainty approaches assume that $\mathbf{z}$ belongs to a set $\mathcal{Z}$, but these approaches make no assumptions about the relative likelihood of different points within that set. The set $\mathcal{Z}$ can be defined in different ways. One way is to define intervals for each component of $\mathcal{Z}$. Another way is to define $\mathcal{Z}$ by a set of inequality constraints, $\mathbf{g}(\mathbf{x}, \mathbf{z}) \leq \mathbf{0}$, similar to what was done for the design space $\mathcal{X}$ in chapter 10.

17.2.1   Minimax

In problems with set-based uncertainty, we often want to minimize the maximum possible value of the objective function. Such a *minimax* approach[8] solves the optimization problem

$$\underset{\mathbf{x} \in \mathcal{X}}{\text{minimize}} \, \underset{\mathbf{z} \in \mathcal{Z}}{\text{maximize}} \, f(\mathbf{x}, \mathbf{z}) \qquad (17.1)$$

In other words, we want to find an $\mathbf{x}$ that minimizes f, assuming the worst-case value for $\mathbf{z}$.

This optimization is equivalent to defining a modified objective function

$$f_{\text{mod}}(\mathbf{x}) = \underset{\mathbf{z} \in \mathcal{Z}}{\text{maximize}} \, f(\mathbf{x}, \mathbf{z}) \qquad (17.2)$$

and then solving

$$\underset{\mathbf{x} \in \mathcal{X}}{\text{minimize}} \, f_{\text{mod}}(\mathbf{x}) \qquad (17.3)$$

Example 17.1 shows this optimization on a univariate problem and illustrates the effect of different levels of uncertainty.

In problems where we have feasibility constraints, our optimization problem becomes

$$\underset{\mathbf{x} \in \mathcal{X}}{\text{minimize}} \, \underset{\mathbf{z} \in \mathcal{Z}}{\text{maximize}} \, f(\mathbf{x}, \mathbf{z}) \text{ subject to } (\mathbf{x}, \mathbf{z}) \in \mathcal{F} \qquad (17.4)$$

Example 17.2 shows the effect of applying minimax on the space of feasible design points when there are constraints.

Consider the objective function

$$f(x,z) = f(x+z) = f(\tilde{x}) = \begin{cases} -\tilde{x} & \text{if } \tilde{x} \le 0 \\ \tilde{x}^2 & \text{otherwise} \end{cases}$$

where $\tilde{x} = x + z$, with a set-based uncertainty region $z \in [-\epsilon, \epsilon]$. The mini-max approach is a minimization problem over the modified objective function $f_{\text{mod}}(x) = \text{maximize}_{z \in [-\epsilon, \epsilon]} f(x, z)$.

Example 17.1. Example of a mini-max approach to optimization under set-based uncertainty.

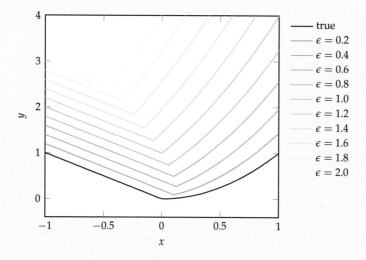

The figure above shows $f_{\text{mod}}(x)$ for several different values of ϵ. The minimum for $\epsilon = 0$ coincides with the minimum of $f(x, 0)$. As ϵ is increased, the minimum first shifts right as x increases faster than x^2 and then shifts left as x^2 increases faster than x. The robust minimizer does not generally coincide with the minimum of $f(x, 0)$.

Consider an uncertain feasible set in the form of a rotated ellipse, where $(\mathbf{x}, z) \in \mathcal{F}$ if and only if $z \in [0, \pi/2]$ and

$$(x_1 \cos z + x_2 \sin z)^2 + (x_1 \sin z - x_2 \cos z)^2 / 16 \leq 1$$

When $z = 0$, the major axis of the ellipse is vertical. Increasing values of z slowly rotates it counter clockwise to horizontal at $z = \pi/2$. The figure below shows the vertical and horizontal ellipses and the set of all points that are feasible for at least one z in blue.

A minimax approach to optimization should consider only design points that are feasible under all values of z. The set of designs that are always feasible are given by the intersection of all ellipses formed by varying z. This set is outlined in red.

Example 17.2. The minimax approach applied to uncertainty in the feasible set.

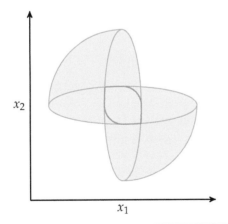

17.2.2   Information-Gap Decision Theory

Instead of assuming the uncertainty set $\mathcal{Z}$ is fixed, an alternative approach known as *information-gap decision theory*[9] parameterizes the uncertainty set by a nonnegative scalar *gap* parameter ϵ. The gap controls the volume of the parameterized set $\mathcal{Z}(\epsilon)$ centered at some nominal value $\bar{\mathbf{z}} = \mathcal{Z}(0)$. One way to define $\mathcal{Z}(\epsilon)$ is as a hypersphere of radius ϵ centered at a nominal point $\bar{\mathbf{z}}$:

$$\mathcal{Z}(\epsilon) = \{\mathbf{z} \mid \|\mathbf{z} - \bar{\mathbf{z}}\|_2 \le \epsilon\} \tag{17.5}$$

Figure 17.2 illustrates this definition in two dimensions.

By parameterizing the uncertainty set, we avoid committing to a particular uncertainty set. Uncertainty sets that are too large sacrifice the quality of the solution, and uncertainty sets that are too small sacrifice robustness. Design points that remain feasible for larger gaps are considered more robust.

In information-gap decision theory, we try to find the design point that allows for the largest gap while preserving feasibility. This design point can be obtained by solving the following optimization problem:

$$\mathbf{x}^* = \arg\max_{\mathbf{x} \in \mathcal{X}} \underset{\epsilon \in [0,\infty)}{\text{maximize}} \begin{cases} \epsilon & \text{if } (\mathbf{x}, \mathbf{z}) \in \mathcal{F} \text{ for all } \mathbf{z} \in \mathcal{Z}(\epsilon) \\ 0 & \text{otherwise} \end{cases} \tag{17.6}$$

This optimization focuses on finding designs that ensure feasibility in the presence of uncertainty. In fact, equation (17.6) does not explicitly include the objective function f. However, we can incorporate the constraint that $f(\mathbf{x}, \mathbf{z})$ be no greater than some threshold $y_{\max}$. Such performance constraints can help us avoid excessive risk aversion. Figure 17.3 and example 17.3 illustrate the application of information-gap decision theory.

17.3   Probabilistic Uncertainty

Models of *probabilistic uncertainty* uses distributions over a set $\mathcal{Z}$. Probabilistic uncertainty models provide more information than set-based uncertainty models, allowing the designer to account for the probability of different outcomes of a design. These distributions can be defined using expert knowledge or learned from data. Given a distribution p over $\mathcal{Z}$, we can infer a distribution over the output of f using methods that will be discussed in chapter 18. This section will

[9] F. M. Hemez and Y. Ben-Haim, "Info-Gap Robustness for the Correlation of Tests and Simulations of a Non-Linear Transient," *Mechanical Systems and Signal Processing*, vol. 18, no. 6, pp. 1443–1467, 2004.

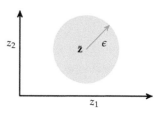

Figure 17.2.   A parametrized uncertainty set $\mathcal{Z}(\epsilon)$ in the form of a hypersphere.

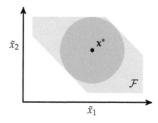

Figure 17.3.   Information-gap decision theory applied to an objective function with additive noise $f(\tilde{\mathbf{x}})$ with $\tilde{\mathbf{x}} = \mathbf{x} + \mathbf{z}$ and a circular uncertainty set

$$\mathcal{Z}(\epsilon) = \{\mathbf{z} \mid \|z\|_2 \le \epsilon\}$$

The design $\mathbf{x}^*$ is optimal under information-gap decision theory as it allows for the largest possible ϵ such that all $\mathbf{x}^* + \mathbf{z}$ are feasible.

Consider the robust optimization of $f(x,z) = \tilde{x}^2 + 6e^{-\tilde{x}^2}$ with $\tilde{x} = x + z$ subject to the constraint $\tilde{x} \in [-2, 2]$ with the uncertainty set $\mathcal{Z}(\epsilon) = [-\epsilon, \epsilon]$.

Example 17.3. We can mitigate excessive risk aversion by applying a constraint on the maximum acceptable objective function value when applying information-gap decision theory.

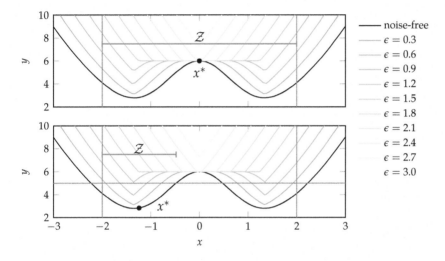

Applying information-gap decision theory to this problem results in a maximally-sized uncertainty set and a design centered in the suboptimal region of the objective function. Applying an additional constraint on the maximum objective function value, $f(x, z) \leq 5$, allows the same approach to find a design with better noise-free performance. The blue lines indicate the worst-case objective function value for a given uncertainty parameter ϵ.

outline five different metrics for converting this distribution into a scalar value given a particular design $\mathbf{x}$. We can then optimize with respect to these metrics.[10]

17.3.1   Expected Value

One way to convert the distribution output by f into a scalar value is to use the *expected value* or *mean*. The expected value is the average output that we can expect when considering all outputs of $f(\mathbf{x}, \mathbf{z})$ for all $\mathbf{z} \in \mathcal{Z}$ and their corresponding probabilities. The expected value as a function of the design point $\mathbf{x}$ is

$$\mathbb{E}_{\mathbf{z} \sim p}[f(\mathbf{x}, \mathbf{z})] = \int_{\mathcal{Z}} f(\mathbf{x}, \mathbf{z}) p(\mathbf{z}) \, d\mathbf{z} \tag{17.7}$$

The expected value does not necessarily correspond to the objective function without noise, as illustrated in example 17.4.

Computing the integral in equation (17.7) analytically may not be possible. One may approximate that value using sampling or a variety of other more sophisticated techniques discussed in chapter 18.

17.3.2   Variance

Besides optimizing with respect to the expected value of the function, we may also be interested in choosing design points whose value is not overly sensitive to uncertainty.[11] Such regions can be quantified using the *variance* of f:

$$\text{Var}[f(\mathbf{x}, \mathbf{z})] = \mathbb{E}_{\mathbf{z} \sim p}\left[\left(f(\mathbf{x}, \mathbf{z}) - \mathbb{E}_{\mathbf{z} \sim p}[f(\mathbf{x}, \mathbf{z})]\right)^2\right] \tag{17.8}$$

$$= \int_{\mathcal{Z}} f(\mathbf{x}, \mathbf{z})^2 p(\mathbf{z}) \, d\mathbf{z} - \mathbb{E}_{\mathbf{z} \sim p}[f(\mathbf{x}, \mathbf{z})]^2 \tag{17.9}$$

[10] Further discussion of various metrics can be found in A. Shapiro, D. Dentcheva, and A. Ruszczyński, *Lectures on Stochastic Programming: Modeling and Theory*, 2nd ed. SIAM, 2014.

[11] Sometimes designers seek plateau-like regions where the output of the objective function is relatively constant, such as producing materials with consistent performance or scheduling trains such that they arrive at a consistent time.

One common model is to apply zero-mean Gaussian noise to the function output, $f(\mathbf{x}, \mathbf{z}) = f(\mathbf{x}) + \mathbf{z}$, as was the case with Gaussian processes in chapter 16. The expected value is equivalent to the noise-free case:

$$\mathbb{E}_{\mathbf{z} \sim \mathcal{N}(\mathbf{0}, \boldsymbol{\Sigma})}[f(\mathbf{x}) + \mathbf{z}] = \mathbb{E}_{\mathbf{z} \sim \mathcal{N}(\mathbf{0}, \boldsymbol{\Sigma})}[f(\mathbf{x})] + \mathbb{E}_{\mathbf{z} \sim \mathcal{N}(\mathbf{0}, \boldsymbol{\Sigma})}[\mathbf{z}] = f(\mathbf{x})$$

It is also common to add noise directly to the design vector, $f(\mathbf{x}, \mathbf{z}) = f(\mathbf{x} + \mathbf{z}) = f(\tilde{\mathbf{x}})$. In such cases the expected value is affected by the variance of zero-mean Gaussian noise.

Consider minimizing the expected value of $f(\tilde{x}) = \sin(2\tilde{x})/\tilde{x}$ with $\tilde{x} = x + z$ for z drawn from a zero-mean Gaussian distribution $\mathcal{N}(0, \nu)$. Increasing the variance increases the effect that the local function landscape has on a design.

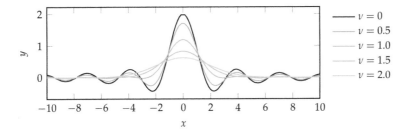

The plot above shows that changing the variance affects the location of the optima.

Example 17.4. The expected value of an uncertain objective function depends on how the uncertainty is incorporated into the objective function.

We call design points with large variance *sensitive* and design points with small variance *robust*. Examples of sensitive and robust points are shown in figure 17.4. We are typically interested in good points as measured by their expected value that are also robust. Managing the trade-off between the expected objective function value and the variance is a multiobjective optimization problem (see example 17.5), and we can use techniques discussed in chapter 12.

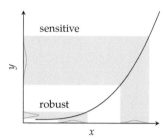

Figure 17.4.   Probabilistic approaches produce probability distributions over the model output. Design points can be sensitive or robust to uncertainty. The blue regions show how the distribution over a normally distributed design is affected by the objective function.

17.3.3   Statistical Feasibility

An alternative metric against which to optimize is *statistical feasibility*. Given $p(\mathbf{z})$, we can compute the probability a design point $\mathbf{x}$ is feasible:

$$P((\mathbf{x}, \mathbf{z}) \in \mathcal{F}) = \int_{\mathcal{Z}} ((\mathbf{x}, \mathbf{z}) \in \mathcal{F}) p(\mathbf{z}) \, d\mathbf{z} \qquad (17.10)$$

This probability can be estimated through sampling. If we are also interested in ensuring that the objective value does not exceed a certain threshold, we can incorporate a constraint $f(\mathbf{x}, \mathbf{z}) \le y_{\max}$ as is done with information-gap decision theory. Unlike the expected value and variance metrics, we want to maximize this metric.

17.3.4   Value at Risk

The *value at risk* (*VaR*) is the best objective value that can be guaranteed with probability α. We can write this definition mathematically in terms of the *cumulative distribution function*, denoted $\Phi(y)$, over the random output of the objective function. The probability that the outcome is less than or equal to y is given by $\Phi(y)$. VaR with confidence α is the minimum value of y such that $\Phi(y) \ge \alpha$. This definition is equivalent to the α *quantile* of a probability distribution. An α close to 1 is sensitive to unfavorable outliers, whereas an α close to 0 is overly optimistic and close to the best possible outcome.

17.3.5   Conditional Value at Risk

The *conditional value at risk* (*CVaR*) is related to the value at risk.[12] CVaR is the expected value of the top $1 - \alpha$ quantile of the probability distribution over the output. This quantity is illustrated in figure 17.5.

[12] The conditional value at risk is also known as the *mean excess loss, mean shortfall*, and *tail value at risk*. R. T. Rockafellar and S. Uryasev, "Optimization of Conditional Value-at-Risk," *Journal of Risk*, vol. 2, pp. 21–42, 2000.

Consider the objective function $f(x,z) = x^2 + z$, with z drawn from a Gamma distribution that depends on x. We can construct a function `dist(x)` that returns a Gamma distribution from the `Distributions.jl` package:

Example 17.5. Considering both the expected value and the variance in optimization under uncertainty.

```
dist(x) = Gamma(2/(1+abs(x)),2)
```

This distribution has mean $4/(1+|x|)$ and variance $8/(1+|x|)$.

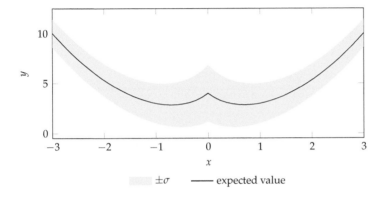

$\pm\sigma$ —— expected value

We can find a robust optimizer that minimizes both the expected value and the variance. Minimizing with respect to the expected value, ignoring the variance, produces two minima at $x \approx \pm 0.695$. Incorporating a penalty for the variance shifts these minima away from the origin. The figure below shows objective functions of the form $\alpha\,\mathbb{E}[y \mid x] + (1-\alpha)\sqrt{\mathrm{Var}[y \mid x]}$ for $\alpha \in [0,1]$ along with their associated minima.

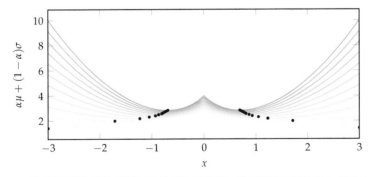

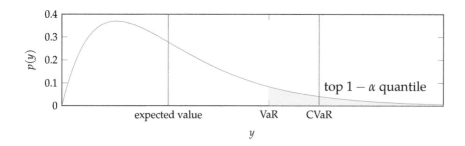

Figure 17.5. CVaR and VaR for a particular level α. CVaR is the expected value of the top $1 - \alpha$ quantile, whereas the VaR is the lowest objective function value over the same quantile.

CVaR has some theoretical and computational advantages over VaR. CVaR is less sensitive to estimation errors in the distribution over the objective output. For example, if the cumulative distribution function is flat in some intervals, then VaR can jump with small changes in α. In addition, VaR does not account for costs beyond the α quantile, which is undesirable if there are rare outliers with very poor objective values.[13]

17.4   Summary

- Uncertainty in the optimization process can arise due to errors in the data, the models, or the optimization method itself.

- Accounting for these sources of uncertainty is important in ensuring robust designs.

- Optimization with respect to set-based uncertainty includes the minimax approach that assumes the worst-case and information-gap decision theory that finds a design robust to a maximally sized uncertainty set.

- Probabilistic approaches typically minimize the expected value, the variance, risk of infeasibility, value at risk, conditional value at risk, or a combination of these.

17.5   Exercises

Exercise 17.1. Suppose we have zero-mean Gaussian noise in the input such that $f(x, z) = f(x + z)$. Consider the three points a, b, and c in the figure below:

[13] For an overview of properties, see G. C. Pflug, "Some Remarks on the Value-at-Risk and the Conditional Value-at-Risk," in *Probabilistic Constrained Optimization: Methodology and Applications*, S. P. Uryasev, ed. Springer, 2000, pp. 272–281. and R. T. Rockafellar and S. Uryasev, "Conditional Value-at-Risk for General Loss Distributions," *Journal of Banking and Finance*, vol. 26, pp. 1443–1471, 2002.

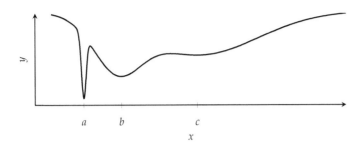

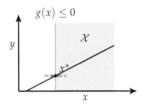

Figure 17.6. Optima with active constraints are often sensitive to uncertainty.

Which design point is best if we are minimizing the expected value minus the standard deviation?

Exercise 17.2. Optima, such as the one depicted in figure 17.6, often lie on a constraint boundary and are thus sensitive to uncertainties that could cause them to become infeasible. One approach to overcome uncertainty with respect to feasibility is to make the constraints more stringent, reducing the size of the feasible region as shown in figure 17.7.

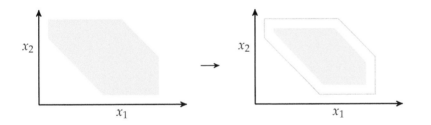

Figure 17.7. Applying more stringent constraints during optimization prevents designs from being too close to the true feasibility boundary.

It is common to rewrite constraints of the form $g(\mathbf{x}) \leq g_{max}$ to $\gamma g(\mathbf{x}) \leq g_{max}$, where $\gamma > 1$ is a *factor of safety*. Optimizing such that the constraint values stay below g_{max}/γ provides an additional safety buffer.

Consider a beam with a square cross section thought to fail when the stresses exceed $\sigma_{max} = 1$. We wish to minimize the cross section $f(x) = x^2$, where x is the cross section length. The stress in the beam is also a function of the cross section length $g(x) = x^{-2}$. Plot the probability that the optimized design does not fail as the factor of safety varies from 1 to 2:

- Uncertainty in maximum stress, $g(x, z) = x^{-2} + z$

- Uncertainty in construction tolerance, $g(x, z) = (x + z)^{-2}$

- Uncertainty in material properties, $g(x, z) = (1 + z)x^{-2}$

where z is zero-mean noise with variance 0.01.

Exercise 17.3. The *six-sigma* method is a special case of statistical feasibility in which a production or industrial process is improved until its assumed Gaussian output violates design requirements only with outliers that exceed six standard deviations. This requirement is fairly demanding, as is illustrated in figure 17.8.

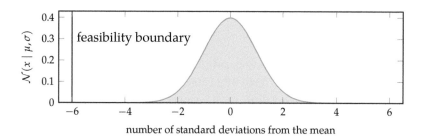

Figure 17.8. Statistical feasibility can be met either by shifting the objective function mean away from the feasibility boundary or by reducing the variance of the objective function.

Consider the optimization problem

$$\underset{\mathbf{x}}{\text{minimize}} \ x_1$$

$$\text{subject to } e^{x_1} \leq x_2 + z \leq 2e^{x_1}$$

with $z \sim \mathcal{N}(0, 1)$. Find the optimal design x^* such that (x, z) is feasible for all $|z| \leq 6$.

18 Uncertainty Propagation

As discussed in the previous chapter, probabilistic approaches to optimization under uncertainty model some of the inputs to the objective function as a probability distribution. This chapter discusses how to propagate known input distributions to estimate quantities associated with the output distribution, such as the mean and variance of the objective function. There are a variety of approaches to *uncertainty propagation*, some based on mathematical concepts such as Monte Carlo, the Taylor series approximation, orthogonal polynomials, and Gaussian processes. These approaches differ in the assumptions they make and the quality of their estimates.

18.1 Sampling Methods

The mean and variance of the objective function at a particular design point can be approximated using *Monte Carlo integration*,[1] which approximates the integral using m samples $\mathbf{z}^{(1)}, \ldots, \mathbf{z}^{(m)}$, from the distribution p over $\mathcal{Z}$. These estimates are also called the *sample mean* and *sample variance*:

[1] Alternatively, quasi Monte Carlo integration can be used to produce estimates with faster convergence as discussed in chapter 13.

$$\mathbb{E}_{\mathbf{z}\sim p}[f(\mathbf{z})] \approx \hat{\mu} = \frac{1}{m}\sum_{i=1}^{m} f(\mathbf{z}^{(i)}) \tag{18.1}$$

$$\text{Var}_{\mathbf{z}\sim p}[f(\mathbf{z})] \approx \hat{\nu} = \left(\frac{1}{m}\sum_{i=1}^{m} f(\mathbf{z}^{(i)})^2\right) - \hat{\mu}^2 \tag{18.2}$$

In the equation above, and for the rest of this chapter, we drop $\mathbf{x}$ from $f(\mathbf{x}, \mathbf{z})$ for notational convenience, but the dependency on $\mathbf{x}$ still exists. For each new design point $\mathbf{x}$ in our optimization process, we recompute the mean and variance.

A desirable property of this sampling-based approach is that p does not need to be known exactly. We can obtain samples directly from simulation or real-world experiments. A potential limitation of this approach is that many samples may be required before there is convergence to a suitable estimate. The variance of the sample mean for a normally distributed f is $\text{Var}[\hat{\mu}] = v/m$, where v is the true variance of f. Thus, doubling the number of samples m tends to halve the variance of the sample mean.

18.2 Taylor Approximation

Another way to estimate $\hat{\mu}$ and $\hat{v}$ is to use the Taylor series approximation for f at a fixed design point $\mathbf{x}$.[2] For the moment, we will assume that the n components of $\mathbf{z}$ are uncorrelated and have finite variance. We will denote the mean of the distribution over $\mathbf{z}$ as $\boldsymbol{\mu}$ and the variances of the individual components of $\mathbf{z}$ as $\mathbf{v}$.[3] The following is the second-order Taylor series approximation of $f(\mathbf{z})$ at the point $\mathbf{z} = \boldsymbol{\mu}$:

$$\hat{f}(\mathbf{z}) = f(\boldsymbol{\mu}) + \sum_{i=1}^{n} \frac{\partial f}{\partial z_i}(z_i - \mu_i) + \frac{1}{2}\sum_{i=1}^{n}\sum_{j=1}^{n}\frac{\partial^2 f}{\partial z_i \partial z_j}(z_i - \mu_i)(z_j - \mu_j) \tag{18.3}$$

> [2] For a derivation of the mean and variance of a general function of n random variables, see H. Benaroya and S. M. Han, *Probability Models in Engineering and Science*. Taylor & Francis, 2005.
>
> [3] If the components of $\mathbf{z}$ are uncorrelated, then the covariance matrix is diagonal and $\mathbf{v}$ would be the vector composed of the diagonal elements.

From this approximation, we can analytically compute estimates of the mean and variance of f:

$$\hat{\mu} = f(\boldsymbol{\mu}) + \frac{1}{2}\sum_{i=1}^{n}\frac{\partial^2 f}{\partial z_i^2}v_i\bigg|_{\mathbf{z}=\boldsymbol{\mu}} \tag{18.4}$$

$$\hat{v} = \sum_{i=1}^{n}\left(\frac{\partial f}{\partial z_i}\right)^2 v_i + \frac{1}{2}\sum_{i=1}^{n}\sum_{j=1}^{n}\left(\frac{\partial^2 f}{\partial z_i \partial z_j}\right)^2 v_i v_j\bigg|_{\mathbf{z}=\boldsymbol{\mu}} \tag{18.5}$$

The higher-order terms can be neglected to obtain a first-order approximation:

$$\hat{\mu} = f(\boldsymbol{\mu}) \qquad\qquad \hat{v} = \sum_{i=1}^{n}\left(\frac{\partial f}{\partial z_i}\right)^2 v_i\bigg|_{\mathbf{z}=\boldsymbol{\mu}} \tag{18.6}$$

We can relax the assumption that the components of $\mathbf{z}$ are uncorrelated, but it makes the mathematics more complex. In practice, it can be easier to transform the random variables so that they are uncorrelated. We can transform a vector of n correlated random variables $\mathbf{c}$ with covariance matrix $\mathbf{C}$ into m uncorrelated random variables $\mathbf{z}$ by multiplying by an orthogonal $m \times n$ matrix $\mathbf{T}$ containing the eigenvectors corresponding to the m largest eigenvalues of $\mathbf{C}$. We have $\mathbf{z} = \mathbf{Tc}$.[4]

The Taylor approximation method is implemented in algorithm 18.1. First- and second-order approximations are compared in example 18.1.

```
using ForwardDiff
function taylor_approx(f, μ, ν, secondorder=false)
    μhat = f(μ)
    ∇ = (z -> ForwardDiff.gradient(f, z))(μ)
    νhat = ∇.^2⋅ν
    if secondorder
        H = (z -> ForwardDiff.hessian(f, z))(μ)
        μhat += (diag(H)⋅ν)/2
        νhat += ν⋅(H.^2*ν)/2
    end
    return (μhat, νhat)
end
```

[4] It is also common to scale the outputs such that the covariance matrix becomes the identity matrix. This process is known as *whitening*. J. H. Friedman, "Exploratory Projection Pursuit," *Journal of the American Statistical Association*, vol. 82, no. 397, pp. 249–266, 1987.

Algorithm 18.1. A method for automatically computing the Taylor approximation of the mean and variance of objective function f at design point x with noise mean vector μ and variance vector ν. The Boolean parameter secondorder controls whether the first- or second-order approximation is computed.

18.3 Polynomial Chaos

Polynomial chaos is a method for fitting a polynomial to evaluations of $f(\mathbf{z})$ and using the resulting surrogate model to estimate the mean and variance. We will begin this section by discussing how polynomial chaos is used in the univariate case. We will then generalize the concept to multivariate functions and show how to obtain estimates of the mean and variance by integrating the function represented by our surrogate model.

18.3.1 Univariate

In one dimension, we approximate $f(z)$ with a surrogate model consisting of k polynomial basis functions, $b_1, \ldots, b_k$:

$$f(z) \approx \hat{f}(z) = \sum_{i=1}^{k} \theta_i b_i(z) \tag{18.7}$$

Consider the objective function $f(x,z) = \sin(x + z_1)\cos(x + z_2)$, where z_1 and z_2 are zero-mean Gaussian noise with variances 0.1 and 0.2, respectively.

The first and second partial derivatives of f with respect to the zs are

$$\frac{\partial f}{\partial z_1} = \cos(x + z_1)\cos(x + z_2) \qquad \frac{\partial^2 f}{\partial z_2^2} = -\sin(x + z_1)\cos(x + z_2)$$

$$\frac{\partial f}{\partial z_2} = -\sin(x + z_1)\sin(x + z_2) \qquad \frac{\partial^2 f}{\partial z_1 \partial z_2} = -\cos(x + z_1)\sin(x + z_2)$$

$$\frac{\partial^2 f}{\partial z_1^2} = -\sin(x + z_1)\cos(x + z_2)$$

which allow us to construct the Taylor approximation:

$$\hat{\mu}(x) = 0.85\sin(x)\cos(x)$$
$$\hat{v}(x) = 0.1\cos^4(x) + 0.2\sin^4(x) + 0.045\sin^2(x)\cos^2(x)$$

We can use `taylor_approx` for a given x using:

```
taylor_approx(z->sin(x+z[1])*cos(x+z[2]), [0,0], [0.1,0.2])
```

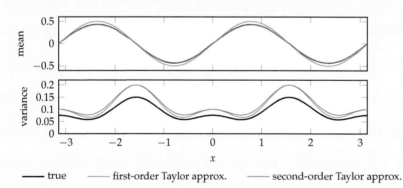

Example 18.1.   The Taylor approximation applied to a univariate design problem with two-dimensional Gaussian noise.

In contrast with the Monte Carlo methods discussed in section 18.1, our samples of z do not have to be randomly drawn from p. In fact, it may be desirable to obtain samples using one of the sampling plans discussed in chapter 13. We will discuss how to obtain the basis coefficients in section 18.3.2.

We can use the surrogate model $\hat{f}$ to estimate the mean:

$$\hat{\mu} = \mathbb{E}\left[\hat{f}\right] \tag{18.8}$$

$$= \int_{\mathcal{Z}} \hat{f}(z)p(z)\,dz \tag{18.9}$$

$$= \int_{\mathcal{Z}} \sum_{i=1}^{k} \theta_i b_i(z)p(z)\,dz \tag{18.10}$$

$$= \sum_{i=1}^{k} \theta_i \int_{\mathcal{Z}} b_i(z)p(z)\,dz \tag{18.11}$$

$$= \theta_1 \int_{\mathcal{Z}} b_1(z)p(z)\,dz + \ldots + \theta_k \int_{\mathcal{Z}} b_k(z)p(z)\,dz \tag{18.12}$$

We can also estimate the variance:

$$\hat{\nu} = \mathbb{E}\left[\left(\hat{f} - \mathbb{E}\left[\hat{f}\right]\right)^2\right] \tag{18.13}$$

$$= \mathbb{E}\left[\hat{f}^2 - 2\hat{f}\,\mathbb{E}\left[\hat{f}\right] + \mathbb{E}\left[\hat{f}\right]^2\right] \tag{18.14}$$

$$= \mathbb{E}\left[\hat{f}^2\right] - \mathbb{E}\left[\hat{f}\right]^2 \tag{18.15}$$

$$= \int_{\mathcal{Z}} \hat{f}(z)^2 p(z)\,dz - \mu^2 \tag{18.16}$$

$$= \int_{\mathcal{Z}} \sum_{i=1}^{k}\sum_{j=1}^{k} \theta_i\theta_j b_i(z)b_j(z)p(z)\,dz - \mu^2 \tag{18.17}$$

$$= \int_{\mathcal{Z}} \left(\sum_{i=1}^{k} \theta_i^2 b_i(z)^2 + 2\sum_{i=2}^{k}\sum_{j=1}^{i-1} \theta_i\theta_j b_i(z)b_j(z)\right)p(z)\,dz - \mu^2 \tag{18.18}$$

$$= \sum_{i=1}^{k} \theta_i^2 \int_{\mathcal{Z}} b_i(z)^2 p(z)\,dz + 2\sum_{i=2}^{k}\sum_{j=1}^{i-1} \theta_i\theta_j \int_{\mathcal{Z}} b_i(z)b_j(z)p(z)\,dz - \mu^2 \tag{18.19}$$

The mean and variance can be efficiently computed if the basis functions are chosen to be *orthogonal* under p. Two basis functions b_i and b_j are orthogonal with respect to a probability density $p(z)$ if

$$\int_{\mathcal{Z}} b_i(z)b_j(z)p(z)\,dz = 0 \quad \text{if } i \neq j \tag{18.20}$$

If the chosen basis functions are all orthogonal to one another and the first basis function is $b_1(z) = 1$, the mean is:

$$\hat{\mu} = \theta_1 \int_{\mathcal{Z}} b_1(z)p(z)\,dz + \theta_2 \int_{\mathcal{Z}} b_2(z)p(z)\,dz + \cdots + \theta_k \int_{\mathcal{Z}} b_k(z)p(z)\,dz \tag{18.21}$$

$$= \theta_1 \int_{\mathcal{Z}} b_1(z)^2 p(z)\,dz + \theta_2 \int_{\mathcal{Z}} b_1(z)b_2(z)p(z)\,dz + \cdots + \theta_k \int_{\mathcal{Z}} b_1(z)b_k(z)p(z)\,dz \tag{18.22}$$

$$= \theta_1 \int_{\mathcal{Z}} p(z)\,dz + 0 + \cdots + 0 \tag{18.23}$$

$$= \theta_1 \tag{18.24}$$

Similarly, the variance is:

$$\hat{v} = \sum_{i=1}^{k} \theta_i^2 \int_{\mathcal{Z}} b_i(z)^2 p(z)\,dz + 2\sum_{i=2}^{k}\sum_{j=1}^{i-1} \theta_i \theta_j \int_{\mathcal{Z}} b_i(z)b_j(z)p(z)\,dz - \mu^2 \tag{18.25}$$

$$= \sum_{i=1}^{k} \theta_i^2 \int_{\mathcal{Z}} b_i(z)^2 p(z)\,dz - \mu^2 \tag{18.26}$$

$$= \theta_1^2 \int_{\mathcal{Z}} b_1(z)^2 p(z)\,dz + \sum_{i=2}^{k} \theta_i^2 \int_{\mathcal{Z}} b_i(z)^2 p(z)\,dz - \theta_1^2 \tag{18.27}$$

$$= \sum_{i=2}^{k} \theta_i^2 \int_{\mathcal{Z}} b_i(z)^2 p(z)\,dz \tag{18.28}$$

The mean thus falls immediately from fitting a surrogate model to the observed data, and the variance can be very efficiently computed given the values $\int_{\mathcal{Z}} b_i(z)^2 p(z)\,dz$ for a choice of basis functions and probability distribution.[5] Example 18.2 uses these procedures to estimate the mean and variance with different sample sizes.

Polynomial chaos approximates the function using kth degree orthogonal polynomial basis functions with $i \in \{1, \ldots, k+1\}$ and $b_1 = 1$. All orthogonal polynomials satisfy the recurrence relation:

$$b_{i+1}(z) = \begin{cases} (z - \alpha_i)b_i(z) & \text{for } i = 1 \\ (z - \alpha_i)b_i(z) - \beta_i b_{i-1}(z) & \text{for } i > 1 \end{cases} \tag{18.29}$$

[5] Integrals of this form can be efficiently computed using Gaussian quadrature, covered in appendix C.8.

Consider optimizing the (unknown) objective function

$$f(x,z) = 1 - e^{-(x+z-1)^2} - 2e^{-(x+z-3)^2}$$

with z known to be drawn from a zero-mean unit-Gaussian distribution.

The objective function, its true expected value, and estimated expected values with different sample counts are plotted below. The estimated expected value is computed using third-order Hermite polynomials.

Example 18.2. Estimating the expected value of an unknown objective function using polynomial chaos.

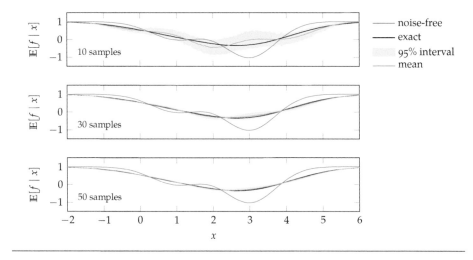

with $b_1(z) = 1$ and weights

$$
\begin{aligned}
\alpha_i &= \frac{\int_{\mathcal{Z}} z \, b_i(z)^2 p(z) \, dz}{\int_{\mathcal{Z}} b_i(z)^2 p(z) \, dz} \\
\beta_i &= \frac{\int_{\mathcal{Z}} b_i(z)^2 p(z) \, dz}{\int_{\mathcal{Z}} b_{i-1}(z)^2 p(z) \, dz}
\end{aligned}
\tag{18.30}
$$

The recurrence relation can be used to generate the basis functions. Each basis function b_i is a polynomial of degree $i - 1$. The basis functions for several common probability distributions are given in table 18.1, can be generated using the methods in algorithm 18.2, and are plotted in figure 18.1. Example 18.3 illustrates the effect the polynomial order has on the estimates of the mean and variance.

Distribution	Domain	Density	Name	Recursive Form	Closed Form
Uniform	$[-1,1]$	$\frac{1}{2}$	Legendre	$\mathrm{Le}_k(x) = \frac{1}{2^k k!} \frac{d^k}{dx^k}\left[(x^2 - 1)^k\right]$	$b_i(x) = \sum_{j=0}^{i-1} \binom{i-1}{j}\binom{-i-2}{j}\left(\frac{1-x}{2}\right)^j$
Exponential	$[0,\infty)$	e^{-x}	Laguerre	$\frac{d}{dx}\mathrm{La}_k(x) = \left(\frac{d}{dx} - 1\right)\mathrm{La}_{k-1}$	$b_i(x) = \sum_{j=0}^{i-1} \binom{i-1}{j}\frac{(-1)^j}{j!}x^j$
Unit Gaussian	$(-\infty,\infty)$	$\frac{1}{\sqrt{2\pi}}e^{-x^2/2}$	Hermite	$\mathrm{H}_k(x) = x\mathrm{H}_{k-1} - \frac{d}{dx}\mathrm{H}_{k-1}$	$b_i(x) = \sum_{j=0}^{\lfloor (i-1)/2 \rfloor} (i-1)! \frac{(-1)^{\frac{i-1}{2}-j}}{(2j)!(\frac{i-1}{2}-j)!}(2x)^{2j}$

Basis functions for arbitrary probability density functions and domains can be constructed both analytically and numerically.[6] The *Stieltjes algorithm*[7] (algorithm 18.3) generates orthogonal polynomials using the recurrence relation in equation (18.29). Example 18.4 shows how the polynomial order affects the estimates of the mean and variance.

Table 18.1. Orthogonal polynomial basis functions for several common probability distributions.

[6] The polynomials can be scaled by a nonzero factor. It is convention to set $b_1(x) = 1$.

[7] T. J. Stieltjes, "Quelques Recherches sur la Théorie des Quadratures Dites Mécaniques," *Annales Scientifiques de l'École Normale Supérieure*, vol. 1, pp. 409–426, 1884. in French. An overview in English is provided by W. Gautschi, *Orthogonal Polynomials: Computation and Approximation*. Oxford University Press, 2004.

18.3.2   Coefficients

The coefficients $\theta_1, \ldots, \theta_k$ in equation (18.7) can be inferred in two different ways. The first way is to fit the values of the samples from $\mathcal{Z}$ using the linear regression method discussed in section 14.3. The second way is to exploit the orthogonality of the basis functions, producing an integration term amenable to Gaussian quadrature.

```julia
using Polynomials
function legendre(i)
    n = i-1
    p = Polynomial([-1,0,1])^n
    for i in 1 : n
        p = derivative(p)
    end
    return p / (2^n * factorial(n))
end
function laguerre(i)
    p = Polynomial([1])
    for j in 2 : i
        p = integrate(derivative(p) - p) + 1
    end
    return p
end
function hermite(i)
    p = Polynomial([1])
    x = Polynomial([0,1])
    for j in 2 : i
        p = x*p - derivative(p)
    end
    return p
end
```

Algorithm 18.2. Methods for constructing polynomial orthogonal basis functions, where i indicates the construction of b_i.

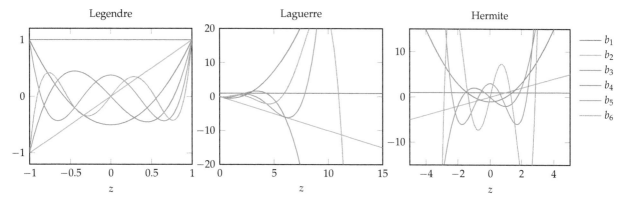

Figure 18.1. Orthogonal basis functions for uniform, exponential, and unit Gaussian distributions.

Consider the function $f(z) = \sin(\pi z)$ with input z drawn from a uniform distribution over the domain $[-1, 1]$. The true mean and variance can be computed analytically:

$$\mu = \int_a^b f(z)p(z)\,dz = \int_{-1}^1 \sin(\pi z)\frac{1}{2}\,dz = 0 \tag{18.31}$$

$$\nu = \int_a^b f(z)^2 p(z)\,dz - \mu^2 = \int_{-1}^1 \sin^2(\pi z)\frac{1}{2}\,dz - 0 = \frac{1}{2} \tag{18.32}$$

Suppose we have five samples of f at $z = \{-1, -0.2, 0.3, 0.7, 0.9\}$. We can fit a Legendre polynomial to the data to obtain our surrogate model $\hat{f}$. Polynomials of different degrees yield:

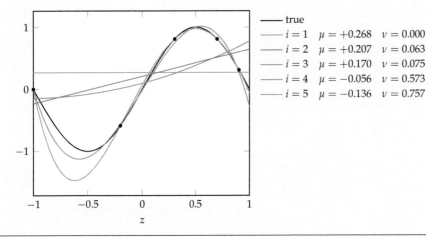

	true
$i = 1$	$\mu = +0.268$ $\nu = 0.000$
$i = 2$	$\mu = +0.207$ $\nu = 0.063$
$i = 3$	$\mu = +0.170$ $\nu = 0.075$
$i = 4$	$\mu = -0.056$ $\nu = 0.573$
$i = 5$	$\mu = -0.136$ $\nu = 0.757$

Example 18.3. Legendre polynomials used to estimate the mean and variance of a function with a uniformly distributed input.

```
using Polynomials
function orthogonal_recurrence(bs, p, dom, ε=1e-6)
    i = length(bs)
    c1 = quadgk(z->z*bs[i](z)^2*p(z), dom..., atol=ε)[1]
    c2 = quadgk(z->  bs[i](z)^2*p(z), dom..., atol=ε)[1]
    α = c1 / c2
    if i > 1
        c3 = quadgk(z->bs[i-1](z)^2*p(z), dom..., atol=ε)[1]
        β = c2 / c3
        return Polynomial([-α, 1])*bs[i] - β*bs[i-1]
    else
        return Polynomial([-α, 1])*bs[i]
    end
end
```

Algorithm 18.3. The Stieltjes algorithm for constructing the next polynomial basis function b_{i+1} according to the orthogonal recurrence relation, where bs contains $\{b_1, \ldots, b_i\}$, p is the probability distribution, and dom is a tuple containing a lower and upper bound for z. The optional parameter ε controls the absolute tolerance of the numerical integration. We make use of the Polynomials.jl package.

We multiply each side of equation (18.7) by the jth basis and our probability density function and integrate:

$$f(z) \approx \sum_{i=1}^{k} \theta_i b_i(z) \tag{18.33}$$

$$\int_{\mathcal{Z}} f(z) b_j(z) p(z)\, dz \approx \int_{\mathcal{Z}} \left(\sum_{i=1}^{k} \theta_i b_i(z) \right) b_j(z) p(z)\, dz \tag{18.34}$$

$$= \sum_{i=1}^{k} \theta_i \int_{\mathcal{Z}} b_i(z) b_j(z) p(z)\, dz \tag{18.35}$$

$$= \theta_j \int_{\mathcal{Z}} b_j(z)^2 p(z)\, dz \tag{18.36}$$

where we made use of the orthogonality property from equation (18.20).

It follows that the jth coefficient is:

$$\theta_j = \frac{\int_{\mathcal{Z}} f(z) b_j(z) p(z)\, dz}{\int_{\mathcal{Z}} b_j(z)^2 p(z)\, dz} \tag{18.37}$$

The denominator of equation (18.37) typically has a known analytic solution or can be inexpensively precomputed. Calculating the coefficient thus primarily requires solving the integral in the numerator, which can be done numerically using Gaussian quadrature.[8]

[8] Gaussian quadrature is implemented in QuadGK.jl via the quadgk function, and is covered in appendix C.8. Quadrature rules can also be obtained using the eigenvalues and eigenvectors of a tri-diagonal matrix formed using the coefficients α_i and β_i from equation (18.30). G. H. Golub and J. H. Welsch, "Calculation of Gauss Quadrature Rules," *Mathematics of Computation*, vol. 23, no. 106, pp. 221–230, 1969.

Consider the function $f(z) = \sin(\pi z)$ with input z drawn from a truncated Gaussian distribution with mean 3 and variance 1 over the domain $[2, 5]$. The true mean and variance are:

$$\mu = \int_a^b f(z)p(z)\,dz = \int_2^5 \sin(\pi z)p(z)\,dz \approx 0.104$$

$$\nu = \int_a^b f(z)^2 p(z)\,dz - \mu^2 = \int_2^5 \sin^2(\pi z)p(z)\,dz - 0.104^2 \approx 0.495$$

where the probability density of the truncated Gaussian is:

$$p(z) = \begin{cases} \frac{\mathcal{N}(z|3,1)}{\int_2^5 \mathcal{N}(\tau|3,1)\,d\tau} & \text{if } z \in [2, 5] \\ 0 & \text{otherwise} \end{cases}$$

Suppose we have five samples of f at $z = \{2.1, 2.5, 3.3, 3.9, 4.7\}$. We can fit orthogonal polynomials to the data to obtain our surrogate model $\hat{f}$. Polynomials of different degrees yield:

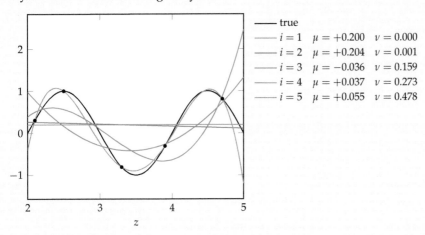

Example 18.4. Legendre polynomials constructed using the Stieltjes method to estimate the mean and variance of a function with a random variable input.

18.3.3 Multivariate

Polynomial chaos can be applied to functions with multiple random inputs. Multivariate basis functions over m variables are constructed as a product over univariate orthogonal polynomials:

$$b_i(\mathbf{z}) = \prod_{j=1}^{m} b_{a_j}(z_j) \tag{18.38}$$

where $\mathbf{a}$ is an assignment vector that assigns the a_jth basis function to the jth random component. This basis function construction is demonstrated in example 18.5.

Consider a three-dimensional polynomial chaos model of which one of the multidimensional basis functions is $b(\mathbf{z}) = b_3(z_1)b_1(z_2)b_3(z_3)$. The corresponding assignment vector is $\mathbf{a} = [3, 1, 3]$.

Example 18.5. Constructing a multivariate polynomial chaos basis function using equation (18.38).

A common method for constructing multivariate basis functions is to generate univariate orthogonal polynomials for each random variable and then to construct a multivariate basis function for every possible combination.[9] This procedure is implemented in algorithm 18.4. Constructing basis functions in this manner assumes that the variables are independent. Interdependence can be resolved using the same transformation discussed in section 18.2.

[9] Here the number of multivariate exponential basis functions grows exponentially in the number of variables.

```
function polynomial_chaos_bases(bases1d)
    bases = []
    for a in Iterators.product(bases1d...)
        push!(bases,
            z -> prod(b(z[i]) for (i,b) in enumerate(a)))
    end
    return bases
end
```

Algorithm 18.4. A method for constructing multivariate basis functions where bases1d contains lists of univariate orthogonal basis functions for each random variable.

A multivariate polynomial chaos approximation with k basis functions is still a linear combination of terms

$$f(\mathbf{z}) \approx \hat{f}(\mathbf{z}) = \sum_{i=1}^{k} \theta_i b_i(\mathbf{z}) \tag{18.39}$$

where the mean and variance can be computed using the equations in section 18.3.1, provided that $b_1(\mathbf{z}) = 1$.

18.4 Bayesian Monte Carlo

Gaussian processes, covered in chapter 16, are probability distributions over functions. They can be used as surrogates for stochastic objective functions. We can incorporate prior information, such as the expected smoothness of the objective function, in a process known as *Bayesian Monte Carlo* or *Bayes-Hermite Quadrature*.

Consider a Gaussian process fit to several points with the same value for the design point $\mathbf{x}$ but different values for the uncertain point $\mathbf{z}$. The Gaussian process obtained is a distribution over functions based on the observed data. When obtaining the expected value through integration, we must consider the expected value of the functions in the probability distribution represented by the Gaussian process $p(\hat{f})$:

$$\mathbb{E}_{\mathbf{z} \sim p}[f] \approx \mathbb{E}_{\hat{f} \sim p(\hat{f})}[\hat{f}] \tag{18.40}$$

$$= \int_{\hat{\mathcal{F}}} \left(\int_{\mathcal{Z}} \hat{f}(\mathbf{z}) p(\mathbf{z}) \, d\mathbf{z} \right) p(\hat{f}) \, d\hat{f} \tag{18.41}$$

$$= \int_{\mathcal{Z}} \left(\int_{\hat{\mathcal{F}}} \hat{f}(\mathbf{z}) p(\hat{f}) \, d\hat{f} \right) p(\mathbf{z}) \, d\mathbf{z} \tag{18.42}$$

$$= \int_{\mathcal{Z}} \hat{\mu}(\mathbf{z}) p(\mathbf{z}) \, d\mathbf{z} \tag{18.43}$$

where $\hat{\mu}(\mathbf{z})$ is the predicted mean under the Gaussian process and $\hat{\mathcal{F}}$ is the space of functions. The variance of the estimate is

$$\text{Var}_{\mathbf{z} \sim p}[f] \approx \text{Var}_{\hat{f} \sim p(\hat{f})}[\hat{f}] \tag{18.44}$$

$$= \int_{\hat{\mathcal{F}}} \left(\int_{\mathcal{Z}} \hat{f}(\mathbf{z}) p(\mathbf{z}) \, d\mathbf{z} - \int_{\mathcal{Z}} \mathbb{E}[\hat{f}(\mathbf{z}')] p(\mathbf{z}') \, d\mathbf{z}' \right)^2 p(\hat{f}) \, d\hat{f} \tag{18.45}$$

$$= \int_{\mathcal{Z}} \int_{\mathcal{Z}} \int_{\hat{\mathcal{F}}} \left[\hat{f}(\mathbf{z}) - \mathbb{E}[\hat{f}(\mathbf{z})] \right] \left[\hat{f}(\mathbf{z}') - \mathbb{E}[\hat{f}(\mathbf{z}')] \right] p(\hat{f}) \, d\hat{f} p(\mathbf{z}) p(\mathbf{z}') \, d\mathbf{z} \, d\mathbf{z}' \tag{18.46}$$

$$= \int_{\mathcal{Z}} \int_{\mathcal{Z}} \text{Cov}(\hat{f}(\mathbf{z}), \hat{f}(\mathbf{z}')) p(\mathbf{z}) p(\mathbf{z}') \, d\mathbf{z} \, d\mathbf{z}' \tag{18.47}$$

where Cov is the posterior covariance under the Gaussian process:

$$\text{Cov}(\hat{f}(\mathbf{z}), \hat{f}(\mathbf{z}')) = k(\mathbf{z}, \mathbf{z}') - k(\mathbf{z}, Z) \mathbf{K}(Z, Z)^{-1} k(Z, \mathbf{z}') \tag{18.48}$$

where Z contains the observed inputs.

Analytic expressions exist for the mean and variance for the special case where $\mathbf{z}$ is Gaussian.[10] Under a Gaussian kernel,

$$k(\mathbf{x}, \mathbf{x}') = \exp\left(-\frac{1}{2}\sum_{i=1}^{n}\frac{(x_i - x_i')^2}{w_i^2}\right) \tag{18.49}$$

the mean for Gaussian uncertainty $\mathbf{z} \sim \mathcal{N}(\mathbf{\mu_z}, \mathbf{\Sigma_z})$ is

$$\mathbb{E}_{\mathbf{z}\sim p}[f] = \mathbf{q}^{\top}\mathbf{K}^{-1}\mathbf{y} \tag{18.50}$$

with

$$q_i = |\mathbf{W}^{-1}\mathbf{\Sigma_z} + \mathbf{I}|^{-1/2}\exp\left(-\frac{1}{2}\left(\mathbf{\mu_z} - \hat{\mu}(\mathbf{z}^{(i)})\right)^{\top}(\mathbf{\Sigma_z} + \mathbf{W})^{-1}\left(\mathbf{\mu_z} - \mathbf{z}^{(i)}\right)\right) \tag{18.51}$$

where $\mathbf{W} = \mathrm{diag}[w_1^2, \ldots, w_n^2]$, and we have constructed our Gaussian process using samples $(\mathbf{z}^{(i)}, y_i)$ for $i \in \{1, \ldots, m\}$.[11]

The variance is

$$\mathrm{Var}_{\mathbf{z}\sim p}[f] = |2\mathbf{W}^{-1}\mathbf{\Sigma_z} + \mathbf{I}|^{-1/2} - \mathbf{q}^{\top}\mathbf{K}^{-1}\mathbf{q} \tag{18.52}$$

Even when the analytic expressions are not available, there are many problems for which numerically evaluating the expectation is sufficiently inexpensive that the Gaussian process approach is better than a Monte Carlo estimation.

Bayesian Monte Carlo is implemented in algorithm 18.5 and is worked out in example 18.6.

[10] It is also required that the covariance function obey the *product correlation rule*, that is, it can be written as the product of a univariate positive-definite function r:

$$k(\mathbf{x}, \mathbf{x}') = \prod_{i=1}^{n} r(x_i - x_i')$$

Analytic results exist for polynomial kernels and mixtures of Gaussians. C. E. Rasmussen and Z. Ghahramani, "Bayesian Monte Carlo," in *Advances in Neural Information Processing Systems (NIPS)*, 2003.

[11] See A. Girard, C. E. Rasmussen, J. Q. Candela, and R. Murray-Smith, "Gaussian Process Priors with Uncertain Inputs—Application to Multiple-Step Ahead Time Series Forecasting," in *Advances in Neural Information Processing Systems (NIPS)*, 2003.

```
function bayesian_monte_carlo(GP, w, μz, Σz)
    W = Matrix(Diagonal(w.^2))
    invK = inv(K(GP.X, GP.X, GP.k))
    q = [exp(-((z-μz)·(inv(W+Σz)*(z-μz)))/2) for z in GP.X]
    q .*= (det(W\Σz + I))^(-0.5)
    μ = q'*invK*GP.y
    ν = (det(2W\Σz + I))^(-0.5) - (q'*invK*q)[1]
    return (μ, ν)
end
```

Algorithm 18.5. A method for obtaining the Bayesian Monte Carlo estimate for the expected value of a function under a Gaussian process GP with a Gaussian kernel with weights w, where the variables are drawn from a normal distribution with mean μz and covariance Σz.

Consider again estimating the expected value and variance of $f(x,z) = \sin(x + z_1)\cos(x + z_2)$, where z_1 and z_2 are zero-mean Gaussian noise with variances 1 and $1/2$, respectively: $\boldsymbol{\mu_z} = [0,0]$ and $\boldsymbol{\Sigma_z} = \text{diag}([1,1/2])$.

We use Bayesian Monte Carlo with a Gaussian kernel with unit weights $\mathbf{w} = [1,1]$ for $x = 0$ with samples $Z = \{[0,0],[1,0],[-1,0],[0,1],[0,-1]\}$.

We compute:

$$\mathbf{W} = \begin{bmatrix} 1 & 0 \\ 0 & 1 \end{bmatrix}$$

$$\mathbf{K} = \begin{bmatrix} 1 & 0.607 & 0.607 & 0.607 & 0.607 \\ 0.607 & 1 & 0.135 & 0.368 & 0.368 \\ 0.607 & 0.135 & 1 & 0.368 & 0.368 \\ 0.607 & 0.368 & 0.368 & 1 & 0.135 \\ 0.607 & 0.368 & 0.368 & 0.135 & 1 \end{bmatrix}$$

$$\mathbf{q} = [0.577, 0.450, 0.450, 0.417, 0.417]$$

$$\mathbb{E}_{\mathbf{z}\sim p}[f] = 0.0$$

$$\text{Var}_{\mathbf{z}\sim p}[f] = 0.327$$

Below we plot the expected value as a function of x using the same approach with ten random samples of $\mathbf{z}$ at each point.

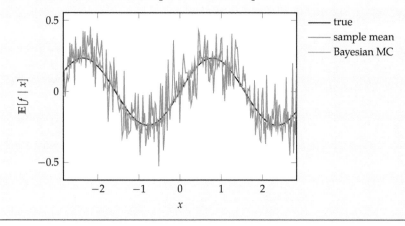

Example 18.6. An example of using Bayesian Monte Carlo to estimate the expected value and variance of a function.

The figure compares the Bayesian Monte Carlo method to the sample mean for estimating the expected value of a function. The same randomly sampled $\mathbf{z}$ values generated for each evaluated x were input to each method.

18.5 Summary

- The expected value and variance of the objective function are useful when optimizing problems involving uncertainty, but computing these quantities reliably can be challenging.

- One of the simplest approaches is to estimate the moments using sampling in a process known as Monte Carlo integration.

- Other approaches, such as the Taylor approximation, use knowledge of the objective function's partial derivatives.

- Polynomial chaos is a powerful uncertainty propagation technique based on orthogonal polynomials.

- Bayesian Monte Carlo uses Gaussian processes to efficiently arrive at the moments with analytic results for Gaussian kernels.

18.6 Exercises

Exercise 18.1. Suppose we draw a sample from a univariate Gaussian distribution. What is the probability that our sample falls within one standard deviation of the mean ($x \in [\mu - \sigma, \mu + \sigma]$)? What is the probability that our sample is less than one standard deviation above the mean ($x < \mu + \sigma$)?

Exercise 18.2. Let $x^{(1)}, x^{(2)}, \ldots, x^{(m)}$ be a random sample of independent, identically distributed values of size m from a distribution with mean μ and variance ν. Show that the variance of the sample mean $\text{Var}(\hat{\mu})$ is ν / m.

Exercise 18.3. Derive the recurrence relation equation (18.29) that is satisfied by all orthogonal polynomials.

Exercise 18.4. Suppose we have fitted a polynomial chaos model of an objective function $f(\mathbf{x}, \mathbf{z})$ for a particular design point $\mathbf{x}$ using m evaluations with $\mathbf{z}^{(1)}, \ldots, \mathbf{z}^{(m)}$. Derive an expression for estimating the partial derivative of the polynomial chaos coefficients with respect to a design component x_i.

Exercise 18.5. Consider an objective function $f(\mathbf{x}, \mathbf{z})$ with design variables $\mathbf{x}$ and random variables $\mathbf{z}$. As discussed in chapter 17, optimization under uncertainty often involves minimizing a linear combination of the estimated mean and variance:

$$f_{\text{mod}}(\mathbf{x}, \mathbf{z}) = \alpha \hat{\mu}(\mathbf{x}) + (1 - \alpha) \hat{v}(\mathbf{x})$$

How can one use polynomial chaos to estimate the gradient of f_{mod} with respect to a design variable $\mathbf{x}$?

19 Discrete Optimization

Previous chapters have focused on optimizing problems involving design variables that are continuous. Many problems, however, have design variables that are naturally discrete, such as manufacturing problems involving mechanical components that come in fixed sizes or navigation problems involving choices between discrete paths. A *discrete optimization* problem has constraints such that the design variables must be chosen from a discrete set. Some discrete optimization problems have infinite design spaces, and others are finite.[1] Even for finite problems, where we could in theory enumerate every possible solution, it is generally not computationally feasible to do so in practice. This chapter discusses both exact and approximate approaches to solving discrete optimization problems that avoid enumeration. Many of the methods covered earlier, such as simulated annealing and genetic programming, can easily be adapted for discrete optimization problems, but we will focus this chapter on categories of techniques we have not yet discussed.

[1] Discrete optimization with a finite design space is sometimes referred to as *combinatorial optimization*. For a review, see B. Korte and J. Vygen, *Combinatorial Optimization: Theory and Algorithms*, 5th ed. Springer, 2012.

Discrete optimization constrains the design to be integral. Consider the problem:

$$\underset{\mathbf{x}}{\text{minimize}} \quad x_1 + x_2$$

$$\text{subject to} \quad \|\mathbf{x}\| \leq 2$$

$$\mathbf{x} \text{ is integral}$$

The optimum in the continuous case is $\mathbf{x}^* = [-\sqrt{2}, -\sqrt{2}]$ with a value of $y = -2\sqrt{2} \approx -2.828$. If x_1 and x_2 are constrained to be integer-valued, then the best we can do is to have $y = -2$ with $\mathbf{x}^* \in \{[-2, 0], [-1, -1], [0, -2]\}$.

Example 19.1. Discrete versions of problems constrain the solution, often resulting in worse solutions than their continuous counterparts.

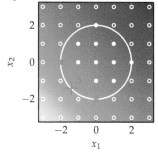

19.1   Integer Programs

An *integer program* is a linear program[2] with integral constraints. By integral constraints, we mean that the design variables must come from the set of integers.[3] Integer programs are sometimes referred to as *integer linear programs* to emphasize the assumption that the objective function and constraints are linear.

An integer program in standard form is expressed as:

$$
\begin{aligned}
\underset{\mathbf{x}}{\text{minimize}} \quad & \mathbf{c}^\top \mathbf{x} \\
\text{subject to} \quad & \mathbf{A}\mathbf{x} \leq \mathbf{b} \\
& \mathbf{x} \geq \mathbf{0} \\
& \mathbf{x} \in \mathbb{Z}^n
\end{aligned}
\tag{19.1}
$$

where $\mathbb{Z}^n$ is the set of n-dimensional integral vectors.

Like linear programs, integer programs are often solved in equality form. Transforming an integer program to equality form often requires adding additional slack variables $\mathbf{s}$ that do not need to be integral. Thus, the equality form for integral programs is:

$$
\begin{aligned}
\underset{\mathbf{x}}{\text{minimize}} \quad & \mathbf{c}^\top \mathbf{x} \\
\text{subject to} \quad & \mathbf{A}\mathbf{x} + \mathbf{s} = \mathbf{b} \\
& \mathbf{x} \geq \mathbf{0} \\
& \mathbf{s} \geq \mathbf{0} \\
& \mathbf{x} \in \mathbb{Z}^n
\end{aligned}
\tag{19.2}
$$

More generally, a *mixed integer program* (algorithm 19.1) includes both continuous and discrete design components. Such a problem, in equality form, is expressed as:

$$
\begin{aligned}
\underset{\mathbf{x}}{\text{minimize}} \quad & \mathbf{c}^\top \mathbf{x} \\
\text{subject to} \quad & \mathbf{A}\mathbf{x} = \mathbf{b} \\
& \mathbf{x} \geq \mathbf{0} \\
& \mathbf{x}_{\mathcal{D}} \in \mathbb{Z}^{\|\mathcal{D}\|}
\end{aligned}
\tag{19.3}
$$

where $\mathcal{D}$ is a set of indices into the design variables that are constrained to be discrete. Here, $\mathbf{x} = [\mathbf{x}_{\mathcal{D}}, \mathbf{x}_{\mathcal{C}}]$, where $\mathbf{x}_{\mathcal{D}}$ represents the vector of discrete design variables and $\mathbf{x}_{\mathcal{C}}$ the vector of continuous design variables.

[2] See chapter 11.

[3] Integer programming is a very mature field, with applications in operations research, communications networks, task scheduling, and other disciplines. Modern solvers, such as Gurobi and CPLEX, can routinely handle problems with millions of variables. There are packages for Julia that provide access to Gurobi, CPLEX, and a variety of other solvers.

```
mutable struct MixedIntegerProgram
    A
    b
    c
    D
end
```

Algorithm 19.1. A mixed integer linear program type that reflects equation (19.3). Here, D is the set of design indices constrained to be discrete.

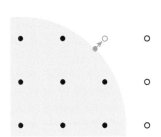

Figure 19.1. Rounding can produce an infeasible design point.

19.2 Rounding

A common approach to discrete optimization is to *relax* the constraint that the design points must come from a discrete set. The advantage of this relaxation is that we can use techniques, like gradient descent or linear programming, that take advantage of the continuous nature of the objective function to direct the search. After a continuous solution is found, the design variables are *rounded* to the nearest feasible discrete design.

There are potential issues with rounding. Rounding might result in an infeasible design point, as shown in figure 19.1. Even if rounding results in a feasible point, it may be far from optimal, as shown in figure 19.2. The addition of the discrete constraint will typically worsen the objective value as illustrated in example 19.1. However, for some problems, we can show the relaxed solution is close to the optimal discrete solution.

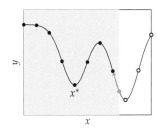

Figure 19.2. The nearest feasible discrete design may be significantly worse than the best feasible discrete design.

We can solve integer programs using rounding by removing the integer constraint, solving the corresponding linear program, or LP, and then rounding the solution to the nearest integer. This method is implemented in algorithm 19.2.

```
relax(MIP) = LinearProgram(MIP.A, MIP.b, MIP.c)
function round_ip(MIP)
    x = minimize_lp(relax(MIP))
    for i in MIP.D
        x[i] = round(Int, x[i])
    end
    return x
end
```

Algorithm 19.2. Methods for relaxing a mixed integer linear program into a linear program and solving a mixed integer linear program by rounding. Both methods accept a mixed integer linear program MIP. The solution obtained by rounding may be suboptimal or infeasible.

We can show that rounding the continuous solution for a constraint $\mathbf{Ax} \leq \mathbf{b}$ when $\mathbf{A}$ is integral is never too far from the optimal integral solution.[4] If $\mathbf{x}_c^*$ is an optimal solution of the LP with $m \times n$ matrix $\mathbf{A}$, then there exists an optimal

[4] W. Cook, A. M. Gerards, A. Schrijver, and É. Tardos, "Sensitivity Theorems in Integer Linear Programming," *Mathematical Programming*, vol. 34, no. 3, pp. 251–264, 1986.

discrete solution $\mathbf{x}_d^*$ with $\left\|\mathbf{x}_c^* - \mathbf{x}_d^*\right\|_\infty$ less than or equal to n times the maximum absolute value of the determinants of the submatrices of $\mathbf{A}$.

The vector $\mathbf{c}$ need not be integral for an LP to have an optimal integral solution because the feasible region is purely determined by $\mathbf{A}$ and $\mathbf{b}$. Some approaches use the dual formulation for the LP, which has a feasible region dependent on $\mathbf{c}$, in which case having an integral $\mathbf{c}$ is also required.

In the special case of *totally unimodular* integer programs, where $\mathbf{A}, \mathbf{b}$, and $\mathbf{c}$ have all integer entries and $\mathbf{A}$ is totally unimodular, the simplex algorithm is guaranteed to return an integer solution. A matrix is totally unimodular if the determinant of every *submatrix*[5] is 0, 1, or −1, and the inverse of a totally unimodular matrix is also integral. In fact, every vertex solution of a totally unimodular integer program is integral. Thus, every $\mathbf{A}\mathbf{x} = \mathbf{b}$ for unimodular $\mathbf{A}$ and integral $\mathbf{b}$ has an integral solution.

[5] A submatrix is a matrix obtained by deleting rows and/or columns of another matrix.

Several matrices and their total unimodularity are discussed in example 19.2. Methods for determining whether a matrix or an integer linear program are totally unimodular are given in algorithm 19.3.

Consider the following matrices:

$$\begin{bmatrix} 1 & 0 & 1 \\ 0 & 0 & 0 \\ 1 & 0 & -1 \end{bmatrix} \quad \begin{bmatrix} 1 & 0 & 1 \\ 0 & 0 & 0 \\ 1 & 0 & 0 \end{bmatrix} \quad \begin{bmatrix} -1 & -1 & 0 & 0 & 0 \\ 1 & 0 & -1 & -1 & 0 \\ 0 & 1 & 1 & 0 & -1 \end{bmatrix}$$

The left matrix is not totally unimodular, as

$$\begin{vmatrix} 1 & 1 \\ 1 & -1 \end{vmatrix} = -2$$

The other two matrices are totally unimodular.

Example 19.2. Examples of totally unimodular matrices.

19.3   *Cutting Planes*

The *cutting plane method* is an exact method for solving mixed integer programs when $\mathbf{A}$ is not totally unimodular.[6] Modern practical methods for solving integer programs use *branch and cut* algorithms[7] that combine the cutting plane method

[6] R. E. Gomory, "An Algorithm for Integer Solutions to Linear Programs," *Recent Advances in Mathematical Programming*, vol. 64, pp. 269–302, 1963.

[7] M. Padberg and G. Rinaldi, "A Branch-and-Cut Algorithm for the Resolution of Large-Scale Symmetric Traveling Salesman Problems," *SIAM Review*, vol. 33, no. 1, pp. 60–100, 1991.

```
isint(x, ϵ=1e-10) = abs(round(x) - x) ≤ ϵ
function is_totally_unimodular(A::Matrix)
    # all entries must be in [0,1,-1]
    if any(a ∉ (0,-1,1) for a in A)
        return false
    end
    # brute force check every subdeterminant
    r,c = size(A)
    for i in 1 : min(r,c)
        for a in subsets(1:r, i)
            for b in subsets(1:c, i)
                B = A[a,b]
                if det(B) ∉ (0,-1,1)
                    return false
                end
            end
        end
    end
    return true
end
function is_totally_unimodular(MIP)
    return is_totally_unimodular(MIP.A) &&
            all(isint, MIP.b) && all(isint, MIP.c)
end
```

Algorithm 19.3. Methods for determining whether matrices A or mixed integer programs MIP are totally unimodular. The method isint returns true if the given value is integral.

with the branch and bound method, discussed in the next section. The cutting plane method works by solving the relaxed LP and then adding linear constraints that result in an optimal solution.

We begin the cutting method with a solution $\mathbf{x}_c^*$ to the relaxed LP, which must be a vertex of $\mathbf{Ax} = \mathbf{b}$. If the $\mathcal{D}$ components in $\mathbf{x}_c^*$ are integral, then it is also an optimal solution to the original mixed integer program, and we are done. As long as the $\mathcal{D}$ components in $\mathbf{x}_c^*$ are not integral, we find a hyperplane with $\mathbf{x}_c^*$ on one side and all feasible discrete solutions on the other. This *cutting plane* is an additional linear constraint to exclude $\mathbf{x}_c^*$. The augmented LP is then solved for a new $\mathbf{x}_c^*$.

Each iteration of algorithm 19.4 introduces cutting planes that make nonintegral components of $\mathbf{x}_c^*$ infeasible while preserving the feasibility of the nearest integral solutions and the rest of the feasible set. The integer program modified with these cutting plane constraints is solved for a new relaxed solution. Figure 19.3 illustrates this process.

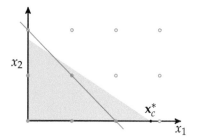 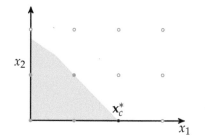

Figure 19.3. The cutting plane method introduces constraints until the solution to the LP is integral. The cutting plane is shown as a red line on the left. The feasible region of the augmented LP is on the right.

We wish to add constraints that cut out nonintegral components of $\mathbf{x}_c^*$. For an LP in equality form with constraint $\mathbf{Ax} = \mathbf{b}$, recall from section 11.2.1 that we can partition a vertex solution $\mathbf{x}_c^*$ to arrive at

$$\mathbf{A}_{\mathcal{B}}\mathbf{x}_{\mathcal{B}}^* + \mathbf{A}_{\mathcal{V}}\mathbf{x}_{\mathcal{V}}^* = \mathbf{b} \tag{19.4}$$

where $\mathbf{x}_{\mathcal{V}}^* = \mathbf{0}$. The nonintegral components of $\mathbf{x}_c^*$ will thus occur only in $\mathbf{x}_{\mathcal{B}}^*$.

We can introduce an additional inequality constraint for each $b \in \mathcal{B}$ such that x_b^* is nonintegral:[8]

$$x_b^* - \lfloor x_b^* \rfloor - \sum_{v \in \mathcal{V}} \left(\bar{A}_{bv} - \lfloor \bar{A}_{bv} \rfloor \right) x_v \leq 0 \tag{19.5}$$

[8] Note that $\lfloor x \rfloor$, or *floor* of x, rounds x down to the nearest integer.

```
frac(x) = modf(x)[1]
function cutting_plane(MIP)
    LP = relax(MIP)
    x, b_inds, v_inds = minimize_lp(LP)
    n_orig = length(x)
    D = copy(MIP.D)
    while !all(isint(x[i]) for i in D)
        AB, AV = LP.A[:,b_inds], LP.A[:,v_inds]
        Abar = AB\AV
        b = 0
        for i in D
            if !isint(x[i])
                b += 1
                A2 = [LP.A zeros(size(LP.A,1));
                        zeros(1,size(LP.A,2)+1)]
                A2[end,end] = 1
                A2[end,v_inds] = (x->floor(x) - x).(Abar[b,:])
                b2 = vcat(LP.b, -frac(x[i]))
                c2 = vcat(LP.c, 0)
                LP = LinearProgram(A2,b2,c2)
            end
        end
        x, b_inds, v_inds = minimize_lp(LP)
    end
    return x[1:n_orig]
end
```

Algorithm 19.4. The cutting plane method solves a given mixed integer program MIP and returns an optimal design vector. An error is thrown if no feasible solution exists. The helper function frac returns the fractional part of a number, and the implementation for minimize_lp, algorithm 11.5, has been adjusted to return the basic and nonbasic indices b_inds and v_inds along with an optimal design x.

where $\bar{\mathbf{A}} = \mathbf{A}_{\mathcal{B}}^{-1}\mathbf{A}_{\mathcal{V}}$. These cutting planes use only the $\mathcal{V}$-components to cut out the nonintegral components of $\mathbf{x}_c^*$.

Introducing a cutting plane constraint cuts out the relaxed solution $\mathbf{x}_c^*$, because all x_v are zero:

$$\underbrace{x_b^* - \lfloor x_b^* \rfloor}_{>0} - \underbrace{\sum_{v \in \mathcal{V}} \left(\bar{A}_{bv} - \lfloor \bar{A}_{bv} \rfloor\right) x_v}_{0} > 0 \tag{19.6}$$

A cutting plane is written in equality form using an additional integral slack variable x_k:

$$x_k + \sum_{v \in \mathcal{V}} \left(\lfloor \bar{A}_{bv} \rfloor - \bar{A}_{bv}\right) x_v = \lfloor x_b^* \rfloor - x_b^* \tag{19.7}$$

Each iteration of algorithm 19.4 thus increases the number of constraints and the number of variables until solving the LP produces an integral solution. Only the components corresponding to the original design variables are returned.

The cutting plane method is used to solve a simple integer linear program in example 19.3.

19.4   Branch and Bound

One method for finding the global optimum of a discrete problem, such as an integer program, is to enumerate all possible solutions. The *branch and bound*[9] method guarantees that an optimal solution is found without having to evaluate all possible solutions. Many commercial integer program solvers use ideas from both the cutting plane method and branch and bound. The method gets its name from the *branch operation* that partitions the solution space[10] and the *bound operation* that computes a lower bound for a partition.

Branch and bound is a general method that can be applied to many kinds of discrete optimization problems, but we will focus here on how it can be used for integer programming. Algorithm 19.5 provides an implementation that uses a *priority queue*, which is a data structure that associates priorities with elements in a collection. We can add an element and its priority value to a priority queue using the enqueue! operation. We can remove the element with the minimum priority value using the dequeue! operation.

[9] A. H. Land and A. G. Doig, "An Automatic Method of Solving Discrete Programming Problems," *Econometrica*, vol. 28, no. 3, pp. 497–520, 1960.

[10] The subsets are typically disjoint, but this is not required. For branch and bound to work, at least one subset must have an optimal solution. D. A. Bader, W. E. Hart, and C. A. Phillips, "Parallel Algorithm Design for Branch and Bound," in *Tutorials on Emerging Methodologies and Applications in Operations Research*, H. J. Greenberg, ed., Kluwer Academic Press, 2004.

Consider the integer program:

$$\underset{\mathbf{x}}{\text{minimize}} \quad 2x_1 + x_2 + 3x_3$$

$$\text{subject to} \quad \begin{bmatrix} 0.5 & -0.5 & 1.0 \\ 2.0 & 0.5 & -1.5 \end{bmatrix} \mathbf{x} = \begin{bmatrix} 2.5 \\ -1.5 \end{bmatrix}$$

$$\mathbf{x} \geq \mathbf{0} \qquad \mathbf{x} \in \mathbb{Z}^3$$

Example 19.3. The cutting plane method used to solve an integer program.

The relaxed solution is $\mathbf{x}^* \approx [0.818, 0, 2.091]$, yielding:

$$\mathbf{A}_{\mathcal{B}} = \begin{bmatrix} 0.5 & 1 \\ 2 & -1.5 \end{bmatrix} \quad \mathbf{A}_V = \begin{bmatrix} -0.5 \\ 0.5 \end{bmatrix} \quad \bar{\mathbf{A}} = \begin{bmatrix} -0.091 \\ -0.455 \end{bmatrix}$$

From equation (19.7), the constraint for x_1 with slack variable x_4 is:

$$x_4 + (\lfloor -0.091 \rfloor - (-0.091))x_2 = \lfloor 0.818 \rfloor - 0.818$$
$$x_4 - 0.909x_2 = -0.818$$

The constraint for x_3 with slack variable x_5 is:

$$x_5 + (\lfloor -0.455 \rfloor - (-0.455))x_2 = \lfloor 2.091 \rfloor - 2.091$$
$$x_5 - 0.545x_2 = -0.091$$

The modified integer program has:

$$\mathbf{A} = \begin{bmatrix} 0.5 & -0.5 & 1 & 0 & 0 \\ 2 & 0.5 & -1.5 & 0 & 0 \\ 0 & -0.909 & 0 & 1 & 0 \\ 0 & -0.545 & 0 & 0 & 1 \end{bmatrix} \quad \mathbf{b} = \begin{bmatrix} 2.5 \\ -1.5 \\ -0.818 \\ -0.091 \end{bmatrix} \quad \mathbf{c} = \begin{bmatrix} 2 \\ 1 \\ 3 \\ 0 \\ 0 \end{bmatrix}$$

Solving the modified LP, we get $\mathbf{x}^* \approx [0.9, 0.9, 2.5, 0.0, 0.4]$. Since this point is not integral, we repeat the procedure with constraints:

$$x_6 - 0.9x_4 = -0.9 \qquad\qquad x_7 - 0.9x_4 = -0.9$$
$$x_8 - 0.5x_4 = -0.5 \qquad\qquad x_9 - 0.4x_4 = -0.4$$

and solve a third LP to obtain: $\mathbf{x}^* = [1, 2, 3, 1, 1, 0, 0, 0, 0]$ with a final solution of $\mathbf{x}_i^* = [1, 2, 3]$.

The algorithm begins with a priority queue containing a single LP relaxation of the original mixed integer program. Associated with that LP is a solution $\mathbf{x}_c^*$ and objective value $y_c = \mathbf{c}^\top \mathbf{x}_c^*$. The objective value serves as a lower bound on the solution and thus serves as the LP's priority in the priority queue. At each iteration of the algorithm, we check whether the priority queue is empty. If it is not empty, we dequeue the LP with the lowest priority value. If the solution associated with that element has the necessary integral components, then we keep track of whether it is the best integral solution found so far.

If the dequeued solution has one or more components in $\mathcal{D}$ that are nonintegral, we choose from $\mathbf{x}_c^*$ such a component that is farthest from an integer value. Suppose this component corresponds to the ith design variable. We *branch* by considering two new LPs, each one created by adding one of the following constraints to the dequeued LP:[11]

$$x_i \leq \lfloor x_{i,c}^* \rfloor \qquad \text{or} \qquad x_i \geq \lceil x_{i,c}^* \rceil \tag{19.8}$$

as shown in figure 19.4. Example 19.4 demonstrates this process.

We compute the solution associated with these two LPs, which provide lower bounds on the value of the original mixed integer program. If either solution lowers the objective value when compared to the best integral solution seen so far, it is placed into the priority queue. Not placing solutions already known to be inferior to the best integral solution seen thus far allows branch and bound to prune the search space. The process continues until the priority queue is empty, and we return the best feasible integral solution. Example 19.5 shows how branch and bound can be applied to a small integer program.

[11] Note that $\lceil x \rceil$, or *ceiling* of x, rounds x up to the nearest integer.

Consider a relaxed solution $\mathbf{x}_c^* = [3, 2.4, 1.2, 5.8]$ for an integer program with $\mathbf{c} = [-1, -2, -3, -4]$. The lower bound is

$$y \geq \mathbf{c}^\top \mathbf{x}_c^* = -34.6$$

We branch on a nonintegral coordinate of $\mathbf{x}_c^*$, typically the one farthest from an integer value. In this case, we choose the first nonintegral coordinate, $x_{2,c}^*$, which is 0.4 from the nearest integer value. We then consider two new LPs, one with $x_2 \leq 2$ as an additional constraint and the other with $x_2 \geq 3$ as an additional constraint.

Example 19.4. An example of a single application of the branching step in branch and bound.

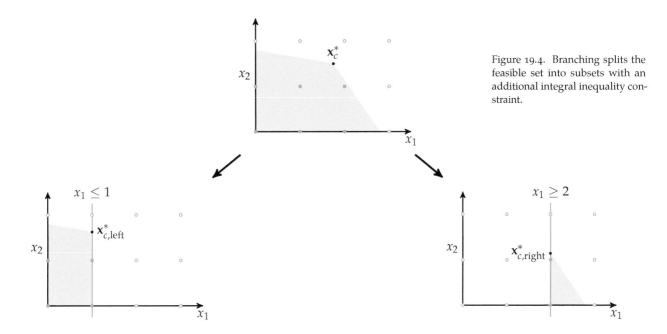

Figure 19.4. Branching splits the feasible set into subsets with an additional integral inequality constraint.

```
function minimize_lp_and_y(LP)
    try
        x = minimize_lp(LP)
        return (x, x·LP.c)
    catch
        return (fill(NaN, length(LP.c)), Inf)
    end
end
function branch_and_bound(MIP)
    LP = relax(MIP)
    x, y = minimize_lp_and_y(LP)
    n = length(x)
    x_best, y_best, Q = deepcopy(x), Inf, PriorityQueue()
    enqueue!(Q, (LP,x,y), y)
    while !isempty(Q)
        LP, x, y = dequeue!(Q)
        if any(isnan.(x)) || all(isint(x[i]) for i in MIP.D)
            if y < y_best
                x_best, y_best = x[1:n], y
            end
        else
            i = argmax([abs(x[i] - round(x[i])) for i in MIP.D])
            # x_i ≤ floor(x_i)
            A, b, c = LP.A, LP.b, LP.c
            A2 = [A zeros(size(A,1));
                  [j==i for j in 1:size(A,2)]' 1]
            b2, c2 = vcat(b, floor(x[i])), vcat(c, 0)
            LP2 = LinearProgram(A2,b2,c2)
            x2, y2 = minimize_lp_and_y(LP2)
            if y2 ≤ y_best
                enqueue!(Q, (LP2,x2,y2), y2)
            end
            # x_i ≥ ceil(x_i)
            A2 = [A zeros(size(A,1));
                  [j==i for j in 1:size(A,2)]' -1]
            b2, c2 = vcat(b, ceil(x[i])), vcat(c, 0)
            LP2 = LinearProgram(A2,b2,c2)
            x2, y2 = minimize_lp_and_y(LP2)
            if y2 ≤ y_best
                enqueue!(Q, (LP2,x2,y2), y2)
            end
        end
    end
    return x_best
end
```

Algorithm 19.5.   The branch and bound algorithm for solving a mixted integer program MIP. The helper method minimize_lp_and_y solves an LP and returns both the solution and its value. An infeasible LP produces a NaN solution and an Inf value. More sophisticated implementations will drop variables whose solutions are known in order to speed computation. The PriorityQueue type is provided by the DataStructures.jl package.

We can use branch and bound to solve the integer program in example 19.3. As before, the relaxed solution is $\mathbf{x}_c^* = [0.818, 0, 2.09]$, with a value of 7.909. We branch on the first component, resulting in two integer programs, one with $x_1 \leq 0$ and one with $x_1 \geq 1$:

Example 19.5. Appying branch and bound to solve an integer programming problem.

$$\mathbf{A}_{\text{left}} = \begin{bmatrix} 0.5 & -0.5 & 1 & 0 \\ 2 & 0.5 & -1.5 & 0 \\ 1 & 0 & 0 & 1 \end{bmatrix} \quad \mathbf{b}_{\text{left}} = \begin{bmatrix} 2.5 \\ -1.5 \\ 0 \end{bmatrix} \quad \mathbf{c}_{\text{left}} = \begin{bmatrix} 2 \\ 1 \\ 3 \\ 0 \end{bmatrix}$$

$$\mathbf{A}_{\text{right}} = \begin{bmatrix} 0.5 & -0.5 & 1 & 0 \\ 2 & 0.5 & -1.5 & 0 \\ 1 & 0 & 0 & -1 \end{bmatrix} \quad \mathbf{b}_{\text{right}} = \begin{bmatrix} 2.5 \\ -1.5 \\ 1 \end{bmatrix} \quad \mathbf{c}_{\text{right}} = \begin{bmatrix} 2 \\ 1 \\ 3 \\ 0 \end{bmatrix}$$

The left LP with $x_1 \leq 0$ is infeasible. The right LP with $x_1 \geq 1$ has a relaxed solution, $\mathbf{x}_c^* = [1, 2, 3, 0]$, and a value of 13. We have thus obtained our integral solution, $\mathbf{x}_i^* = [1, 2, 3]$.

19.5   Dynamic Programming

Dynamic programming[12] is a technique that can be applied to problems with *optimal substructure* and *overlapping subproblems*. A problem has optimal substructure if an optimal solution can be constructed from optimal solutions of its subproblems. Figure 19.5 shows an example.

[12] The term dynamic programming was chosen by Richard Bellman to reflect the time-varying aspect of the problems he applied it to and to avoid the sometimes negative connotations words like *research* and *mathematics* had. He wrote, "I thought dynamic programming was a good name. It was something not even a Congressman could object to. So I used it as an umbrella for my activities." R. Bellman, *Eye of the Hurricane: An Autobiography*. World Scientific, 1984. p. 159.

Figure 19.5. Shortest path problems have optimal substructure because if the shortest path from any a to c passes through b, then the subpaths $a \rightarrow b$ and $b \rightarrow c$ are both shortest paths.

A problem with overlapping subproblems solved recursively will encounter the same subproblem many times. Instead of enumerating exponentially many potential solutions, dynamic programming either stores subproblem solutions, and thereby avoids having to recompute them, or recursively builds the optimal solution in a single pass. Problems with recurrence relations often have overlapping subproblems. Figure 19.6 shows an example.

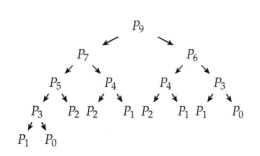 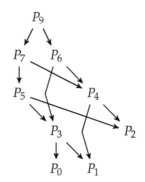

Figure 19.6. We can compute the nth term of the *Padovan sequence*, $P_n = P_{n-2} + P_{n-3}$, with $P_0 = P_1 = P_2 = 1$ by recursing through all subterms (left). A more efficient approach is to compute subterms once and reuse their values in subsequent calculations by exploiting the problem's overlapping substructure (right).

Dynamic programming can be implemented either top-down or bottom-up, as demonstrated in algorithm 19.6. The top-down approach begins with the desired problem and recurses down to smaller and smaller subproblems. Subproblem solutions are stored so that when we are given a new subproblem, we can either retrieve the computed solution or solve and store it for future use.[13] The bottom-up approach starts by solving the smaller subproblems and uses their solutions to obtain solutions to larger problems.

[13] Storing subproblem solutions in this manner is called *memoization*.

```
function padovan_topdown(n, P=Dict())
    if !haskey(P, n)
        P[n] = n < 3 ? 1 :
               padovan_topdown(n-2,P) + padovan_topdown(n-3,P)
    end
    return P[n]
end
function padovan_bottomup(n)
    P = Dict(0=>1,1=>1,2=>1)
    for i in 3 : n
        P[i] = P[i-2] + P[i-3]
    end
    return P[n]
end
```

Algorithm 19.6. Computing the Padovan sequence using dynamic programming, with both the top-down and bottom-up approaches.

The *knapsack problem* is a well-known combinatorial optimization problem that often arises in resource allocation.[14] Suppose we are packing our knapsack for a trip, but we have limited space and want to pack the most valuable items. There are several variations of the knapsack problem. In the 0-1 knapsack problem, we have the following optimization problem:

[14] The knapsack problem is an integer program with a single constraint, but it can be efficiently solved using dynamic programming.

$$\underset{\mathbf{x}}{\text{minimize}} \quad -\sum_{i=1}^{n} v_i x_i$$

$$\text{subject to} \quad \sum_{i=1}^{n} w_i x_i \leq w_{\max} \tag{19.9}$$

$$x_i \in \{0,1\} \text{ for all } i \text{ in } \{1,\ldots,n\}$$

We have n items, with the ith item having integral weight $w_i > 0$ and value v_i. The design vector $\mathbf{x}$ consists of binary values that indicate whether an item is packed. The total weight cannot exceed our integral capacity $w_{\max}$, and we seek to maximize the total value of packed items.

There are 2^n possible design vectors, which makes direct enumeration for large n intractable. However, we can use dynamic programming. The 0-1 knapsack problem has optimal substructure and overlapping subproblems. Consider having solved knapsack problems with n items and several capacities up to $w_{\max}$. The solution to a larger knapsack problem with one additional item with weight w_{n+1} and capacity $w_{\max}$ will either include or not include the new item:

- If it is not worth including the new item, the solution will have the same value as a knapsack with $n - 1$ items and capacity $w_{\max}$.

- If it is worth including the new item, the solution will have the value of a knapsack with $n - 1$ items and capacity $w_{\max} - w_{n+1}$ plus the value of the new item.

The recurrence relation is:

$$\text{knapsack}(i, w_{\max}) = \begin{cases} 0 & \text{if } i = 0 \\ \text{knapsack}(i-1, w_{\max}) & \text{if } w_i > w_{\max} \\ \max \begin{cases} \text{knapsack}(i-1, w_{\max}) & \text{(discard new item)} \\ \text{knapsack}(i-1, w_{\max} - w_i) + v_i & \text{(include new item)} \end{cases} & \text{otherwise} \end{cases} \tag{19.10}$$

The 0-1 knapsack problem can be solved using the implementation in algorithm 19.7.

```
function knapsack(v, w, w_max)
    n = length(v)
    y = Dict((0,j) => 0.0 for j in 0:w_max)
    for i in 1 : n
        for j in 0 : w_max
            y[i,j] = w[i] > j ? y[i-1,j] :
                    max(y[i-1,j],
                        y[i-1,j-w[i]] + v[i])
        end
    end

    # recover solution
    x, j = falses(n), w_max
    for i in n: -1 : 1
        if w[i] ≤ j && y[i,j] - y[i-1, j-w[i]] == v[i]
            # the ith element is in the knapsack
            x[i] = true
            j -= w[i]
        end
    end
    return x
end
```

Algorithm 19.7. A method for solving the 0-1 knapsack problem with
item values v, integral item weights
w, and integral capacity w_max. Re-
covering the design vector from
the cached solutions requires additional iteration.

19.6 Ant Colony Optimization

Ant colony optimization[15] is a stochastic method for optimizing paths through
graphs. This method was inspired by some ant species that wander randomly
in search of food, leaving *pheromone* trails as they go. Other ants that stumble
upon a pheromone trail are likely to start following it, thereby reinforcing the
trail's scent. Pheromones slowly evaporate over time, causing unused trails to
fade. Short paths, with stronger pheromones, are traveled more often and thus
attract more ants. Thus, short paths create positive feedback that lead other ants
to follow and further reinforce the shorter path.

Basic shortest path problems, such as the shortest paths found by ants between
the ant hill and sources of food, can be efficiently solved using dynamic pro-

[15] M. Dorigo, V. Maniezzo, and A.
Colorni, "Ant System: Optimization by a Colony of Cooperating
Agents," *IEEE Transactions on Systems, Man, and Cybernetics, Part B
(Cybernetics)*, vol. 26, no. 1, pp. 29–
41, 1996.

gramming. Ant colony optimization has been used to find near-optimal solutions to the *traveling salesman problem*, a much more difficult problem in which we want to find the shortest path that passes through each node exactly once. Ant colony optimization has also been used to route multiple vehicles, find optimal locations for factories, and fold proteins.[16] The algorithm is stochastic in nature and is thus resistant to changes to the problem over time, such as traffic delays changing effective edge lengths in the graph or networking issues that remove edges entirely.

Ants move stochastically based on the attractiveness of the edges available to them. The attractiveness of transition $i \rightarrow j$ depends on the pheromone level and an optional prior factor:

$$A(i \rightarrow j) = \tau(i \rightarrow j)^\alpha \eta(i \rightarrow j)^\beta \qquad (19.11)$$

where α and β are exponents for the pheromone level τ and prior factor η, respectively.[17] For problems involving shortest paths, we can set the prior factor to the inverse edge length $\ell(i \rightarrow j)$ to encourage the traversal of shorter paths: $\eta(i \rightarrow j) = 1/\ell(i \rightarrow j)$. A method for computing the edge attractiveness is given in algorithm 19.8.

Suppose an ant is at node i and can transition to any of the nodes $j \in \mathcal{J}$. The set of successor nodes $\mathcal{J}$ contains all valid outgoing neighbors.[18] Sometimes edges are excluded, such as in the traveling salesman problem where ants are prevented from visiting the same node twice. It follows that $\mathcal{J}$ is dependent on both i and the ant's history.

```
function edge_attractiveness(graph, τ, η; α=1, β=5)
    A = Dict()
    for i in 1 : nv(graph)
        neighbors = outneighbors(graph, i)
        for j in neighbors
            v = τ[(i,j)]^α * η[(i,j)]^β
            A[(i,j)] = v
        end
    end
    return A
end
```

[16] These and other applications are discussed in these references: M. Manfrin, "Ant Colony Optimization for the Vehicle Routing Problem," Ph.D. dissertation, Université Libre de Bruxelles, 2004. T. Stützle, "MAX-MIN Ant System for Quadratic Assignment Problems," Technical University Darmstadt, Tech. Rep., 1997. and A. Shmygelska, R. Aguirre-Hernández, and H. H. Hoos, "An Ant Colony Algorithm for the 2D HP Protein Folding Problem," in *International Workshop on Ant Algorithms (ANTS)*, 2002.

[17] Dorigo, Maniezzo, and Colorni recommend $\alpha = 1$ and $\beta = 5$.

[18] The *outgoing neighbors* of a node i are all nodes j such that $i \rightarrow j$ is in the graph. In an undirected graph, the neighbors and the outgoing neighbors are identical.

Algorithm 19.8. A method for computing the edge attractiveness table given graph `graph`, pheromone levels τ, prior edge weights η, pheromone exponent α, and prior exponent β.

The probability of edge transition $i \rightarrow j$ is:

$$P(i \rightarrow j) = \frac{A(i \rightarrow j)}{\sum_{j' \in \mathcal{J}} A(i \rightarrow j')} \qquad (19.12)$$

Ants affect subsequent generations of ants by depositing pheromones. There are several methods for modeling pheromone deposition. A common approach is to deposit pheromones establishing after a complete path.[19] Ants that do not find a path do not deposit pheromones. For shortest path problems, a successful ant that has established a path of length ℓ deposits $1/\ell$ pheromones on each edge it traversed.

[19] M. Dorigo, G. Di Caro, and L. M. Gambardella, "Ant Algorithms for Discrete Optimization," *Artificial Life*, vol. 5, no. 2, pp. 137–172, 1999.

```
import StatsBase: Weights, sample
function run_ant(G, lengths, τ, A, x_best, y_best)
    x = [1]
    while length(x) < nv(G)
        i = x[end]
        neighbors = setdiff(outneighbors(G, i), x)
        if isempty(neighbors) # ant got stuck
            return (x_best, y_best)
        end

        as = [A[(i,j)] for j in neighbors]
        push!(x, neighbors[sample(Weights(as))])
    end

    l = sum(lengths[(x[i-1],x[i])] for i in 2:length(x))
    for i in 2 : length(x)
        τ[(x[i-1],x[i])] += 1/l
    end
    if l < y_best
        return (x, l)
    else
        return (x_best, y_best)
    end
end
```

Algorithm 19.9. A method for simulating a single ant on a traveling salesman problem in which the ant starts at the first node and attempts to visit each node exactly once. Pheromone levels are increased at the end of a successful tour. The parameters are the graph G, edge lengths lengths, pheromone levels τ, edge attractiveness A, the best solution found thus far x_best, and its value y_best.

Ant colony optimization also models pheromone evaporation, which naturally occurs in the real world. Modeling evaporation helps prevent the algorithm from prematurely converging to a single, potentially suboptimal, solution. Pheromone evaporation is executed at the end of each iteration after all ant simulations have been completed. Evaporation decreases the pheromone level of each transition by a factor of $1 - \rho$, with $\rho \in [0, 1]$.[20]

[20] It is common to use $\rho = 1/2$.

For m ants at iteration k, the effective pheromone update is

$$\tau(i \rightarrow j)^{(k+1)} = (1 - \rho)\tau(i \rightarrow j)^{(k)} + \sum_{a=1}^{m} \frac{1}{\ell^{(a)}} \left((i \rightarrow j) \in \mathcal{P}^{(a)} \right) \qquad (19.13)$$

where $\ell^{(a)}$ is the path length and $\mathcal{P}^{(a)}$ is the set of edges traversed by ant a.

Ant colony optimization is implemented in algorithm 19.10, with individual ant simulations using algorithm 19.9. Figure 19.7 visualizes ant colony optimization used to solve a traveling salesman problem.

```
function ant_colony_optimization(G, lengths;
    m = 1000, k_max=100, α=1.0, β=5.0, ρ=0.5,
    η = Dict((e.src,e.dst)=>1/lengths[(e.src,e.dst)]
            for e in edges(G)))
    τ = Dict((e.src,e.dst)=>1.0 for e in edges(G))
    x_best, y_best = [], Inf
    for k in 1 : k_max
        A = edge_attractiveness(G, τ, η, α=α, β=β)
        for (e,v) in τ
            τ[e] = (1-ρ)*v
        end
        for ant in 1 : m
            x_best,y_best = run_ant(G,lengths,τ,A,x_best,y_best)
        end
    end
    return x_best
end
```

Algorithm 19.10. Ant colony optimization, which takes a directed or undirected graph G from LightGraphs.jl and a dictionary of edge tuples to path lengths lengths. Ants start at the first node in the graph. Optional parameters include the number of ants per iteration m, the number of iterations k_max, the pheromone exponent α, the prior exponent β, the evaporation scalar ρ, and a dictionary of prior edge weights η.

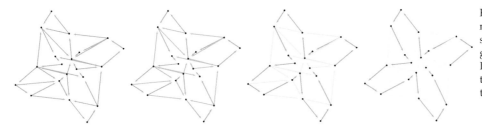

Figure 19.7. Ant colony optimization used to solve a traveling salesman problem on a directed graph using 50 ants per iteration. Path lengths are the Euclidean distances. Color opacity corresponds to pheromone level.

19.7 Summary

- Discrete optimization problems require that the design variables be chosen from discrete sets.

- Relaxation, in which the continuous version of the discrete problem is solved, is by itself an unreliable technique for finding an optimal discrete solution but is central to more sophisticated algorithms.

- Many combinatorial optimization problems can be framed as an integer program, which is a linear program with integer constraints.

- Both the cutting plane and branch and bound methods can be used to solve integer programs efficiently and exactly. The branch and bound method is quite general and can be applied to a wide variety of discrete optimization problems.

- Dynamic programming is a powerful technique that exploits optimal overlapping substructure in some problems.

- Ant colony optimization is a nature-inspired algorithm that can be used for optimizing paths in graphs.

19.8 Exercises

Exercise 19.1. A *Boolean satisfiability problem*, often abbreviated SAT, requires determining whether a Boolean design exists that causes a Boolean-valued objective function to output true. SAT problems were the first to be proven to belong to the difficult class of NP-complete problems.[21] This means that SAT is at least as difficult as all other problems whose solutions can be verified in polynomial time. Consider the Boolean objective function:

$$f(\mathbf{x}) = x_1 \wedge (x_2 \vee \neg x_3) \wedge (\neg x_1 \vee \neg x_2)$$

Find an optimal design using enumeration. How many designs must be considered for an n-dimensional design vector in the worst case?

Exercise 19.2. Formulate the problem in exercise 19.1 as an integer linear program. Can any Boolean satisfiability problem be formulated as an integer linear program?

[21] S. Cook, "The Complexity of Theorem-Proving Procedures," in *ACM Symposium on Theory of Computing*, 1971.

Exercise 19.3. Why are we interested in totally unimodular matrices? Furthermore, why does every totally unimodular matrix contain only entries that are 0, 1, or −1?

Exercise 19.4. This chapter solved the 0-1 knapsack problem using dynamic programming. Show how to apply branch and bound to the 0-1 knapsack problem, and use your approach to solve the knapsack problem with values $\mathbf{v} = [9, 4, 2, 3, 5, 3]$, and weights $\mathbf{w} = [7, 8, 4, 5, 9, 4]$ with capacity $w_{max} = 20$.

20 Expression Optimization

Previous chapters discussed optimization over a fixed set of design variables. For many problems, the number of variables is unknown, such as in the optimization of graphical structures or computer programs. Designs in these contexts can be represented by expressions that belong to a grammar. This chapter discusses ways to make the search of optimal designs more efficient by accounting for the grammatical structure of the design space.

20.1 Grammars

An expression can be represented by a tree of *symbols*. For example, the mathematical expression $x + \ln 2$ can be represented using the tree in figure 20.1 consisting of the symbols $+$, x, $\ln$, and 2. *Grammars* specify constraints on the space of possible expressions.

A grammar is represented by a set of *production rules*. These rules involve symbols as well as *types*. A type can be interpreted as a set of expression trees. A production rule represents a possible expansion of type to an expression involving symbols or types. If a rule expands only to symbols, then it is called *terminal* because it cannot be expanded further. An example of a nonterminal rule is $\mathbb{R} \mapsto \mathbb{R} + \mathbb{R}$, which means that the type $\mathbb{R}$ can consist of elements of the set $\mathbb{R}$ added to elements in the set $\mathbb{R}$.[1]

We can generate an expression from a grammar by starting with a *start type* and then recursively applying different production rules. We stop when the tree contains only symbols. Figure 20.2 illustrates this process for the expression $x + \ln 2$. An application to natural language expressions is shown in example 20.1.

The number of possible expressions allowed by a grammar can be infinite. Example 20.2 shows a grammar that allows for infinitely many valid expressions.

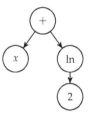

Figure 20.1. The expression $x + \ln 2$ represented as a tree.

[1] This chapter focuses on *context-free grammars*, but other forms exist. See L. Kallmeyer, *Parsing Beyond Context-Free Grammars*. Springer, 2010.

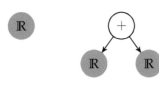
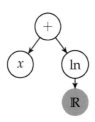
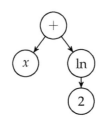

Figure 20.2. Using the production rules

$$\mathbb{R} \mapsto \mathbb{R} + \mathbb{R}$$
$$\mathbb{R} \mapsto x$$
$$\mathbb{R} \mapsto \ln(\mathbb{R})$$
$$\mathbb{R} \mapsto 2$$

to generate $x + \ln 2$. Blue nodes are unexpanded types.

Consider a grammar that allows for the generation of simple English statements:

$$\mathbb{S} \mapsto \mathbb{N}\,\mathbb{V}$$
$$\mathbb{V} \mapsto \mathbb{V}\,\mathbb{A}$$
$$\mathbb{A} \mapsto \text{rapidly} \mid \text{efficiently}$$
$$\mathbb{N} \mapsto \text{Alice} \mid \text{Bob} \mid \text{Mykel} \mid \text{Tim}$$
$$\mathbb{V} \mapsto \text{runs} \mid \text{reads} \mid \text{writes}$$

The types $\mathbb{S}$, $\mathbb{N}$, $\mathbb{V}$, and $\mathbb{A}$ correspond to statements, nouns, verbs, and adverbs, respectively. An expression is generated by starting with the type $\mathbb{S}$ and iteratively replacing types:

$$\mathbb{S}$$
$$\mathbb{N}\,\mathbb{V}$$
$$\text{Mykel}\,\mathbb{V}\,\mathbb{A}$$
$$\text{Mykel writes rapidly}$$

Not all terminal symbol categories must be used. For instance, the statement "Alice runs" can also be generated.

Example 20.1. A grammar for producing simple English statements. Using | on the right-hand side of an expression is shorthand for *or*. Thus, the rule

$$\mathbb{A} \mapsto \text{rapidly} \mid \text{efficiently}$$

is equivalent to having two rules, $\mathbb{A} \mapsto$ rapidly and $\mathbb{A} \mapsto$ efficiently.

Expression optimization often constrains expressions to a maximum depth or penalizes expressions based on their depth or node count. Even if the grammar allows a finite number of expressions, the space is often too vast to search exhaustively. Hence, there is a need for algorithms that efficiently search the space of possible expressions for one that optimizes an objective function.

Consider a four-function calculator grammar that applies addition, subtraction, multiplication, and division to the ten digits:

$$\mathbb{R} \mapsto \mathbb{R} + \mathbb{R}$$
$$\mathbb{R} \mapsto \mathbb{R} - \mathbb{R}$$
$$\mathbb{R} \mapsto \mathbb{R} \times \mathbb{R}$$
$$\mathbb{R} \mapsto \mathbb{R} \ / \ \mathbb{R}$$
$$\mathbb{R} \mapsto 0 \mid 1 \mid 2 \mid 3 \mid 4 \mid 5 \mid 6 \mid 7 \mid 8 \mid 9$$

An infinite number of expressions can be generated because the nonterminal $\mathbb{R}$ can always be expanded into one of the calculator operations.

Many expressions will produce the same value. Addition and multiplication operators are commutative, meaning that the order does not matter. For example, $a + b$ is the same as $b + a$. These operations are also associative, meaning the order in which multiple operations of the same type occur do not matter. For example, $a \times b \times c$ is the same as $c \times b \times a$. Other operations preserve values, like adding zero or multiplying by one.

Not all expressions under this grammar are mathematically valid. For example, division by zero is undefined. Removing zero as a terminal symbol is insufficient to prevent this error because zero can be constructed using other operations, such as $1 - 1$. Such exceptions are often handled by the objective function, which can catch exceptions and penalize them.

Example 20.2. Some of the challenges associated with grammars, as illustrated with a four-function calculator grammar.

The expression optimization routines covered in the chapter use `ExprRules.jl`. Grammars can be defined using the `grammar` macro by listing the production rules, as shown in example 20.3.

Many of the expression optimization algorithms involve manipulating components of an expression tree in a way that preserves the way the types were expanded. A `RuleNode` object tracks which production rules were applied when

We may define a grammar using the `grammar` macro. The nonterminals are on the left of the equal sign, and the expressions with terminals and nonterminals are the on the right. The package includes some syntax to represent grammars more compactly.

Example 20.3. Example of defining a grammar using the `ExprRules.jl` package.

```
using ExprRules
grammar = @grammar begin
    R = x            # reference a variable
    R = R * A        # multiple children
    R = f(R)         # call a function
    R = _(randn())   # random variable generated on node creation
    R = 1 | 2 | 3    # equivalent to R = 1, R = 2, and R = 3
    R = |(4:6)       # equivalent to R = 4, R = 5, and R = 6
    A = 7            # rules for different return types
end;
```

doing an expansion. Calling `rand` with a specified starting type will generate a random expression represented by a `RuleNode`. Calling `sample` will select a random `RuleNode` from an existing `RuleNode` tree. Nodes are evaluated using `eval`.

The method `return_type` returns the node's return type as a symbol, `isterminal` returns whether the symbol is terminal, `child_types` returns the list of nonterminal symbols associated with the node's production rule, and `nchildren` returns the number of children. These four methods each take as input the grammar and the node. The number of nodes in an expression tree is obtained using `length(node)`, and the depth is obtained using `depth(node)`.

A third type, `NodeLoc`, is used to refer to a node's location in the expression tree. Subtrees manipulation often requires `NodeLocs`.

```
loc = sample(NodeLoc, node); # uniformly sample a node loc
loc = sample(NodeLoc, node, :R, grammar); # sample a node loc of type R
subtree = get(node, loc);
```

20.2 Genetic Programming

Genetic algorithms (see chapter 9) use chromosomes that encode design points in a sequential format. *Genetic programming*[2] represents individuals using trees

[2] J. R. Koza, *Genetic Programming: On the Programming of Computers by Means of Natural Selection.* MIT Press, 1992.

instead (figure 20.3), which are better at representing mathematical functions, programs, decision trees, and other hierarchical structures.

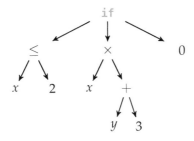

Figure 20.3. A tree representation of the Julia method:
`x ≤ 2 ? x*(y+3) : 0`

Similar to genetic algorithms, genetic programs are initialized randomly and support crossover and mutation. In *tree crossover* (figure 20.4), two parent trees are mixed to form a child tree. A random node is chosen in each parent, and the subtree at the chosen node in the first parent is replaced with the subtree at the chosen node of the second parent. Tree crossover works on parents with different sizes and shapes, allowing arbitrary trees to mix. In some cases one must ensure that replacement nodes have certain types, such as Boolean values input into the condition of an `if` statement.[3] Tree crossover is implemented in algorithm 20.1.

[3] This book focuses only on genetic operations that adhere to the constraints of the grammar. Sometimes genetic programming with this restriction is referred to as *strongly typed genetic programming*, as discussed in D. J. Montana, "Strongly Typed Genetic Programming," *Evolutionary Computation*, vol. 3, no. 2, pp. 199–230, 1995.

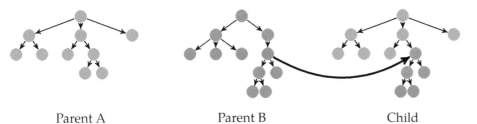

Parent A   Parent B   Child

Figure 20.4. Tree crossover is used to combine two parent trees to produce a child tree.

Tree crossover tends to produce trees with greater depth than the parent trees. Each generation tends to increase in complexity, which often results in overly complicated solutions and slower runtimes. We encourage *parsimony*, or simplicity, in the solution, by introducing a small bias in the objective function value based on a tree's depth or node count.

Applying *tree mutation* (figure 20.5) starts by choosing a random node in the tree. The subtree rooted at that node is deleted, and a new random subtree is generated to replace the old subtree. In contrast to mutation in binary chromo-

```julia
struct TreeCrossover <: CrossoverMethod
    grammar
    max_depth
end
function crossover(C::TreeCrossover, a, b)
    child = deepcopy(a)
    crosspoint = sample(b)
    typ = return_type(C.grammar, crosspoint.ind)
    d_subtree = depth(crosspoint)
    d_max = C.max_depth + 1 - d_subtree
    if d_max > 0 && contains_returntype(child,C.grammar,typ,d_max)
        loc = sample(NodeLoc, child, typ, C.grammar, d_max)
        insert!(child, loc, deepcopy(crosspoint))
    end
    child
end
```

Algorithm 20.1. Tree crossover implemented for a and b of type RuleNode from ExprRules.jl. The TreeCrossover struct contains a rule set grammar and a maximum depth max_depth.

somes, tree mutation can typically occur at most once, often with a low probability around 1%. Tree mutation is implemented in algorithm 20.2.

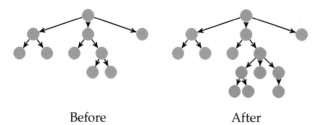

Before   After

Figure 20.5. Tree mutation deletes a random subtree and generates a new one to replace it.

```
struct TreeMutation <: MutationMethod
    grammar
    p
end
function mutate(M::TreeMutation, a)
    child = deepcopy(a)
    if rand() < M.p
        loc = sample(NodeLoc, child)
        typ = return_type(M.grammar, get(child, loc).ind)
        subtree = rand(RuleNode, M.grammar, typ)
        insert!(child, loc, subtree)
    end
    return child
end
```

Algorithm 20.2. Tree mutation implemented for an individual a of type RuleNode from ExprRules.jl. The TreeMutation struct contains a rule set grammar and a mutation probability p.

Tree permutation (figure 20.6) is a second form of genetic mutation. Here, the children of a randomly chosen node are randomly permuted. Tree permutation alone is typically not sufficient to introduce new genetic material into a population and is often combined with tree mutation. Tree permutation is implemented in algorithm 20.3.

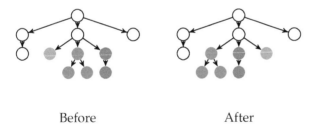

Before   After

Figure 20.6. Tree permutation permutes the children of a randomly chosen node.

The implementation of genetic programming is otherwise identical to that of genetic algorithms. More care must typically be taken in implementing the crossover and mutation routines, particularly when determining what sorts of nodes can be generated and that only syntactically correct trees are produced. Genetic programming is used to generate an expression that approximates π in example 20.4.

```
struct TreePermutation <: MutationMethod
    grammar
    p
end
function mutate(M::TreePermutation, a)
    child = deepcopy(a)
    if rand() < M.p
        node = sample(child)
        n = length(node.children)
        types = child_types(M.grammar, node)
        for i in 1 : n-1
            c = 1
            for k in i+1 : n
                if types[k] == types[i] &&
                    rand() < 1/(c+=1)

                    node.children[i], node.children[k] =
                        node.children[k], node.children[i]
                end
            end
        end
    end
    return child
end
```

Algorithm 20.3. Tree permutation implemented for an individual a of type RuleNode from ExprRules.jl, where p is the mutation probability.

Consider approximating π using only operations on a four-function calcula-tor. We can solve this problem using genetic programming where nodes can be any of the elementary operations: add, subtract, multiply, and divide, and the digits $1 - 9$.

We use ExprRules.jl to specify our grammar:

```
grammar = @grammar begin
    R = |(1:9)
    R = R + R
    R = R - R
    R = R / R
    R = R * R
end
```

Example 20.4. Using genetic pro-gramming to estimate π using only digits and the four principle arith-metic operations.

We construct an objective function and penalize large trees:

```
function f(node)
    value = eval(node, grammar)
    if isinf(value) || isnan(value)
        return Inf
    end
    Δ = abs(value - π)
    return log(Δ) + length(node)/1e3
end
```

We finally run our genetic program using a call to the genetic_algorithm function from section 9.2:

```
population = [rand(RuleNode, grammar, :R) for i in 1:1000]
best_tree = genetic_algorithm(f, population, 30,
                              TruncationSelection(50),
                              TreeCrossover(grammar, 10),
                              TreeMutation(grammar, 0.25))
```

The best performing tree is shown on the right. It evaluates to 3.141586, which matches π to four decimal places.

20.3   Grammatical Evolution

Grammatical evolution[4] operates on an integer array instead of a tree, allowing the same techniques employed in genetic algorithms to be applied. Unlike genetic algorithms, the chromosomes in grammatical evolution encode expressions based on a grammar. Grammatical evolution was inspired by genetic material, which is inherently serial like the chromosomes used in genetic algorithms. [5]

In grammatical evolution, designs are integer arrays much like the chromosomes used in genetic algorithms. Each integer is unbounded because indexing is performed using modular arithmetic. The integer array can be translated into an expression tree by parsing it from left to right.

We begin with a starting symbol and a grammar. Suppose n rules in the grammar can be applied to the starting symbol. The jth rule is applied, where $j = i \bmod_1 n$ and i is the first integer in the integer array.[6]

We then consider the rules applicable to the resulting expression and use similar modular arithmetic based on the second integer in the array to select which rule to apply. This process is repeated until no rules can be applied and the phenotype is complete.[7] The decoding process is implemented in algorithm 20.4 and is worked through in example 20.5.

It is possible for the integer array to be too short, thereby causing the translation process to run past the length of the array. Rather than producing an invalid individual and penalizing it in the objective function, the process wraps around to the beginning of the array instead. This wrap-around effect means that the same decision can be read several times during the transcription process. Transcription can result in infinite loops, which can be prevented by a maximum depth.

Genetic operations work directly on the integer design array. We can adopt the operations used on real-valued chromosomes and apply them to the integer-valued chromosomes. The only change is that mutation must preserve real values. A mutation method for integer-valued chromosomes using zero-mean Gaussian perturbations is implemented in algorithm 20.5.

Grammatical evolution uses two additional genetic operators. The first, *gene duplication*, occurs naturally as an error in DNA replication and repair. Gene duplication can allow new genetic material to be generated and can store a second copy of a useful gene to reduce the chance of a lethal mutation removing the gene from the gene pool. Gene duplication chooses a random interval of genes in the

[4] C. Ryan, J. J. Collins, and M. O. Neill, "Grammatical Evolution: Evolving Programs for an Arbitrary Language," in *European Conference on Genetic Programming*, 1998.

[5] Our serial DNA is read and used to construct complicated protein structures. DNA is often referred to as the genotype—the object on which genetic operations are performed. The protein structure is the phenotype—the object encoded by the genotype whose performance is evaluated. The grammatical evolution literature often refers to the integer design vector as the genotype and the resulting expression as the phenotype.

[6] We use $x \bmod_1 n$ to refer to the 1-index modulus:

$$((x - 1) \bmod n) + 1$$

This type of modulus is useful with 1-based indexing. The corresponding Julia function is mod1.

[7] No genetic information is read when there is only a single applicable rule.

```
struct DecodedExpression
    node
    n_rules_applied
end
function decode(x, grammar, sym, c_max=1000, c=0)
    node, c = _decode(x, grammar, sym, c_max, c)
    DecodedExpression(node, c)
end
function _decode(x, grammar, typ, c_max, c)
    types = grammar[typ]
    if length(types) > 1
        g = x[mod1(c+=1, length(x))]
        rule = types[mod1(g, length(types))]
    else
        rule = types[1]
    end
    node = RuleNode(rule)
    childtypes = child_types(grammar, node)
    if !isempty(childtypes) && c < c_max
        for ctyp in childtypes
            cnode, c = _decode(x, grammar, ctyp, c_max, c)
            push!(node.children, cnode)
        end
    end
    return (node, c)
end
```

Algorithm 20.4. A method for decoding an integer design vector to produce an expression, where x is a vector of integers, grammar is a Grammar, and sym is the root symbol. The counter c is used during the recursion process and the parameter c_max is an upper limit on the maximum number of rule applications, to prevent an infinite loop. The method returns a DecodedExpression, which contains the expression tree and the number of rules applied during the decoding process.

Consider a grammar for real-valued strings:

$$\mathbb{R} \mapsto \mathbb{D}\,\mathbb{D}'\,\mathbb{P}\,\mathbb{E}$$

$$\mathbb{D}' \mapsto \mathbb{D}\,\mathbb{D}' \mid \epsilon$$

$$\mathbb{P} \mapsto .\,\mathbb{D}\,\mathbb{D}' \mid \epsilon$$

$$\mathbb{E} \mapsto \mathbb{E}\,\mathbb{S}\,\mathbb{D}\,\mathbb{D}' \mid \epsilon$$

$$\mathbb{S} \mapsto + \mid - \mid \epsilon$$

$$\mathbb{D} \mapsto 0 \mid 1 \mid 2 \mid 3 \mid 4 \mid 5 \mid 6 \mid 7 \mid 8 \mid 9$$

where $\mathbb{R}$ is a real value, $\mathbb{D}$ is a terminal decimal, $\mathbb{D}'$ is a nonterminal decimal, $\mathbb{P}$ is the decimal part, $\mathbb{E}$ is the exponent, and $\mathbb{S}$ is the sign. Any ϵ values produce empty strings.

Suppose our design is $[205, 52, 4, 27, 10, 59, 6]$ and we have the starting symbol $\mathbb{R}$. There is only one applicable rule, so we do not use any genetic information and we replace $\mathbb{R}$ with $\mathbb{D}\mathbb{D}'\mathbb{P}\mathbb{E}$.

Next we must replace $\mathbb{D}$. There are 10 options. We select $205 \bmod_1 10 = 5$, and thus obtain $4\mathbb{D}'\mathbb{P}\mathbb{E}$

Next we replace $\mathbb{D}'$, which has two options. We select index $52 \bmod_1 2 = 2$, which corresponds to ϵ.

Continuing in this manner we produce the string *4E+8*.

The above grammar can be implemented in ExprRules using:

```
grammar = @grammar begin
    R  =  D * De * P * E
    De =  D * De | ""
    P  =  "." * D * De | ""
    E  =  "E" * S * D * De | ""
    S  =  "+" | "-" | ""
    D  =  "0"|"1"|"2"|"3"|"4"|"5"|"6"|"7"|"8"|"9"
end
```

and can be evaluated using:

```
x = [205, 52, 4, 27, 10, 59, 6]
str = eval(decode(x, grammar, :R).node, grammar)
```

Example 20.5. The process by which an integer design vector in grammatical evolution is decoded into an expression.

Our implementation is depth-first. If 52 were instead 51, the rule $\mathbb{D}' \mapsto \mathbb{D}\,\mathbb{D}'$ would be applied, followed by selecting a rule for the new $\mathbb{D}$, eventually resulting in *43950.950E+8*.

```
struct IntegerGaussianMutation <: MutationMethod
    σ
end
function mutate(M::IntegerGaussianMutation, child)
    return child + round.(Int, randn(length(child)).*M.σ)
end
```

Algorithm 20.5. The Gaussian mutation method modified to preserve integer values for integer-valued chromosomes. Each value is perturbed by a zero-mean Gaussian random value with standard deviation σ and then rounded to the nearest integer.

chromosome to duplicate. A copy of the selected interval is appended to the back of the chromosome. Duplication is implemented in algorithm 20.6.

```
struct GeneDuplication <: MutationMethod
end
function mutate(M::GeneDuplication, child)
    n = length(child)
    i, j = rand(1:n), rand(1:n)
    interval = min(i,j) : max(i,j)
    return vcat(child, deepcopy(child[interval]))
end
```

Algorithm 20.6. The gene duplication method used in grammatical evolution.

The second genetic operation, *pruning*, tackles a problem encountered during crossover. As illustrated in figure 20.7, crossover will select a crossover point at random in each chromosome and construct a new chromosome using the left side of the first and the right side of the second chromosome. Unlike genetic algorithms, the trailing entries in chromosomes of grammatical evolution may not be used; during parsing, once the tree is complete, the remaining entries are ignored. The more unused entries, the more likely it is that the crossover point lies in the inactive region, thus not providing new beneficial material. An individual is pruned with a specified probability, and, if pruned, its chromosome is truncated such that only active genes are retained. Pruning is implemented in algorithm 20.7 and is visualized in figure 20.8.

Like genetic programming, grammatical evolution can use the genetic algorithm method.[8] We can construct a MutationMethod that applies multiple mutation methods in order to use pruning, duplication, and standard mutation approaches in tandem. Such a method is implemented in algorithm 20.8.

Grammatical evolution suffers from two primary drawbacks. First, it is difficult to tell whether the chromosome is feasible without decoding it into an expression.

[8] Genotype to phenotype mapping would occur in the objective function.

Figure 20.7. Crossover applied to chromosomes in grammatical evolution may not affect the active genes in the front of the chromosome. The child shown here inherits all of the active shaded genes from parent *a* so it will effectively act as an identical expression. Pruning was developed to overcome this issue.

```
struct GenePruning <: MutationMethod
    p
    grammar
    typ
end
function mutate(M::GenePruning, child)
    if rand() < M.p
        c = decode(child, M.grammar, M.typ).n_rules_applied
        if c < length(child)
            child = child[1:c]
        end
    end
    return child
end
```

Algorithm 20.7. The gene pruning method used in grammatical evolution.

Figure 20.8. Pruning truncates the chromosome such that only active genes remain.

```
struct MultiMutate <: MutationMethod
    Ms
end
function mutate(M::MultiMutate, child)
    for m in M.Ms
        child = mutate(m, child)
    end
    return child
end
```

Algorithm 20.8. A MutationMethod for applying all mutation methods stored in the vector Ms.

Second, a small change in the chromosome may produce a large change in the corresponding expression.

20.4 Probabilistic Grammars

A *probabilistic grammar*[9] adds a weight to each rule in the genetic program's grammar. When sampling from all applicable rules for a given node, we select a rule stochastically according to the relative weights. The probability of an expression is the product of the probabilities of sampling each rule. Algorithm 20.9 implements the probability calculation. Example 20.6 demonstrates sampling an expression from a probabilistic grammar and computes its likelihood.

[9] T. L. Booth and R. A. Thompson, "Applying Probability Measures to Abstract Languages," *IEEE Transactions on Computers*, vol. C-22, no. 5, pp. 442–450, 1973.

```
struct ProbabilisticGrammar
    grammar
    ws
end
function probability(probgram, node)
    typ = return_type(probgram.grammar, node)
    i = findfirst(isequal(node.ind), probgram.grammar[typ])
    prob = probgram.ws[typ][i] / sum(probgram.ws[typ])
    for (i,c) in enumerate(node.children)
        prob *= probability(probgram, c)
    end
    return prob
end
```

Algorithm 20.9. A method for computing the probability of an expression based on a probabilistic grammar, where probgram is a probabilistic grammar consisting of a grammar grammar and a mapping of types to weights for all applicable rules ws, and node is a RuleNode expression.

Optimization using a probabilistic grammar improves its weights with each iteration using elite samples from a population. At each iteration, a population of expressions is sampled and their objective function values are computed. Some number of the best expressions are considered the elite samples and can be used to update the weights. A new set of weights is generated for the probabilistic grammar, where the weight $w_i^\mathbb{T}$ for the ith production rule applicable to return type $\mathbb{T}$ is set to the number of times the production rule was used in generating the elite samples. This update procedure is implemented in algorithm 20.10.

The probabilistic grammars above can be extended to more complicated probability distributions that consider other factors, such as the depth in the expression or local dependencies among siblings in subtrees. One approach is to use Bayesian networks.[10]

[10] P. K. Wong, L. Y. Lo, M. L. Wong, and K. S. Leung, "Grammar-Based Genetic Programming with Bayesian Network," in *IEEE Congress on Evolutionary Computation (CEC)*, 2014.

Consider a probabilistic grammar for strings composed entirely of "a"s:

$$\begin{aligned}
\mathbb{A} &\mapsto \mathtt{a}\,\mathbb{A} & w_1^{\mathbb{A}} &= 1 \\
&\mapsto \mathtt{a}\,\mathbb{B}\,\mathtt{a}\,\mathbb{A} & w_2^{\mathbb{A}} &= 3 \\
&\mapsto \epsilon & w_3^{\mathbb{A}} &= 2 \\
\mathbb{B} &\mapsto \mathtt{a}\,\mathbb{B} & w_1^{\mathbb{B}} &= 4 \\
&\mapsto \epsilon & w_1^{\mathbb{B}} &= 1
\end{aligned}$$

where we have a set of weights $\mathbf{w}$ for each parent type and ϵ is an empty string.

Suppose we generate an expression starting with the type $\mathbb{A}$. The probability distribution over the three possible rules is:

$$\begin{aligned}
P(\mathbb{A} \mapsto \mathtt{a}\,\mathbb{A}) &= 1/(1+3+2) &= 1/6 \\
P(\mathbb{A} \mapsto \mathtt{a}\,\mathbb{B}\,\mathtt{a}\,\mathbb{A}) &= 3/(1+3+2) &= 1/2 \\
P(\mathbb{A} \mapsto \epsilon) &= 2/(1+3+2) &= 1/3
\end{aligned}$$

Suppose we sample the second rule and obtain a $\mathbb{B}$ a $\mathbb{A}$.

Next we sample a rule to apply to $\mathbb{B}$. The probability distribution over the two possible rules is:

$$\begin{aligned}
P(\mathbb{B} \mapsto \mathtt{a}\,\mathbb{B}) &= 4/(4+1) &= 4/5 \\
P(\mathbb{B} \mapsto \epsilon) &= 1/(4+1) &= 1/5
\end{aligned}$$

Suppose we sample the second rule and obtain a ϵ a $\mathbb{A}$.

Next we sample a rule to apply to $\mathbb{A}$. Suppose we sample $\mathbb{A} \mapsto \epsilon$ to obtain a ϵ a ϵ, which produces the "a"-string "aa". The probability of the sequence of rules applied to produce "aa" under the probabilistic grammar is:

$$P(\mathbb{A} \mapsto \mathtt{a}\,\mathbb{B}\,\mathtt{a}\,\mathbb{A})P(\mathbb{B} \mapsto \epsilon)P(\mathbb{A} \mapsto \epsilon) = \frac{1}{2} \cdot \frac{1}{5} \cdot \frac{1}{3} = \frac{1}{30}$$

Note that this is not the same as the probability of obtaining "aa", as other sequences of production rules could also have produced it.

Example 20.6. Sampling an expression from a probabilistic grammar and computing the expression's likelihood.

```
function _update!(probgram, x)
    grammar = probgram.grammar
    typ = return_type(grammar, x)
    i = findfirst(isequal(x.ind), grammar[typ])
    probgram.ws[typ][i] += 1
    for c in x.children
        _update!(probgram, c)
    end
    return probgram
end
function update!(probgram, Xs)
    for w in values(probgram.ws)
        fill!(w,0)
    end
    for x in Xs
        _update!(probgram, x)
    end
    return probgram
end
```

Algorithm 20.10. A method for applying a learning update to a probabilistic grammar probgram based on an elite sample of expressions Xs.

20.5   Probabilistic Prototype Trees

The *probabilistic prototype tree*[11] is a different approach that learns a distribution for every node in the expression tree. Each node in a probabilistic prototype tree contains a probability vector representing a categorical distribution over the grammar's production rules. The probability vectors are updated to reflect knowledge gained from successive generations of expressions. The maximum number of children for a node is the maximum number of nonterminals among rules in the grammar.[12]

Probability vectors are randomly initialized when a node is created. Random probability vectors can be drawn from a *Dirichlet distribution*.[13] The original implementation initializes terminals to a scalar value of 0.6 and nonterminals to 0.4. In order to handle strongly-typed grammars we maintain a probability vector for applicable rules to each parent type. Algorithm 20.11 defines a node type and implements this initialization method.

Expressions are sampled using the probability vectors in the probabilistic prototype tree. A rule in a node is drawn from the categorical distribution defined by the node's probability vector for the required return type, normalizing the

[11] R. Salustowicz and J. Schmidhuber, "Probabilistic Incremental Program Evolution," *Evolutionary Computation*, vol. 5, no. 2, pp. 123–141, 1997.

[12] The *arity* of a function is the number of arguments. The arity of a grammar rule, which can be viewed as a function, is the number of nonterminals in the rule.

[13] The Dirichlet distribution is often used to represent a distribution over discrete distributions. D. Barber, *Bayesian Reasoning and Machine Learning*. Cambridge University Press, 2012.

```
struct PPTNode
    ps
    children
end
function PPTNode(grammar;
    w_terminal = 0.6,
    w_nonterm = 1-w_terminal,
    )

    ps = Dict(typ => normalize!([isterminal(grammar, i) ?
                    w_terminal : w_nonterm
                    for i in grammar[typ]], 1)
            for typ in nonterminals(grammar))
    PPTNode(ps, PPTNode[])
end
function get_child(ppt::PPTNode, grammar, i)
    if i > length(ppt.children)
        push!(ppt.children, PPTNode(grammar))
    end
    return ppt.children[i]
end
```

Algorithm 20.11. A probabilistic prototype tree node type and associated initialization function where ps is a dictionary mapping a symbol corresponding to a return type to a probability vector over applicable rules, and children is a list of PPTNodes. The method get_child will automatically expand the tree when attempting to access a nonexistent child.

associated probability vector values to obtain a valid probability distribution. The tree is traversed in depth-first order. This sampling procedure is implemented in algorithm 20.12 and visualized in figure 20.9.

Learning can use information either from an entire sampled population or from elite samples. Let the best expression in the current generation be x_{best} and the best expression found so far be x_{elite}. The node probabilities are updated to increase the likelihood of generating x_{best}.[14]

The probability of generating x_{best} is the product of the probabilities of choosing each rule in x_{best} when traversing through the probabilistic prototype tree. We compute a target probability for $P(x_{\text{best}})$:

$$P_{\text{target}} = P(x_{\text{best}}) + (1 - P(x_{\text{best}})) \cdot \alpha \cdot \frac{\epsilon - y_{\text{elite}}}{\epsilon - y_{\text{best}}} \tag{20.1}$$

[14] The original probabilistic prototype tree implementation will periodically increase the likelihood of generating x_{elite}.

where α and ϵ are positive constants. The fraction on the right-hand side produces larger steps toward expressions with better objective function values. The target probability can be calculated using algorithm 20.13.

```
function rand(ppt, grammar, typ)
    rules = grammar[typ]
    rule_index = sample(rules, Weights(ppt.ps[typ]))
    ctypes = child_types(grammar, rule_index)

    arr = Vector{RuleNode}(undef, length(ctypes))
    node = iseval(grammar, rule_index) ?
        RuleNode(rule_index, eval(grammar, rule_index), arr) :
        RuleNode(rule_index, arr)

    for (i,typ) in enumerate(ctypes)
        node.children[i] =
            rand(get_child(ppt, grammar, i), grammar, typ)
    end
    return node
end
```

Algorithm 20.12. A method for sampling an expression from a probabilistic prototype tree. The tree is expanded as needed.

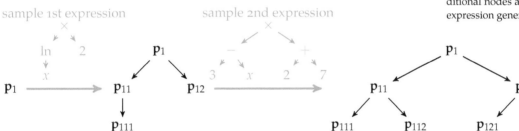

Figure 20.9. A probabilistic prototype tree initially contains only the root node but expands as additional nodes are needed during expression generation.

```
function probability(ppt, grammar, expr)
    typ = return_type(grammar, expr)
    i = findfirst(isequal(expr.ind), grammar[typ])
    p = ppt.ps[typ][i]
    for (i,c) in enumerate(expr.children)
        p *= probability(get_child(ppt, grammar, i), grammar, c)
    end
    return p
end
function p_target(ppt, grammar, x_best, y_best, y_elite, α, ϵ)
    p_best = probability(ppt, grammar, x_best)
    return p_best + (1-p_best)*α*(ϵ - y_elite)/(ϵ - y_best)
end
```

Algorithm 20.13. Methods for computing the probability of an expression and the target probability, where ppt is the root node of the probabilistic prototype tree, grammar is the grammar, expr and x_best are RuleNode expressions, y_best and y_elite are scalar objective function values, and α and ϵ are scalar parameters.

The target probability is used to adjust the probability vectors in the probabilistic prototype tree. The probabilities associated with the chosen nodes are increased iteratively until the target probability is exceeded:

$$P\left(x_{\text{best}}^{(i)}\right) \leftarrow P\left(x_{\text{best}}^{(i)}\right) + c \cdot \alpha \cdot \left(1 - P\left(x_{\text{best}}^{(i)}\right)\right) \text{ for all } i \text{ in } \{1, 2, \ldots\} \quad (20.2)$$

where $x_{\text{best}}^{(i)}$ is the ith rule applied in expression x_{best} and c is a scalar.[15]

15 A recommended value is $c = 0.1$.

The adapted probability vectors are then renormalized to 1 by downscaling the values of all nonincreased vector components proportionally to their current value. The probability vector $\mathbf{p}$, where the ith component was increased, is adjusted according to:

$$p_j \leftarrow p_j \frac{1 - p_i}{\|\mathbf{p}\|_1 - p_i} \text{ for } j \neq i \quad (20.3)$$

The learning update is implemented in algorithm 20.14.

```
function _update!(ppt, grammar, x, c, α)
    typ = return_type(grammar, x)
    i = findfirst(isequal(x.ind), grammar[typ])
    p = ppt.ps[typ]
    p[i] += c*α*(1-p[i])
    psum = sum(p)
    for j in 1 : length(p)
        if j != i
            p[j] *= (1- p[i])/(psum - p[i])
        end
    end
    for (pptchild,xchild) in zip(ppt.children, x.children)
        _update!(pptchild, grammar, xchild, c, α)
    end
    return ppt
end
function update!(ppt, grammar, x_best, y_best, y_elite, α, c, ϵ)
    p_targ = p_target(ppt, grammar, x_best, y_best, y_elite, α, ϵ)
    while probability(ppt, grammar, x_best) < p_targ
        _update!(ppt, grammar, x_best, c, α)
    end
    return ppt
end
```

Algorithm 20.14. A method for applying a learning update to a probabilistic prototype tree with root ppt, grammar grammar, best expression x_best with objective function value y_best, elite objective function value y_elite, learning rate α, learning rate multiplier c, and parameter ϵ.

In addition to population-based learning, probabilistic prototype trees can also explore the design space via mutations. The tree is mutated to explore the region around $\mathbf{x}_{\text{best}}$. Let $\mathbf{p}$ be a probability vector in a node that was accessed when generating $\mathbf{x}_{\text{best}}$. Each component in $\mathbf{p}$ is mutated with a probability proportional to the problem size:

$$\frac{p_{\text{mutation}}}{\#\mathbf{p}\sqrt{\#\mathbf{x}_{\text{best}}}} \tag{20.4}$$

where p_{mutation} is a mutation parameter, $\#\mathbf{p}$ is the number of components in $\mathbf{p}$, and $\#\mathbf{x}_{\text{best}}$ is the number of rules applied in $\mathbf{x}_{\text{best}}$. A component i, selected for mutation, is adjusted according to:

$$p_i \leftarrow p_i + \beta \cdot (1 - p_i) \tag{20.5}$$

where β controls the amount of mutation. Small probabilities undergo larger mutations than do larger probabilities. All mutated probability vectors must be renormalized. Mutation is implemented in algorithm 20.15 and visualized in figure 20.10.

```
function mutate!(ppt, grammar, x_best, p_mutation, β;
    sqrtlen = sqrt(length(x_best)),
    )
    typ = return_type(grammar, x_best)
    p = ppt.ps[typ]
    prob = p_mutation/(length(p)*sqrtlen)
    for i in 1 : length(p)
        if rand() < prob
            p[i] += β*(1-p[i])
        end
    end
    normalize!(p, 1)
    for (pptchild,xchild) in zip(ppt.children, x_best.children)
        mutate!(pptchild, grammar, xchild, p_mutation, β,
                sqrtlen=sqrtlen)
    end
    return ppt
end
```

Algorithm 20.15. A method for mutating a probabilistic prototype tree with root ppt, grammar grammar, best expression x_best, mutation parameter p_mutation, and mutation rate β.

Finally, subtrees in the probabilistic prototype tree are pruned in order to remove stale portions of the tree. A child node is removed if its parent contains a probability component above a specified threshold such that, when chosen, causes the child to be irrelevant. This is always the case for a terminal and may

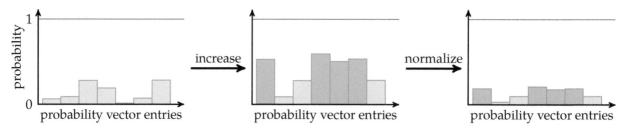

Figure 20.10. Mutating a probability vector in a probabilistic prototype tree with $\beta = 0.5$. The mutated components were increased according to equation (20.5) and the resulting probability vector was renormalized. Notice how smaller probabilities receive greater increases.

be the case for a nonterminal. Pruning is implemented in algorithm 20.16 and demonstrated in example 20.7.

20.6   Summary

- Expression optimization allows for optimizing tree structures that, under a grammar, can express sophisticated programs, structures, and other designs lacking a fixed size.

- Grammars define the rules used to construct expressions.

- Genetic programming adapts genetic algorithms to perform mutation and crossover on expression trees.

- Grammatical evolution operates on an integer array that can be decoded into an expression tree.

- Probabilistic grammars learn which rules are best to generate, and probabilistic prototype trees learn probabilities for every iteration of the expression rule generation process.

```
function prune!(ppt, grammar; p_threshold=0.99)
    kmax, pmax = :None, 0.0
    for (k, p) in ppt.ps
        pmax´ = maximum(p)
        if pmax´ > pmax
            kmax, pmax = k, pmax´
        end
    end
    if pmax > p_threshold
        i = argmax(ppt.ps[kmax])
        if isterminal(grammar, i)
            empty!(ppt.children[kmax])
        else
            max_arity_for_rule = maximum(nchildren(grammar, r) for
                                         r in grammar[kmax])
            while length(ppt.children) > max_arity_for_rule
                pop!(ppt.children)
            end
        end
    end
    return ppt
end
```

Algorithm 20.16. A method for pruning a probabilistic prototype tree with root ppt, grammar grammar, and pruning probability threshold p_treshold.

Consider a node with a probability vector over the rule set:

$$\mathbb{R} \mapsto \mathbb{R} + \mathbb{R}$$
$$\mathbb{R} \mapsto \ln(\mathbb{R})$$
$$\mathbb{R} \mapsto 2 \mid x$$
$$\mathbb{R} \mapsto \mathbb{S}$$

If the probability of selecting 2 or x grows large, then any children in the probabilistic prototype tree are unlikely to be needed and can be pruned. Similarly, if the probability of choosing $\mathbb{S}$ grows large, any children with return type $\mathbb{R}$ are unneeded and can be pruned.

Example 20.7. An example of when pruning for probabilistic prototype trees should be applied.

20.7   Exercises

Exercise 20.1. How many expression trees can be generated using the following grammar and the starting set $\{\mathbb{R}, \mathbb{I}, \mathbb{F}\}$?

$$\mathbb{R} \mapsto \mathbb{I} \mid \mathbb{F}$$
$$\mathbb{I} \mapsto 1 \mid 2$$
$$\mathbb{F} \mapsto \pi$$

Exercise 20.2. The number of expression trees up to height h that can be generated under a grammar grows super-exponentially. As a reference, calculate the number of expressions of height h can be generated using the grammar:[16]

$$\mathbb{N} \mapsto \{\mathbb{N}, \mathbb{N}\} \mid \{\mathbb{N}, \} \mid \{, \mathbb{N}\} \mid \{\} \tag{20.6}$$

[16] Let an empty expression have height 0, the expression $\{\}$ have height 1, and so on.

Exercise 20.3. Define a grammar which can generate any nonnegative integer.

Exercise 20.4. How do expression optimization methods handle divide-by-zero values or other exceptions encountered when generating random subtrees?

Exercise 20.5. Consider an arithmetic grammar such as:

$$\mathbb{R} \mapsto x \mid y \mid z \mid \mathbb{R} + \mathbb{R} \mid \mathbb{R} - \mathbb{R} \mid \mathbb{R} \times \mathbb{R} \mid \mathbb{R}/\mathbb{R} \mid \ln \mathbb{R} \mid \sin \mathbb{R} \mid \cos \mathbb{R}$$

Suppose the variables x, y, and z each have units, and the output is expected to be in particular units. How might such a grammar be modified to respect units?

Exercise 20.6. Consider the grammar

$$\mathbb{S} \mapsto \mathbb{NP} \; \mathbb{VP}$$
$$\mathbb{NP} \mapsto \mathbb{ADJ} \; \mathbb{NP} \mid \mathbb{ADJ} \; \mathbb{N}$$
$$\mathbb{VP} \mapsto \mathbb{V} \; \mathbb{ADV}$$
$$\mathbb{ADJ} \mapsto a \mid the \mid big \mid little \mid blue \mid red$$
$$\mathbb{N} \mapsto mouse \mid cat \mid dog \mid pony$$
$$\mathbb{V} \mapsto ran \mid sat \mid slept \mid ate$$
$$\mathbb{ADV} \mapsto quietly \mid quickly \mid soundly \mid happily$$

What is the phenotype corresponding to the genotype $[2, 10, 19, 0, 6]$ and the starting symbol $\mathbb{S}$?

Exercise 20.7. Use genetic programming to evolve the gear ratios for a clock. Assume all gears are restricted to have radii selected from $\mathcal{R} = \{10, 25, 30, 50, 60, 100\}$. Each gear can either be attached to its parent's axle, thereby sharing the same rotation period, or be interlocked on its parent's rim, thereby having a rotation period depending on the parent's rotation period and on the gear ratio as shown in figure 20.11.

The clock can also contain hands, which are mounted on the axle of a parent gear. Assume the root gear turns with a period of $t_{\text{root}} = 0.1\,\text{s}$ and has a radius of 25. The objective is to produce a clock with a second, minute, and hour hand.

Score each individual according to:

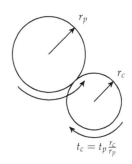

Figure 20.11. The rotation period t_c of a child gear attached to the rim of a parent gear depends on the rotation period of the parent gear, t_p and the ratio of the gears' radii.

$$\left(\underset{\text{hands}}{\text{minimize}}\,(1 - t_{\text{hand}})^2\right) + \left(\underset{\text{hands}}{\text{minimize}}\,(60 - t_{\text{hand}})^2\right) + \left(\underset{\text{hands}}{\text{minimize}}\,(3600 - t_{\text{hand}})^2\right) + \#\text{nodes} \cdot 10^{-3}$$

where t_{hand} is the rotation period of a particular hand in seconds and #nodes is the number of nodes in the expression tree. Ignore rotation direction.

Exercise 20.8. The four 4s puzzle[17] is a mathematical challenge in which we use four 4 digits and mathematical operations to generate expressions for each of the integers from 0 to 100. For example, the first two integers can be produced by $4 + 4 - 4 - 4$ and $44/44$, respectively. Complete the four 4s puzzle.

[17] W. W. R. Ball, *Mathematical Recreations and Essays*. Macmillan, 1892.

Exercise 20.9. Consider the probabilistic grammar

$$\mathbb{R} \mapsto \mathbb{R} + \mathbb{R} \mid \mathbb{R} \times \mathbb{R} \mid \mathbb{F} \mid \mathbb{I} \quad w_{\mathbb{R}} = [1, 1, 5, 5]$$
$$\mathbb{F} \mapsto 1.5 \mid \infty \qquad\qquad\qquad p_{\mathbb{F}} = [4, 3]$$
$$\mathbb{I} \mapsto 1 \mid 2 \mid 3 \qquad\qquad\qquad p_{\mathbb{I}} = [1, 1, 1]$$

What is the generation probability of the expression $1.5 + 2$?

Exercise 20.10. What is the probabilistic grammar from the previous question after clearing the counts and applying a learning update on $1.5 + 2$?

21 *Multidisciplinary Optimization*

Multidisciplinary design optimization (MDO) involves solving optimization problems spanning across disciplines. Many real-world problems involve complicated interactions between several disciplines, and optimizing disciplines individually may not lead to an optimal solution. This chapter discusses a variety of techniques for taking advantage of the structure of MDO problems to reduce the effort required for finding good designs.[1]

21.1 *Disciplinary Analyses*

There are many different *disciplinary analyses* that might factor into a design. For example, the design of a rocket might involve analysis from disciplines such as structures, aerodynamics, and controls. The different disciplines have their own analytical tools, such as finite element analysis. Often these disciplinary analyses tend to be quite sophisticated and computationally expensive. In addition, disciplinary analyses are often tightly coupled with each other, where one discipline may require the output of another's disciplinary analysis. Resolving these interdependencies can be nontrivial.

In an MDO setting, we still have a set of design variables as before, but we also keep track of the outputs, or *response variables*, of each disciplinary analysis.[2] We write the response variables of the ith disciplinary analysis as $\mathbf{y}^{(i)}$. In general, the ith disciplinary analysis F_i can depend on the design variables or the response variables from any other discipline:

$$\mathbf{y}^{(i)} \leftarrow F_i\left(\mathbf{x}, \mathbf{y}^{(1)}, \ldots, \mathbf{y}^{(i-1)}, \mathbf{y}^{(i+1)}, \ldots, \mathbf{y}^{(m)}\right) \tag{21.1}$$

where m is the total number of disciplines. The inputs to a computational fluid dynamics analysis for an aircraft may include the deflections of the wing, which

[1] An extensive survey is provided by J. R. R. A. Martins and A. B. Lambe, "Multidisciplinary Design Optimization: A Survey of Architectures," *AIAA Journal*, vol. 51, no. 9, pp. 2049–2075, 2013. Further discussion can be found in J. Sobieszczanski-Sobieski, A. Morris, and M. van Tooren, *Multidisciplinary Design Optimization Supported by Knowledge Based Engineering*. Wiley, 2015. See also N. M. Alexandrov and M. Y. Hussaini, eds., *Multidisciplinary Design Optimization: State of the Art*. SIAM, 1997.

[2] A disciplinary analysis can provide inputs for other disciplines, the objective function, or the constraints. In addition, it can also provide gradient information for the optimizer.

come from a structural analysis that requires the forces from computational fluid dynamics. An important part of formulating MDO problems is taking into account such dependencies between analyses.

In order to make reasoning about disciplinary analyses easier, we introduce the concept of an *assignment*. An assignment $\mathcal{A}$ is a set of variable names and their corresponding values relevant to a multidisciplinary design optimization problem. To access a variable v, we write $\mathcal{A}[v]$.

A disciplinary analysis is a function that takes an assignment and uses the design point and response variables from other analyses to overwrite the response variable for its discipline:

$$\mathcal{A}' \leftarrow F_i(\mathcal{A}) \tag{21.2}$$

where $F_i(\mathcal{A})$ updates $\mathbf{y}^{(i)}$ in $\mathcal{A}$ to produce $\mathcal{A}'$.

Assignments can be represented in code using dictionaries.[3] Each variable is assigned a name of type `String`. Variables are not restricted to floating-point numbers but can include other objects, such as vectors. Example 21.1 shows an implementation using a dictionary.

[3] A *dictionary*, also called an *associative array*, is a common data structure that allows indexing by keys rather than integers. See appendix A.1.7.

Consider an optimization with one design variable x and two disciplines. Suppose the first disciplinary analysis F_1 computes a response variable $y^{(1)} = f_1(x, y^{(2)})$ and the second disciplinary analysis F_2 computes a response variable $y^{(2)} = f_2(x, y^{(1)})$.

This problem can be implemented as:

```
function F1(A)
    A["y1"] = f1(A["x"], A["y2"])
    return A
end
function F2(A)
    A["y2"] = f2(A["x"], A["y1"])
    return A
end
```

The assignment may be initialized with guesses for $y^{(1)}$ and $y^{(2)}$, and a known input for x. For example:

```
A = Dict("x"=>1, "y1"=>2, "y2"=>3)
```

Example 21.1. Basic code syntax for the assignment-based representation of multidisciplinary design optimization problems.

21.2 *Interdisciplinary Compatibility*

Evaluating the objective function value and feasibility of a design point $\mathbf{x}$ requires obtaining values for the response variables that satisfy *interdisciplinary compatibility*, which means that the response variables must be consistent with the disciplinary analyses. Interdisciplinary compatibility holds for a particular assignment if the assignment is unchanged under all disciplinary analyses:

$$F_i(\mathcal{A}) = \mathcal{A} \text{ for } i \in \{1, \ldots, m\} \tag{21.3}$$

Running any analysis will produce the same values. Finding an assignment that satisfies interdisciplinary compatibility is called a *multidisciplinary analysis*.

System optimization for a single discipline requires an optimizer to select design variables and query the disciplinary analysis in order to evaluate the constraints and the objective function, as shown in figure 21.1. Single-discipline optimization does not require that we consider disciplinary coupling.

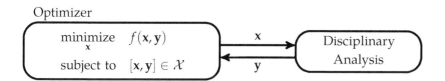

Figure 21.1. Optimization diagram for a single discipline. Gradients may or may not be computed.

System optimization for multiple disciplines can introduce dependencies, in which case coupling becomes an issue. A diagram for two coupled disciplines is given in figure 21.2. Applying conventional optimization to this problem is less straightforward because interdisciplinary compatibility must be established.

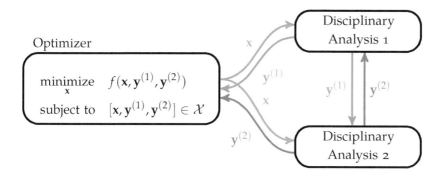

Figure 21.2. Optimization diagram for a two-discipline analysis with interdisciplinary coupling.

If a multidisciplinary analysis does not have a *dependency cycle*,[4] then solving for interdisciplinary compatibility is straightforward. We say discipline i depends on discipline j if i requires any of j's outputs. This dependency relation can be used to form a *dependency graph*, where each node corresponds to a discipline and an edge $j \rightarrow i$ is included if discipline i depends on j. Figure 21.3 shows examples of dependency graphs involving two disciplines with and without cycles.

[4] A dependency cycle arises when disciplines depend on each other.

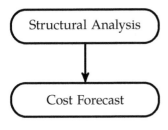

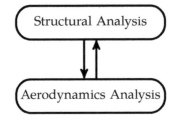

Figure 21.3. Cyclic and acyclic dependency graphs.

An acyclic dependency graph. An evaluation ordering can be specified such that the required inputs for each discipline are available from previously evaluated disciplines.

A cyclic dependency graph. The structural analysis depends on the aerodynamics analysis and vice versa.

If the dependency graph has no cycles, then there always exists an order of evaluation that, if followed, ensures that the necessary disciplinary analyses are evaluated before the disciplinary analyses that depend on them. Such an ordering is called a *topological ordering* and can be found using a topological sorting method such as *Kahn's algorithm*.[5] The reordering of analyses is illustrated in figure 21.4.

If the dependency graph has cycles, then no topological ordering exists. To address cycles, we can use the *Gauss-Seidel method* (algorithm 21.1), which attempts to resolve the multidisciplinary analysis by iterating until convergence.[6] The Gauss-Seidel algorithm is sensitive to the ordering of the disciplines as illustrated by example 21.2. A poor ordering can prevent or slow convergence. The best orderings are those with minimal feedback connections.[7]

It can be advantageous to merge disciplines into a new disciplinary analysis—to group conceptually related analyses, simultaneously evaluate tightly coupled analyses, or more efficiently apply some of the architectures discussed in this chapter. Disciplinary analyses can be merged to form a new analysis whose response variables consist of the response variables of the merged disciplines. The form of the new analysis depends on the disciplinary interdependencies. If

[5] A. B. Kahn, "Topological Sorting of Large Networks," *Communications of the ACM*, vol. 5, no. 11, pp. 558–562, 1962.

[6] The Gauss-Seidel algorithm can also be written to execute analyses in parallel.

[7] In some cases, disciplines can be separated into different clusters that are independent of each other. Each connected cluster can be solved using its own, smaller multidisciplinary analysis.

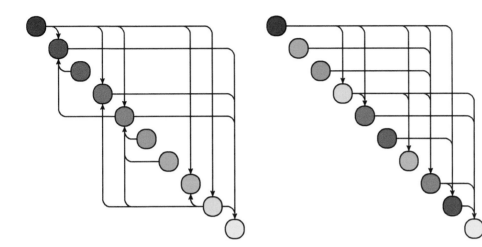

Figure 21.4. A topological sort can be used to reorder the disciplinary analyses to remove feedback connections.

```
function gauss_seidel!(Fs, A; k_max=100, ε=1e-4)
    k, converged = 0, false
    while !converged && k ≤ k_max
        k += 1
        A_old = deepcopy(A)
        for F in Fs
            F(A)
        end
        converged = all(isapprox(A[v], A_old[v], rtol=ε)
                        for v in keys(A))
    end
    return (A, converged)
end
```

Algorithm 21.1. The Gauss-Seidel algorithm for conducting a multidisciplinary analysis. Here, Fs is a vector of disciplinary analysis functions that take and modify an assignment, A. There are two optional arguments: the maximum number of iterations k_max and the relative error tolerance ε. The method returns the modified assignment and whether it converged.

Consider a multidisciplinary design optimization problem with one design variable x and three disciplines, each with one response variable:

$$y^{(1)} \leftarrow F_1(x, y^{(2)}, y^{(3)}) = y^{(2)} - x$$
$$y^{(2)} \leftarrow F_2(x, y^{(1)}, y^{(3)}) = \sin(y^{(1)} + y^{(3)})$$
$$y^{(3)} \leftarrow F_3(x, y^{(1)}, y^{(2)}) = \cos(x + y^{(1)} + y^{(2)})$$

The disciplinary analyses can be implemented as:

```
function F1(A)
    A["y1"] = A["y2"] - A["x"]
    return A
end
function F2(A)
    A["y2"] = sin(A["y1"] + A["y3"])
    return A
end
function F3(A)
    A["y3"] = cos(A["x"] + A["y2"] + A["y1"])
    return A
end
```

Consider running a multidisciplinary analysis for $x = 1$, having initialized our assignment with all 1's:

```
A = Dict("x"=>1, "y1"=>1, "y2"=>1, "y3"=>1)
```

Running the Gauss-Seidel algorithm with the ordering F_1, F_2, F_3 converges, but running with F_1, F_3, F_2 does not.

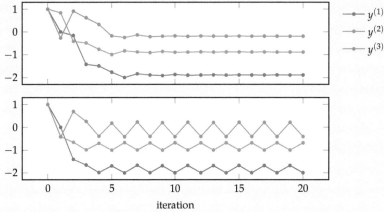

Example 21.2. An example that illustrates the importance of choosing an appropriate ordering when running a multidisciplinary analysis.

the merged disciplines are acyclic then an ordering exists in which the analyses can be serially executed. If the merged disciplines are cyclic, then the new analysis must internally run a multidisciplinary analysis to achieve compatibility.

21.3 Architectures

Multidisciplinary design optimization problems can be written:

$$
\begin{aligned}
&\underset{\mathbf{x}}{\text{minimize}} && f(\mathcal{A}) \\
&\text{subject to} && \mathcal{A} \in \mathcal{X} \\
& && F_i(\mathcal{A}) = \mathcal{A} \text{ for each discipline } i \in \{1, \ldots, m\}
\end{aligned}
\tag{21.4}
$$

where the objective function f and feasible set $\mathcal{X}$ depend on both the design and response variables. The design variables in the assignment are specified by the optimizer. The condition $F_i(\mathcal{A}) = \mathcal{A}$ ensures that the ith discipline is consistent with the values in $\mathcal{A}$. This last condition enforces interdisciplinary compatibility.

There are several challenges associated with optimizing multidisciplinary problems. The interdependence of disciplinary analyses causes the ordering of analyses to matter and often makes parallelization difficult or impossible. There is a trade-off between an optimizer that directly controls all variables and incorporating suboptimizers[8] that leverage discipline-specific expertise to optimize values locally. In addition, there is a trade-off between the expense of running disciplinary analyses and the expense of globally optimizing too many variables. Finally, every architecture must enforce interdisciplinary compatibility in the final solution.

The remainder of this chapter discusses a variety of different optimization architectures for addressing these challenges. These architectures are demonstrated using the hypothetical ride-sharing problem introduced in example 21.3.

[8] A *suboptimizer* is an optimization routine called within another optimization routine.

21.4 Multidisciplinary Design Feasible

The *multidisciplinary design feasible* architecture structures the MDO problem such that standard optimization algorithms can be directly applied to optimize the design variables. A multidisciplinary design analysis is run for any given design point to obtain compatible response values.

Consider a ride-sharing company developing a self-driving fleet. This hypothetical company is simultaneously designing the vehicle, its sensor package, a routing strategy, and a pricing scheme. These portions of the design are referred to as **v**, **s**, **r**, and **p**, respectively, each of which contains numerous design variables. The vehicle, for instance, may include parameters governing the structural geometry, engine and drive train, battery capacity, and passenger capacity.

The objective of the ride-sharing company is to maximize profit. The profit depends on a large-scale simulation of the routing algorithm and passenger demand, which, in turn, depends on response variables from an autonomy analysis of the vehicle and its sensor package. Several analyses extract additional information. The performance of the routing algorithm depends on the demand generated by the pricing scheme and the demand generated by the pricing scheme depends on performance of the routing algorithm. The vehicle range and fuel efficiency depends on the weight, drag, and power consumption of the sensor package. The sensor package requires vehicle geometry and performance information to meet the necessary safety requirements. A dependency diagram is presented below.

Example 21.3. A ride-sharing problem used throughout this chapter to demonstrate optimization architectures.

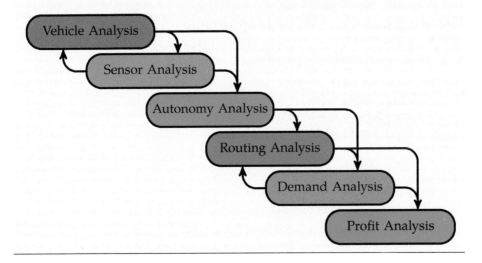

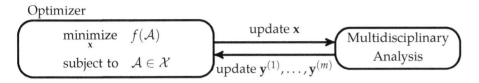

Optimizer

$$\begin{array}{ll} \underset{\mathbf{x}}{\text{minimize}} & f(\mathcal{A}) \\ \text{subject to} & \mathcal{A} \in \mathcal{X} \end{array}$$

update $\mathbf{x}$

update $\mathbf{y}^{(1)}, \ldots, \mathbf{y}^{(m)}$

Multidisciplinary Analysis

Figure 21.5. The multidisciplinary design feasible architecture. The optimizer chooses design points $\mathbf{x}$, and the multidisciplinary analysis computes a consistent assignment $\mathcal{A}$. The structure is similar to that of single-discipline optimization.

$$\begin{array}{ll} \underset{\mathbf{x}}{\text{minimize}} & f\left(\mathbf{x}, \mathbf{y}^{(1)}, \ldots, \mathbf{y}^{(m)}\right) \\ \text{subject to} & \left[\mathbf{x}, \mathbf{y}^{(1)}, \ldots, \mathbf{y}^{(m)}\right] \in \mathcal{X} \end{array} \longrightarrow \begin{array}{ll} \underset{\mathbf{x}}{\text{minimize}} & f(\text{MDA}(\mathbf{x})) \\ \text{subject to} & \text{MDA}(\mathbf{x}) \in \mathcal{X} \end{array}$$

Figure 21.6. Formulating an MDO problem into a typical optimization problem using multidisciplinary design analyses, where $\text{MDA}(\mathbf{x})$ returns a multidisciplinary compatible assignment.

An architecture diagram is given in figure 21.5. It consists of two blocks, the optimizer and the multidisciplinary analysis. The optimizer is the method used for selecting design points with the goal of minimizing the objective function. The optimizer calls the multidisciplinary analysis block by passing it a design point $\mathbf{x}$ and receives a compatible assignment $\mathcal{A}$. If interdisciplinary compatibility is not possible, the multidisciplinary analysis block informs the optimizer and such design points are treated as infeasible. Figure 21.6 shows how an MDO problem can be transformed into a typical optimization problem using multidisciplinary design analyses.

The primary advantages of the multidisciplinary design feasible architecture are its conceptual simplicity and that it is guaranteed to maintain interdisciplinary compatibility at each step in the optimization. Its name reflects the fact that multidisciplinary design analyses are run at every design evaluation, ensuring that the system-level optimizer only considers feasible designs.

The primary disadvantage is that multidisciplinary design analyses are expensive to run, typically requiring several iterations over all analyses. Iterative Gauss-Seidel methods may be slow to converge or may not converge at all, depending on the initialization of the response variables and the ordering of the disciplinary analyses.

Lumping the analyses together makes it necessary for all local variables—typically only relevant to a particular discipline—to be considered by the analysis as a whole. Many practical problems have a very large number of local design variables, such as mesh control points in aerodynamics, element dimensions in structures, component placements in electrical engineering, and neural network weights in machine learning. Multidisciplinary design feasible optimization requires that the system optimizer specify all of these values across all disciplines while satisfying all constraints.

The multidisciplinary design feasible architecture is applied to the ride-sharing problem in example 21.4.

The multidisciplinary design feasible architecture can be applied to the ride-sharing problem. The architectural diagram is shown below.

Optimizer

$$\underset{\mathbf{v},\mathbf{s},\mathbf{r},\mathbf{p}}{\text{minimize}} \quad f\left(\mathbf{v},\mathbf{s},\mathbf{r},\mathbf{p},\mathbf{y}^{(v)},\mathbf{y}^{(s)},\mathbf{y}^{(a)},\mathbf{y}^{(r)},\mathbf{y}^{(d)},\mathbf{y}^{(p)}\right)$$

$$\text{subject to} \quad \left[\mathbf{v},\mathbf{s},\mathbf{r},\mathbf{p},\mathbf{y}^{(v)},\mathbf{y}^{(s)},\mathbf{y}^{(a)},\mathbf{y}^{(r)},\mathbf{y}^{(d)},\mathbf{y}^{(p)}\right] \in \mathcal{X}$$

update $\mathbf{y}^{(v)},\mathbf{y}^{(s)},\mathbf{y}^{(a)},\mathbf{y}^{(r)},\mathbf{y}^{(d)},\mathbf{y}^{(p)}$ ⬆ ⬇ update $\mathbf{v},\mathbf{s},\mathbf{r},\mathbf{p}$

Multidisciplinary Analysis

Example 21.4. The multidisciplinary design feasible architecture applied to the ride-sharing problem. A multidisciplinary analysis over all response variables must be completed for every candidate design point. This tends to be very computationally intensive.

21.5 Sequential Optimization

The *sequential optimization* architecture (figure 21.7) is an architecture that can leverage discipline-specific tools and experience to optimize subproblems but can lead to suboptimal solutions. This architecture is included to demonstrate the limitations of a naive approach and to serve as a baseline against which other architectures can be compared.

A *subproblem* is an optimization procedure conducted at every iteration of an overarching optimization process. Sometimes design variables can be removed from the outer optimization procedure, the *system-level optimizer*, and can be more efficiently optimized in subproblems.

The design variables for the ith discipline can be partitioned according to $\mathbf{x}^{(i)} = [\mathbf{x}_g^{(i)},\mathbf{x}_\ell^{(i)}]$, where $\mathbf{x}_g^{(i)}$ are *global design variables* shared with other disciplines and $\mathbf{x}_\ell^{(i)}$ are *local design variables* used only by the associated disciplinary subproblem.[9] The response variables can be similarly partitioned into the global response variables $\mathbf{y}_g^{(i)}$ and the local response variables $\mathbf{y}_\ell^{(i)}$. Disciplinary autonomy is

[9] The vehicle subproblem in the ride-sharing problem may include global design variables such as the vehicle capacity and range that affect other disciplines but may also include local design variables such as the seating configuration that do not impact other disciplines.

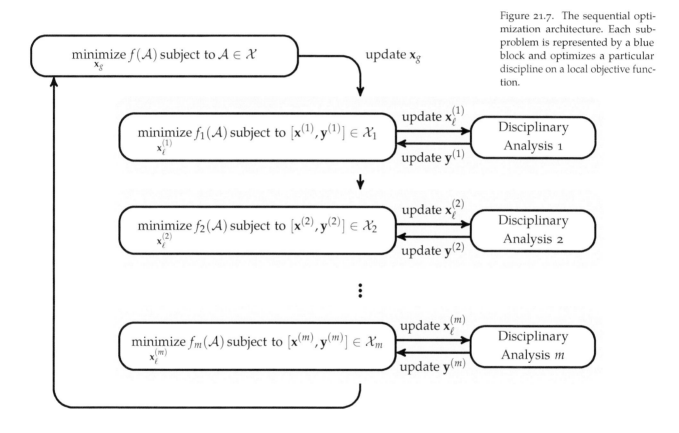

Figure 21.7. The sequential optimization architecture. Each sub-problem is represented by a blue block and optimizes a particular discipline on a local objective function.

achieved by optimizing the local variables in their own disciplinary optimizers. A local objective function f_i must be chosen such that optimizing it also benefits the global objective. A top-level optimizer is responsible for optimizing the global design variables $\mathbf{x}_g$ with respect to the original objective function. An instantiation of $\mathbf{x}_g$ is evaluated through sequential optimizations; each subproblem is optimized one after the other, passing its results to the next subproblem until all have been evaluated.

Sequential optimization takes advantage of the locality of disciplines; that many variables are unique to a particular discipline and do not need to be shared across discipline boundaries. Sequential optimization harnesses each discipline's proficiency at solving its discipline-specific problem. The subproblem optimizers have complete control over their discipline-specific design variables to meet local design objectives and constraints.

Except in special cases, sequential optimization does not lead to an optimal solution of the original problem for the same reason that Gauss-Seidel is not guaranteed to converge. The solution is sensitive to the local objective functions, and finding suitable local objective functions is often a challenge. Sequential optimization does not support parallel execution, and interdisciplinary compatibility is enforced through iteration and does not always converge.

Example 21.5 applies sequential optimization to the ride-sharing problem.

21.6   Individual Discipline Feasible

The *individual discipline feasible* (IDF) architecture removes the need to run expensive multidisciplinary design analyses and allows disciplinary analyses to be executed in parallel. It loses the guarantee that interdisciplinary compatibility is maintained throughout its execution, with eventual agreement enforced through equality constraints in the optimizer. Compatibility is not enforced in multidisciplinary analyses but rather by the optimizer itself.

IDF introduces *coupling variables* to the design space. For each discipline, an additional vector $\mathbf{c}^{(i)}$ is added to the optimization problem to act as aliases for the response variables $\mathbf{y}^{(i)}$. The response variables are unknown until they are computed by their respective domain analyses; inclusion of the coupling variables allows the optimizer to provide these estimates to multiple disciplines simultaneously when running analyses in parallel. Equality between the cou-

The sequential optimization architecture can optimize some variables locally. Figure 21.8 shows the result of applying sequential optimization to the ride-sharing problem.

The design variables for the vehicle, sensor system, routing algorithm, and pricing scheme are split into local discipline-specific variables and top-level global variables. For example, the vehicle subproblem can optimize local vehicle parameters $\mathbf{v}_\ell$ such as drive train components, whereas parameters like vehicle capacity that are used by other analyses are controlled globally in $\mathbf{v}_g$.

The tight coupling between the vehicle and sensor systems is poorly handled by the sequential optimization architecture. While changes made by the vehicle subproblem are immediately addressed by the sensor subproblem, the effect of the sensor subproblem on the vehicle subproblem is not addressed until the next iteration.

Not all analyses require their own subproblems. The profit analysis is assumed not to have any local design variables and can thus be executed without needing a subproblem block.

Example 21.5. Sequential optimization for the ride-sharing problem.

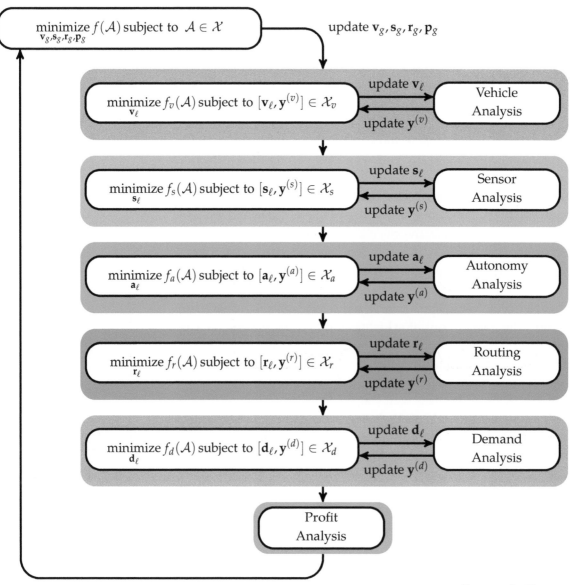

Figure 21.8. The sequential optimization architecture applied to the ride-sharing problem.

pling and response variables is typically reached through iteration. Equality is an optimization constraint, $\mathbf{c}^{(i)} = \mathbf{y}^{(i)}$, for each discipline.

Figure 21.9 shows the general IDF architecture. The system-level optimizer operates on the coupling variables and uses these to populate an assignment that is copied to the disciplinary analysis in each iteration:

$$\mathcal{A}[\mathbf{x}, \mathbf{y}^{(1)}, \ldots, \mathbf{y}^{(m)}] \leftarrow [\mathbf{x}, \mathbf{c}^{(1)}, \ldots, \mathbf{c}^{(m)}] \qquad (21.5)$$

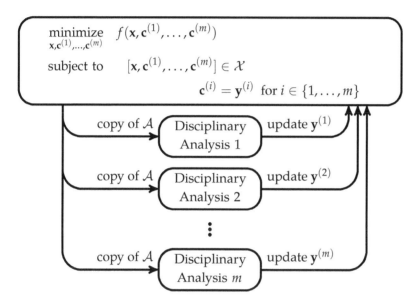

Figure 21.9. The individual discipline feasible architecture allows disciplinary analyses to be run in parallel. This chapter assumes that disciplinary analyses mutate their inputs, so copies of the system level optimizer's assignment are passed to each disciplinary analysis.

Despite the added freedom to execute analyses in parallel, IDF suffers from the shortcoming that it cannot leverage domain-specific optimization procedures in the same way as sequential optimization as optimization is top-level only. Furthermore, the optimizer must satisfy additional equality constraints and has more variables to optimize. IDF can have difficulties with gradient-based optimization since the chosen search direction must take constraints into account as shown in figure 21.10. Changes in the design variables must not cause the coupling variables to become infeasible with respect to the disciplinary analyses. Evaluating the gradients of the objective and constraint function is very costly when the disciplinary analyses are expensive.

The individual discipline feasible architecture is applied to the ride-sharing problem in figure 21.11.

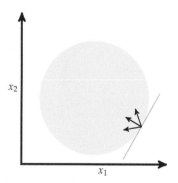

Figure 21.10. The search direction for a point on a constraint boundary must lead into the feasible set.

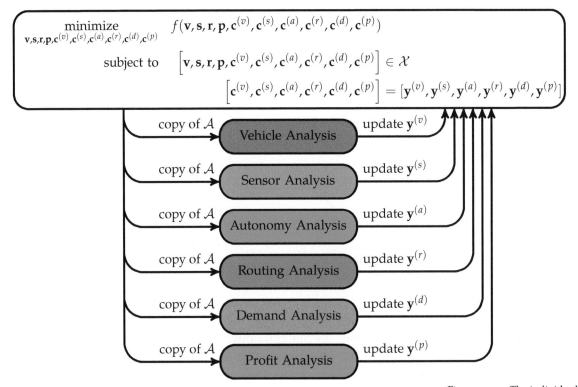

$$\underset{\mathbf{v},\mathbf{s},\mathbf{r},\mathbf{p},\mathbf{c}^{(v)},\mathbf{c}^{(s)},\mathbf{c}^{(a)},\mathbf{c}^{(r)},\mathbf{c}^{(d)},\mathbf{c}^{(p)}}{\text{minimize}} \quad f\left(\mathbf{v},\mathbf{s},\mathbf{r},\mathbf{p},\mathbf{c}^{(v)},\mathbf{c}^{(s)},\mathbf{c}^{(a)},\mathbf{c}^{(r)},\mathbf{c}^{(d)},\mathbf{c}^{(p)}\right)$$

$$\text{subject to} \quad \left[\mathbf{v},\mathbf{s},\mathbf{r},\mathbf{p},\mathbf{c}^{(v)},\mathbf{c}^{(s)},\mathbf{c}^{(a)},\mathbf{c}^{(r)},\mathbf{c}^{(d)},\mathbf{c}^{(p)}\right] \in \mathcal{X}$$

$$\left[\mathbf{c}^{(v)},\mathbf{c}^{(s)},\mathbf{c}^{(a)},\mathbf{c}^{(r)},\mathbf{c}^{(d)},\mathbf{c}^{(p)}\right] = \left[\mathbf{y}^{(v)},\mathbf{y}^{(s)},\mathbf{y}^{(a)},\mathbf{y}^{(r)},\mathbf{y}^{(d)},\mathbf{y}^{(p)}\right]$$

copy of $\mathcal{A}$ → Vehicle Analysis → update $\mathbf{y}^{(v)}$

copy of $\mathcal{A}$ → Sensor Analysis → update $\mathbf{y}^{(s)}$

copy of $\mathcal{A}$ → Autonomy Analysis → update $\mathbf{y}^{(a)}$

copy of $\mathcal{A}$ → Routing Analysis → update $\mathbf{y}^{(r)}$

copy of $\mathcal{A}$ → Demand Analysis → update $\mathbf{y}^{(d)}$

copy of $\mathcal{A}$ → Profit Analysis → update $\mathbf{y}^{(p)}$

Figure 21.11. The individual discipline feasible architecture applied to the ride-sharing problem. The individual design feasible architecture allows for parallel execution of analyses, but the system-level optimizer must optimize a large number of variables.

21.7 Collaborative Optimization

The *collaborative optimization* architecture (figure 21.12) breaks a problem into disciplinary subproblems that have full control over their local design variables and discipline-specific constraints. Subproblems can be solved using discipline-specific tools and can be optimized in parallel.

The *i*th subproblem has the form:

$$
\begin{aligned}
\underset{\mathbf{x}^{(i)}}{\text{minimize}} \quad & f_i(\mathbf{x}^{(i)}, \mathbf{y}^{(i)}) \\
\text{subject to} \quad & [\mathbf{x}^{(i)}, \mathbf{y}^{(i)}] \in \mathcal{X}_i
\end{aligned}
\tag{21.6}
$$

with $\mathbf{x}^{(i)}$ containing a subset of the design variables $\mathbf{x}$ and response variables $\mathbf{y}^{(i)}$. The constraint ensures that the solution satisfies discipline-specific constraints.

Interdisciplinary compatibility requires that the global variables $\mathbf{x}_g^{(i)}$ and $\mathbf{y}_g^{(i)}$ agree between all disciplines. We define a set of coupling variables $\mathcal{A}_g$ that includes variables corresponding to all design and response variables that are global in at least one subproblem. Agreement is enforced by constraining each $\mathbf{x}_g^{(i)}$ and $\mathbf{y}_g^{(i)}$ to match its corresponding coupling variables:

$$
\mathbf{x}_g^{(i)} = \mathcal{A}_g[\mathbf{x}_g^{(i)}] \quad \text{and} \quad \mathbf{y}_g^{(i)} = \mathcal{A}_g[\mathbf{y}_g^{(i)}]
\tag{21.7}
$$

where $\mathcal{A}_g[\mathbf{x}_g^{(i)}]$ and $\mathcal{A}_g[\mathbf{y}_g^{(i)}]$ are the coupling variables corresponding to the global design and response variables in the *i*th discipline. This constraint is enforced using the subproblem objective function:

$$
f_i = \left\| \mathbf{x}_g^{(i)} - \mathcal{A}_g[\mathbf{x}_g^{(i)}] \right\|_2^2 + \left\| \mathbf{y}_g^{(i)} - \mathcal{A}_g[\mathbf{y}_g^{(i)}] \right\|_2^2
\tag{21.8}
$$

Each subproblem thus seeks feasible solutions that minimally deviate from the coupling variables.

The subproblems are managed by a system-level optimizer that is responsible for optimizing the coupling variables $\mathcal{A}_g$ to minimize the objective function. Evaluating an instance of the coupling variables requires running each disciplinary subproblem, typically in parallel.

Disciplinary subproblems may deviate from the coupling variables during the optimization process. This discrepancy occurs when two or more disciplines disagree on a variable or when subproblem constraints prevent matching the target values set by the system-level optimizer. The top-level constraint that $f_i = 0$ for each discipline ensures that coupling is eventually attained.

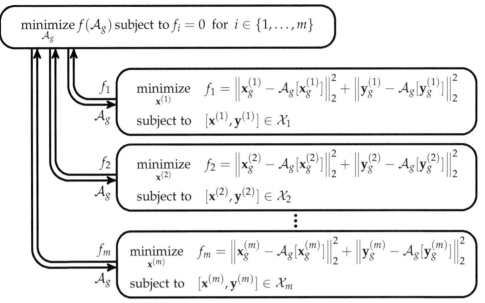

Figure 21.12. Design architecture for collaborative optimization.

The primary advantages of collaborative optimization stem from its ability to isolate some design variables into disciplinary subproblems. Collaborative optimization is readily applicable to real-world multidisciplinary problem solving, as each discipline is typically well segregated, and thus largely unaffected by small decisions made in other disciplines. The decentralized formulation allows traditional discipline optimization methods to be applied, allowing problem designers to leverage existing tools and methodologies.

Collaborative optimization requires optimizing over the coupling variables, which includes both design and response variables. Collaborative optimization does not perform well in problems with high coupling because the additional coupling variables can outweigh the benefits of local optimization.

Collaborative optimization is a *distributed architecture* that decomposes a single optimization problem into a smaller set of optimization problems that have the same solution when their solutions are combined. Distributed architectures have the advantage of reduced solving times, as subproblems can be optimized in parallel.

Collaborative optimization is applied to the ride-sharing problem in example 21.6.

The collaborative optimization architecture can be applied to the vehicle routing problem by producing six different disciplinary subproblems. Unfortunately, having six different subproblems requires any variables shared across disciplines to be optimized at the global level.

Figure 21.13 shows two disciplinary subproblems obtained by grouping the vehicle, sensor, and autonomy disciplines into a transport subproblem and the routing, demand, and profit disciplines into a network subproblem. The disciplines grouped into each subproblem are tightly coupled. Having only two subproblems significantly reduces the number of global variables considered by the system-level optimizer because presumably very few design variables are directly used by both the transport and network subproblems.

The subproblems are each multidisciplinary optimization problems, themselves amenable to optimization using the techniques covered in this chapter. We can, for example, use sequential optimization within the transport subproblem. We can also add another instance of collaborative optimization within the network subproblem.

Example 21.6. Applying collaborative optimization to the ride-sharing problem.

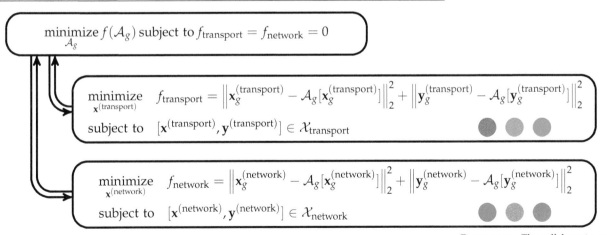

Figure 21.13. The collaborative optimization architecture applied to the ride-sharing problem. The colored circles correspond to the disciplinary analyses contained within each subproblem.

21.8   Simultaneous Analysis and Design

The *simultaneous analysis and design* (SAND) architecture avoids the central challenge of coordinating between multiple disciplinary analyses by having the optimizer conduct the analyses. Instead of running an analysis $F_i(\mathcal{A})$ to obtain a residual, SAND optimizes both the design and response variables subject to a constraint $F_i(\mathcal{A}) = \mathcal{A}$. The optimizer is responsible for simultaneously optimizing the design variables and finding the corresponding response variables.

Any disciplinary analysis can be transformed into *residual form*. The residual $r_i(\mathcal{A})$ is used to indicate whether an assignment $\mathcal{A}$ is compatible with the ith discipline. If $F_i(\mathcal{A}) = \mathcal{A}$, then $r_i(\mathcal{A}) = 0$; otherwise, $r_i(\mathcal{A}) \neq 0$. We can obtain a residual form using the disciplinary analysis:

$$r_i(\mathcal{A}) = \left\| F_i(\mathcal{A}) - \mathcal{A}[\mathbf{y}^{(i)}] \right\| \tag{21.9}$$

though this is typically inefficient, as demonstrated in example 21.7.

Consider a disciplinary analysis that solves the equation $\mathbf{A}\mathbf{y} = \mathbf{x}$. The analysis is $F(\mathbf{x}) = \mathbf{A}^{-1}\mathbf{x}$, which requires an expensive matrix inversion. We can construct a residual form using equation (21.9):

$$r_1(\mathbf{x}, \mathbf{y}) = \|F(\mathbf{x}) - \mathbf{y}\| = \left\| \mathbf{A}^{-1}\mathbf{x} - \mathbf{y} \right\|$$

Alternatively, we can use the original constraint to construct a more efficient residual form:

$$r_2(\mathbf{x}, \mathbf{y}) = \|\mathbf{A}\mathbf{y} - \mathbf{x}\|$$

Example 21.7. Evaluating a disciplinary analysis in a residual is typically counter-productive. The analysis must typically perform additional work to solve the problem whereas a cleaner residual form can more efficiently verify whether the inputs are compatible.

The residual form of a discipline consists of the set of disciplinary equations that are solved by the disciplinary analysis.[10] It is often much easier to evaluate a residual than to run a disciplinary analysis. In SAND, figure 21.14, the analysis effort is the responsibility of the optimizer.

[10] In aerodynamics, these may include the Navier-Stokes equations. In structural engineering, these may include the elasticity equations. In electrical engineering, these may include the differential equations for current flow.

$$\boxed{\text{minimize}_{\mathcal{A}} \, f(\mathcal{A}) \text{ subject to } \mathcal{A} \in \mathcal{X}, \; r_i(\mathcal{A}) = 0 \text{ for each discipline}}$$

Figure 21.14. Simultaneous analysis and design places the entire burden on the optimizer. It uses disciplinary residuals rather than disciplinary analyses.

SAND can explore regions of the design space that are infeasible with respect to the residual equations, as shown in figure 21.15. Exploring infeasible regions can allow us to traverse the design space more easily and find solutions in feasible regions disconnected from the feasible region of the starting design point. SAND suffers from having to simultaneously optimize a very large number of variables for which derivatives and other discipline-specific expertise are not available. Furthermore, SAND gains much of its value from residuals that can be computed more efficiently than can disciplinary analyses. Use of SAND in real-world applications is often limited by the inability to modify existing disciplinary analysis code to produce an efficient residual form.

SAND is applied to the ride-sharing problem in example 21.8.

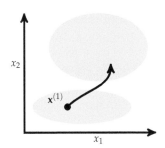

Figure 21.15. SAND can explore regions of the design space that are infeasible and potentially bridge the gap between feasible subsets.

Example 21.8. The simultaneous analysis and design architecture applied to the ride-sharing problem.

Applying the simultaneous analysis and design architecture to the ride-sharing problem requires disciplinary residuals. These can potentially depend on all design and response variables. The architecture requires that the optimizer optimize all of the design variables and all of the response variables.

$$
\begin{aligned}
\underset{\mathbf{v},\mathbf{s},\mathbf{r},\mathbf{p},\mathbf{y}^{(v)},\mathbf{y}^{(s)},\mathbf{y}^{(a)},\mathbf{y}^{(r)},\mathbf{y}^{(d)},\mathbf{y}^{(p)}}{\text{minimize}} \quad & f\left(\mathbf{v},\mathbf{s},\mathbf{r},\mathbf{p},\mathbf{y}^{(v)},\mathbf{y}^{(s)},\mathbf{y}^{(a)},\mathbf{y}^{(r)},\mathbf{y}^{(d)},\mathbf{y}^{(p)}\right) \\
\text{subject to} \quad & \left[\mathbf{v},\mathbf{s},\mathbf{r},\mathbf{p},\mathbf{y}^{(v)},\mathbf{y}^{(s)},\mathbf{y}^{(a)},\mathbf{y}^{(r)},\mathbf{y}^{(d)},\mathbf{y}^{(p)}\right] \in \mathcal{X} \\
& r_v(\mathbf{v},\mathbf{s},\mathbf{r},\mathbf{p},\mathbf{y}^{(v)},\mathbf{y}^{(s)},\mathbf{y}^{(a)},\mathbf{y}^{(r)},\mathbf{y}^{(d)},\mathbf{y}^{(p)}) = 0 \\
& r_s(\mathbf{v},\mathbf{s},\mathbf{r},\mathbf{p},\mathbf{y}^{(v)},\mathbf{y}^{(s)},\mathbf{y}^{(a)},\mathbf{y}^{(r)},\mathbf{y}^{(d)},\mathbf{y}^{(p)}) = 0 \\
& r_a(\mathbf{v},\mathbf{s},\mathbf{r},\mathbf{p},\mathbf{y}^{(v)},\mathbf{y}^{(s)},\mathbf{y}^{(a)},\mathbf{y}^{(r)},\mathbf{y}^{(d)},\mathbf{y}^{(p)}) = 0 \\
& r_r(\mathbf{v},\mathbf{s},\mathbf{r},\mathbf{p},\mathbf{y}^{(v)},\mathbf{y}^{(s)},\mathbf{y}^{(a)},\mathbf{y}^{(r)},\mathbf{y}^{(d)},\mathbf{y}^{(p)}) = 0 \\
& r_d(\mathbf{v},\mathbf{s},\mathbf{r},\mathbf{p},\mathbf{y}^{(v)},\mathbf{y}^{(s)},\mathbf{y}^{(a)},\mathbf{y}^{(r)},\mathbf{y}^{(d)},\mathbf{y}^{(p)}) = 0 \\
& r_p(\mathbf{v},\mathbf{s},\mathbf{r},\mathbf{p},\mathbf{y}^{(v)},\mathbf{y}^{(s)},\mathbf{y}^{(a)},\mathbf{y}^{(r)},\mathbf{y}^{(d)},\mathbf{y}^{(p)}) = 0
\end{aligned}
$$

21.9 Summary

- Multidisciplinary design optimization requires reasoning about multiple disciplines and achieving agreement between coupled variables.

- Disciplinary analyses can often be ordered to minimize dependency cycles.

- Multidisciplinary design problems can be structured in different architectures that take advantage of problem features to improve the optimization process.

- The multidisciplinary design feasible architecture maintains feasibility and compatibility through the use of slow and potentially nonconvergent multidisciplinary design analyses.

- Sequential optimization allows each discipline to optimize its discipline-specific variables but does not always yield optimal designs.

- The individual discipline feasible architecture allows parallel execution of analyses at the expense of adding coupling variables to the global optimizer.

- Collaborative optimization incorporates suboptimizers that can leverage domain specialization to optimize some variables locally.

- The simultaneous analysis and design architecture replaces design analyses with residuals, allowing the optimizer to find compatible solutions but cannot directly use disciplinary solution techniques.

21.10   *Exercises*

Exercise 21.1. Provide an example of a practical engineering problem that is multidisciplinary.

Exercise 21.2. Provide an abstract example of a multidisciplinary problem where the order of the analyses is important.

Exercise 21.3. What is one advantage of the individual discipline feasible architecture over the multidisciplinary design feasible and sequential optimization architectures?

Exercise 21.4. Consider applying multidisciplinary design analysis to minimizing the weight of a wing whose deformation and loading are computed by separate disciplines. We will use a simplified version of the problem, representing the wing as a horizontally-mounted pendulum supported by a torsional spring.

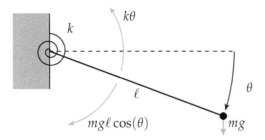

The objective is to minimize the spring stiffness, k, such that the pendulum's displacement does not exceed a target threshold. The pendulum length ℓ, pendulum point mass m, and gravitational constant g are fixed.

We use two simplified analyses in place of more sophisticated analyses used to compute deformations and loadings of aircraft wings. Assuming the pendulum is rigid, the loading moment M is equal to $mg\ell\cos(\theta)$. The torsional spring resists deformation such that the pendulum's angular displacement θ is M/k.

Formulate the spring-pendulum problem under the multidisciplinary design feasible architecture, and then solve it according to that architecture for:

$$m = 1\,\text{kg}, \ell = 1\,\text{m}, g = 9.81\,\text{m}\,\text{s}^{-2}, \theta_{\max} = 10°$$

Exercise 21.5. Formulate the spring-pendulum problem under the individual design feasible architecture.

Exercise 21.6. Formulate the spring-pendulum under the collaborative optimization architecture. Present the two disciplinary optimization problems and the system-level optimization problem.

A *Julia*

Julia is a scientific programming language that is free and open source.[1] It is a relatively new language that borrows inspiration from languages like Python, MATLAB, and R. It was selected for use in this book because it is sufficiently high level[2] so that the algorithms can be compactly expressed and readable while also being fast. This book is compatible with Julia version 1.0. This appendix introduces the necessary concepts for understanding the code included in the text.

[1] Julia may be obtained from http://julialang.org.

[2] In contrast with languages like C++, Julia does not require programmers to worry about memory management and other lower-level details.

A.1 *Types*

Julia has a variety of basic types that can represent data such as truth values, numbers, strings, arrays, tuples, and dictionaries. Users can also define their own types. This section explains how to use some of the basic types and define new types.

A.1.1 *Booleans*

The *Boolean* type in Julia, written `Bool`, includes the values `true` and `false`. We can assign these values to variables. Variable names can be any string of characters, including Unicode, with a few restrictions.

```
done = false
α = false
```

The left-hand side of the equal sign is the variable name, and the right hand side is the value.

We can make assignments in the Julia console. The console will return a response to the expression being evaluated.

```
julia> x = true
true
julia> y = false
false
julia> typeof(x)
Bool
```

The standard Boolean operators are supported.

```
julia> !x      # not
false
julia> x && y # and
false
julia> x || y # or
true
```

The # symbol indicates that the rest of the line is a comment and should not be evaluated.

A.1.2   Numbers

Julia supports integer and floating point numbers as shown here

```
julia> typeof(42)
Int64
julia> typeof(42.0)
Float64
```

Here, Int64 denotes a 64-bit integer, and Float64 denotes a 64-bit floating point value.[3] We can also perform the standard mathematical operations:

[3] On 32-bit machines, an integer literal like 42 is interpreted as an Int32.

```
julia> x = 4
4
julia> y = 2
2
julia> x + y
6
julia> x - y
2
julia> x * y
8
julia> x / y
Error: UndefVarError: Grisu not defined
julia> x ^ y
16
```

```
julia> x % y # x modulo y
0
```

Note that the result of x / y is a `Float64`, even when x and y are integers. We can also perform these operations at the same time as an assignment. For example, x += 1 is shorthand for x = x + 1.

We can also make comparisons:

```
julia> 3 > 4
false
julia> 3 >= 4
false
julia> 3 ≥ 4    # unicode also works
false
julia> 3 < 4
true
julia> 3 <= 4
true
julia> 3 ≤ 4    # unicode also works
true
julia> 3 == 4
false
julia> 3 < 4 < 5
true
```

A.1.3 Strings

A *string* is an array of characters. Strings are not used very much in this textbook except for reporting certain errors. An object of type `String` can be constructed using " characters. For example:

```
julia> x = "optimal"
"optimal"
julia> typeof(x)
String
```

A.1.4 Vectors

A *vector* is a one-dimensional array that stores a sequence of values. We can construct a vector using square brackets, separating elements by commas. Semicolons in these examples suppress the output.

```
julia> x = [];                    # empty vector
julia> x = trues(3);              # Boolean vector containing three trues
julia> x = ones(3);               # vector of three ones
julia> x = zeros(3);              # vector of three zeros
julia> x = rand(3);               # vector of three random numbers between 0 and 1
julia> x = [3, 1, 4];             # vector of integers
julia> x = [3.1415, 1.618, 2.7182]; # vector of floats
```

An *array comprehension* can be used to create vectors. Below, we use the print function so that the output is printed horizontally.[4]

[4] Print statements were used for compactness and are not needed.

```
julia> print([sin(x) for x = 1:5])
[Error: UndefVarError: Grisu not defined
```

We can inspect the type of vectors:

```
julia> typeof([3, 1, 4])              # 1-dimensional array of Int64s
Vector{Int64} (alias for Array{Int64, 1})
julia> typeof([3.1415, 1.618, 2.7182]) # 1-dimensional array of Float64s
Vector{Float64} (alias for Array{Float64, 1})
```

We index into vectors using square brackets.

```
julia> x[1]       # first element is indexed by 1
Error: UndefVarError: Grisu not defined
julia> x[3]       # third element
Error: UndefVarError: Grisu not defined
julia> x[end]     # use end to reference the end of the array
Error: UndefVarError: Grisu not defined
julia> x[end - 1] # this returns the second to last element
Error: UndefVarError: Grisu not defined
```

We can pull out a range of elements from an array. Ranges are specified using a colon notation.

```
julia> x = [1, 1, 2, 3, 5, 8, 13];
julia> print(x[1:3])      # pull out the first three elements
[1, 1, 2]
julia> print(x[1:2:end])  # pull out every other element
[1, 2, 5, 13]
julia> print(x[end:-1:1]) # pull out all the elements in reverse order
[13, 8, 5, 3, 2, 1, 1]
```

We can perform a variety of different operations on arrays. The exclamation mark at the end of function names is often used to indicate that the function mutates (i.e., changes) the input.

```
julia> print([x, x])              # concatenation
[[1, 1, 2, 3, 5, 8, 13], [1, 1, 2, 3, 5, 8, 13]]
julia> length(x)
7
julia> print(push!(x, -1))        # add an element to the end
[1, 1, 2, 3, 5, 8, 13, -1]
julia> pop!(x)                    # remove an element from the end
-1
julia> print(append!(x, [2, 3])) # append y to the end of x
[1, 1, 2, 3, 5, 8, 13, 2, 3]
julia> print(sort!(x))            # sort the elements in the vector
[1, 1, 2, 2, 3, 3, 5, 8, 13]
julia> x[1] = 2; print(x)         # change the first element to 2
[2, 1, 2, 2, 3, 3, 5, 8, 13]
julia> x = [1, 2];
julia> y = [3, 4];
julia> print(x + y)               # add vectors
[4, 6]
julia> print(3x - [1, 2])         # multiply by a scalar and subtract
[2, 4]
julia> print(dot(x, y))           # dot product
11
julia> print(x·y)                 # dot product using unicode character
11
```

It is often useful to apply various functions elementwise to vectors.

```
julia> print(x .* y)    # elementwise multiplication
[3, 8]
julia> print(x .^ 2)    # elementwise squaring
[1, 4]
julia> print(sin.(x))   # elementwise application of sin
[Error: UndefVarError: Grisu not defined
julia> print(sqrt.(x))  # elementwise application of sqrt
[Error: UndefVarError: Grisu not defined
```

A.1.5   Matrices

A *matrix* is a two-dimensional array. Like a vector, it is constructed using square brackets. We use spaces to delimit elements in the same row and semicolons to delimit rows. We can also index into the matrix and output submatrices using ranges.

```
julia> X = [1 2 3; 4 5 6; 7 8 9; 10 11 12];
julia> typeof(X)           # a 2-dimensional array of Int64s
Matrix{Int64} (alias for Array{Int64, 2})
julia> X[2]                # second element using column-major ordering
4
julia> X[3,2]              # element in third row and second column
8
julia> print(X[1,:])       # extract the first row
[1, 2, 3]
julia> print(X[:,2])       # extract the second column
[2, 5, 8, 11]
julia> print(X[:,1:2])     # extract the first two columns
[1 2; 4 5; 7 8; 10 11]
julia> print(X[1:2,1:2])   # extract a 2x2 matrix from the top left of x
[1 2; 4 5]
```

We can also construct a variety of special matrices and use array comprehensions:

```
julia> print(Matrix(1.0I, 3, 3))          # 3x3 identity matrix
[Error: UndefVarError: Grisu not defined
julia> print(Matrix(Diagonal([3, 2, 1]))) # 3x3 diagonal matrix with 3, 2, 1 on diagonal
[3 0 0; 0 2 0; 0 0 1]
julia> print(rand(3,2))                    # 3x2 random matrix
[Error: UndefVarError: Grisu not defined
julia> print(zeros(3,2))                   # 3x2 matrix of zeros
[Error: UndefVarError: Grisu not defined
julia> print([sin(x + y) for x = 1:3, y = 1:2]) # array comprehension
[Error: UndefVarError: Grisu not defined
```

Matrix operations include the following:

```
julia> print(X')          # complex conjugate transpose
[1 4 7 10; 2 5 8 11; 3 6 9 12]
julia> print(3X .+ 2)  # multiplying by scalar and adding scalar
[5 8 11; 14 17 20; 23 26 29; 32 35 38]
julia> X = [1 3; 3 1]; # create an invertible matrix
julia> print(inv(X))    # inversion
[Error: UndefVarError: Grisu not defined
julia> det(X)           # determinant
Error: UndefVarError: Grisu not defined
julia> print([X X])      # horizontal concatenation
[1 3 1 3; 3 1 3 1]
julia> print([X; X])     # vertical concatenation
[1 3; 3 1; 1 3; 3 1]
```

```
julia> print(sin.(X))  # elementwise application of sin
[Error: UndefVarError: Grisu not defined
```

A.1.6 Tuples

A *tuple* is an ordered list of values, potentially of different types. They are constructed with parentheses. They are similar to arrays, but they cannot be mutated.

```
julia> x = (1,) # a single element tuple indicated by the trailing comma
(1,)
julia> x = (1, 0, [1, 2],  2.5029, 4.6692) # third element is a vector
Error: UndefVarError: Grisu not defined
julia> x[2]
0
julia> x[end]
Error: UndefVarError: Grisu not defined
julia> x[4:end]
Error: UndefVarError: Grisu not defined
julia> length(x)
5
```

A.1.7 Dictionaries

A *dictionary* is a collection of key-value pairs. Key-value pairs are indicated with a double arrow operator. We can index into a dictionary using square brackets as with arrays and tuples.

```
julia> x = Dict(); # empty dictionary
julia> x[3] = 4 # associate value 4 with key 3
4
julia> x = Dict(3=>4, 5=>1) # create a dictionary with two key-value pairs
Dict{Int64, Int64} with 2 entries:
  5 => 1
  3 => 4
julia> x[5]          # return value associated with key 5
1
julia> haskey(x, 3) # check whether dictionary has key 3
true
julia> haskey(x, 4) # check whether dictionary has key 4
false
```

A.1.8   Composite Types

A *composite type* is a collection of named fields. By default, an instance of a composite type is immutable (i.e., it cannot change). We use the `struct` keyword and then give the new type a name and list the names of the fields.

```
struct A
    a
    b
end
```

Adding the keyword `mutable` makes it so that an instance can change.

```
mutable struct B
    a
    b
end
```

Composite types are constructed using parentheses, between which we pass in values for the different fields. For example,

```
x = A(1.414, 1.732)
```

The double-colon operator can be used to annotate the types for the fields.

```
struct A
    a::Int64
    b::Float64
end
```

This annotation requires that we pass in an `Int64` for the first field and a `Float64` for the second field. For compactness, this text does not use type annotations, but it is at the expense of performance. Type annotations allow Julia to improve runtime performance because the compiler can optimize the underlying code for specific types.

A.1.9   Abstract Types

So far we have discussed *concrete types*, which are types that we can construct. However, concrete types are only part of the type hierarchy. There are also *abstract types*, which are supertypes of concrete types and other abstract types.

We can explore the type hierarchy of the `Float64` type shown in figure A.1 using the `supertype` and `subtypes` functions.

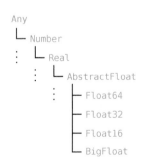

Figure A.1. The type hierarchy for the `Float64` type.

```
julia> supertype(Float64)
AbstractFloat
julia> supertype(AbstractFloat)
Real
julia> supertype(Real)
Number
julia> supertype(Number)
Any
julia> supertype(Any)            # Any is at the top of the hierarchy
Any
julia> subtypes(AbstractFloat) # different types of AbstractFloats
4-element Vector{Any}:
 BigFloat
 Float16
 Float32
 Float64
julia> subtypes(Float64)         # Float64 does not have any subtypes
Type[]
```

We can define our own abstract types.

```
abstract type C end
abstract type D <: C end # D is an abstract subtype of C
struct E <: D # E is composite type that is a subtype of D
    a
end
```

A.1.10 Parametric Types

Julia supports *parametric types*, which are types that take parameters. We have already seen a parametric type with our dictionary example.

```
julia> x = Dict(3=>4, 5=>1)
Dict{Int64, Int64} with 2 entries:
  5 => 1
  3 => 4
```

This constructs a Dict{Int64,Int64}. The parameters to the parametric type are listed within braces and delimited by commas. For the dictionary type, the first parameter specifies the key type, and the second parameter specifies the value type. Julia was able to infer this based on the input, but we could have specified it explicitly.

```
julia> x = Dict{Int64,Int64}(3=>4, 5=>1)
Dict{Int64, Int64} with 2 entries:
  5 => 1
  3 => 4
```

It is possible to define our own parametric types, but we do not do that in this text.

A.2 Functions

A *function* is an object that maps a tuple of argument values to a return value. This section discusses how to define and work with functions.

A.2.1 Named Functions

One way to define a *named function* is to use the `function` keyword, followed by the name of the function and a tuple of names of arguments.

```
function f(x, y)
    return x + y
end
```

We can also define functions compactly using assignment form.

```
julia> f(x, y) = x + y;
julia> f(3, 0.1415)
3.1415
```

A.2.2 Anonymous Functions

An *anonymous function* is not given a name, though it can be assigned to a named variable. One way to define an anonymous function is to use the arrow operator.

```
julia> h = x -> x^2 + 1 # assign anonymous function to a variable
#1 (generic function with 1 method)
julia> g(f, a, b) = [f(a), f(b)]; # applies function f to a and b and returns array
julia> g(h, 5, 10)
2-element Vector{Int64}:
  26
 101
julia> g(x->sin(x)+1, 10, 20)
2-element Vector{Float64}:
 0.4559788891106302
 1.9129452507276277
```

A.2.3   Optional Arguments

We can specify optional arguments by setting default values.

```julia
julia> f(x = 10) = x^2;
julia> f()
100
julia> f(3)
9
julia> f(x, y, z = 1) = x*y + z;
julia> f(1, 2, 3)
5
julia> f(1, 2)
3
```

A.2.4   Keyword Arguments

Functions with keyword arguments are defined using a semicolon.

```julia
julia> f(; x = 0) = x + 1;
julia> f()
1
julia> f(x = 10)
11
julia> f(x, y = 10; z = 2) = (x + y)*z;
julia> f(1)
22
julia> f(2, z = 3)
36
julia> f(2, 3)
10
julia> f(2, 3, z = 1)
5
```

A.2.5   Function Overloading

The types of the arguments passed to a function can be specified using the double colon operator. If multiple functions of the same name are provided, Julia will execute the appropriate function.

```
julia> f(x::Int64) = x + 10;
julia> f(x::Float64) = x + 3.1415;
julia> f(1)
11
julia> f(1.0)
4.141500000000001
julia> f(1.3)
4.4415000000000004
```

The implementation of the most specific function will be used.

```
julia> f(x) = 5;
julia> f(x::Float64) = 3.1415;
julia> f([3, 2, 1])
5
julia> f(0.00787499699)
3.1415
```

A.3 Control Flow

We can control the flow of our programs using conditional evaluation and loops. This section provides some of the syntax used in the book.

A.3.1 Conditional Evaluation

Conditional evaluation will check the value of a Boolean expression and then evaluate the appropriate block of code. One of the most common ways to do this is with an `if` statement.

```
if x < y
    # run this if x < y
elseif x > y
    # run this if x > y
else
    # run this if x == y
end
```

We can also use the *ternary operator* with its question mark and colon syntax. It checks the Boolean expression before the question mark. If the expression evaluates to true, then it returns what comes before the colon; otherwise it returns what comes after the colon.

```
julia> f(x) = x > 0 ? x : 0;
julia> f(-10)
0
julia> f(10)
10
```

A.3.2   Loops

A *loop* allows for repeated evaluation of expressions. One type of loop is the while loop. It repeatedly evaluates a block of expressions until the specified condition after the `while` keyword is met. The following example will sum the values in array x

```
x = [1, 2, 3, 4, 6, 8, 11, 13, 16, 18]
s = 0
while x != []
    s += pop!(x)
end
```

Another type of loop is the for loop. It uses the `for` keyword. The following example will also sum over the values in the array x but will not modify x.

```
x = [1, 2, 3, 4, 6, 8, 11, 13, 16, 18]
s = 0
for i = 1:length(x)
    s += x[i]
end
```

The = can be substituted with `in` or ∈. The following code block is equivalent.

```
x = [1, 2, 3, 4, 6, 8, 11, 13, 16, 18]
s = 0
for y in x
    s += y
end
```

A.4   Packages

A *package* is a collection of Julia code and possibly other external libraries that can be imported to provide additional functionality. Julia has a built-in package manager. A list of registered packages can be found at `https://pkg.julialang.org`. To add a registered package like `Distributions.jl`, we can run:

```
Pkg.add("Distributions")
```

To update packages, we use:

```
Pkg.update()
```

To use a package, we use the keyword `using`:

```
using Distributions
```

Several code blocks in this text specify a package import with `using`. Some code blocks make use of functions that are not explicitly imported. For instance, the `var` function is provided by `Statistics.jl`, and the golden ratio φ is defined in `Base.MathConstants.jl`. Other excluded packages are `InteractiveUtils.jl`, `LinearAlgebra.jl`, `QuadGK.jl`, `Random.jl`, and `StatsBase.jl`.

B   Test Functions

Researchers in optimization use several *test functions* to evaluate optimization algorithms. This section covers several test functions used throughout this book.

B.1   Ackley's Function

Ackley's function (figure B.1) is used to test a method's susceptibility to getting stuck in local minima. It is comprised of two primary components—a sinusoidal component that produces a multitude of local minima and an exponential bell curve centered at the origin, which establishes the function's global minimum.

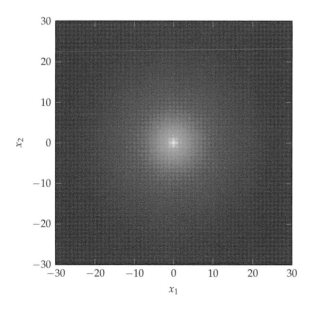

Figure B.1. The two-dimensional version of Ackley's function. The global minimum is at the origin.

Ackley's function is defined for any number of dimensions d:

$$f(\mathbf{x}) = -a\exp\left(-b\sqrt{\frac{1}{d}\sum_{i=1}^{d}x_i^2}\right) - \exp\left(\frac{1}{d}\sum_{i=1}^{d}\cos(cx_i)\right) + a + \exp(1) \quad \text{(B.1)}$$

with a global minimum at the origin with an optimal value of zero. Typically, $a = 20$, $b = 0.2$, and $c = 2\pi$. Ackley's function is implemented in algorithm B.1.

```
function ackley(x, a=20, b=0.2, c=2π)
    d = length(x)
    return -a*exp(-b*sqrt(sum(x.^2)/d)) -
           exp(sum(cos.(c*xi) for xi in x)/d) + a + exp(1)
end
```

Algorithm B.1. Ackley's function with d-dimensional input vector x and three optional parameters.

B.2   Booth's Function

Booth's function (figure B.2) is a two-dimensional quadratic function.

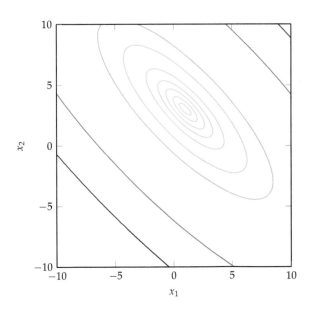

Figure B.2. Booth's function with a global minimum at $[1, 3]$.

Its equation is given by

$$f(\mathbf{x}) = (x_1 + 2x_2 - 7)^2 + (2x_1 + x_2 - 5)^2 \qquad \text{(B.2)}$$

with a global minimum at $[1, 3]$ with an optimal value of zero. It is implemented in algorithm B.2.

```
booth(x) = (x[1]+2x[2]-7)^2 + (2x[1]+x[2]-5)^2
```

Algorithm B.2. Booth's function with two-dimensional input vector x.

B.3 Branin Function

The *Branin function* (figure B.3) is a two-dimensional function,

$$f(\mathbf{x}) = a(x_2 - bx_1^2 + cx_1 - r)^2 + s(1 - t)\cos(x_1) + s \qquad \text{(B.3)}$$

with recommended values $a = 1$, $b = 5.1/(4\pi^2)$, $c = 5/\pi$, $r = 6$, $s = 10$, and $t = 1/(8\pi)$.

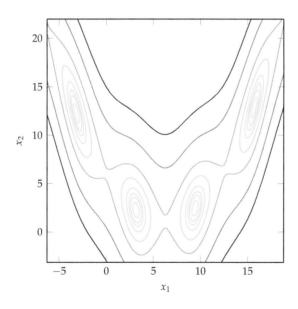

Figure B.3. The Branin function, with four global minima.

It has no local minima aside from global minima with $x_1 = \pi + 2\pi m$ for integral m. Four of these minima are:

$$\left\{ \begin{bmatrix} -\pi \\ 12.275 \end{bmatrix}, \begin{bmatrix} \pi \\ 2.275 \end{bmatrix}, \begin{bmatrix} 3\pi \\ 2.475 \end{bmatrix}, \begin{bmatrix} 5\pi \\ 12.875 \end{bmatrix} \right\} \tag{B.4}$$

with $f(\mathbf{x}^*) \approx 0.397887$. It is implemented in algorithm B.3.

```
function branin(x; a=1, b=5.1/(4π^2), c=5/π, r=6, s=10, t=1/(8π))
    return a*(x[2]-b*x[1]^2+c*x[1]-r)^2 + s*(1-t)*cos(x[1]) + s
end
```

Algorithm B.3. The Branin function with two-dimensional input vector x and six optional parameters.

B.4   Flower Function

The *flower function* (figure B.4) is a two-dimensional test function whose contour function has flower-like petals originating from the origin.

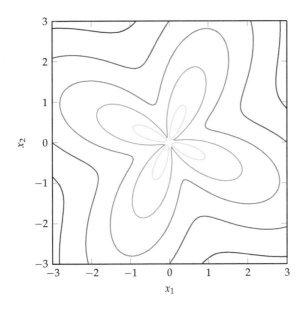

Figure B.4. The flower function.

The equation is

$$f(\mathbf{x}) = a\|\mathbf{x}\| + b\sin(c\tan^{-1}(x_2, x_1)) \tag{B.5}$$

with its parameters typically set to $a = 1$, $b = 1$, and $c = 4$.

The flower function is minimized near the origin but does not have a global minimum due to atan being undefined at $[0, 0]$. It is implemented in algorithm B.4.

```
function flower(x; a=1, b=1, c=4)
    return a*norm(x) + b*sin(c*atan(x[2], x[1]))
end
```

Algorithm B.4. The flower function with two-dimensional input vector x and three optional parameters.

B.5   Michalewicz Function

The *Michalewicz function* (figure B.5) is a d-dimensional optimization function with several steep valleys.

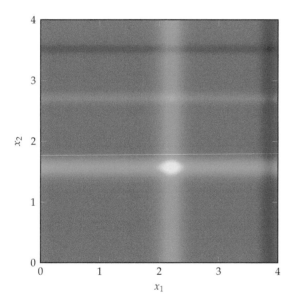

Figure B.5. The Michalewicz function.

Its equation is

$$f(\mathbf{x}) = -\sum_{i=1}^{d} \sin(x_i) \sin^{2m}\left(\frac{ix_i^2}{\pi}\right) \tag{B.6}$$

where the parameter m, typically 10, controls the steepness. The global minimum depends on the number of dimensions. In two dimensions the minimum is at approximately $[2.20, 1.57]$ with $f(\mathbf{x}^*) = -1.8011$. It is implemented in algorithm B.5.

```
function michalewicz(x; m=10)
    return -sum(sin(v)*sin(i*v^2/π)^(2m) for
                (i,v) in enumerate(x))
end
```

Algorithm B.5. The Michalewicz function with input vector x and optional steepness parameter m.

B.6 Rosenbrock's Banana Function

The *Rosenbrock function* (figure B.6), also called Rosenbrock's valley or Rosenbrock's banana function, is a well-known unconstrained test function developed by Rosenbrock in 1960.[1] It has a global minimum inside a long, curved valley. Most optimization algorithms have no problem finding the valley but have difficulties traversing along the valley to the global minimum.

[1] H. H. Rosenbrock, "An Automatic Method for Finding the Greatest or Least Value of a Function," *The Computer Journal*, vol. 3, no. 3, pp. 175–184, 1960.

Figure B.6. The Rosenbrock function with $a = 1$ and $b = 5$. The global minimum is at $[1, 1]$.

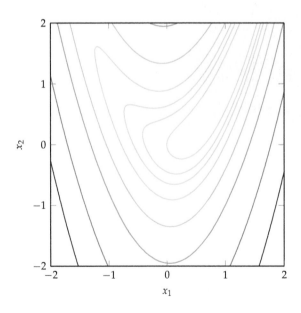

The Rosenbrock function is

$$f(\mathbf{x}) = (a - x_1)^2 + b(x_2 - x_1^2)^2 \qquad (B.7)$$

with a global minimum at $[a, a^2]$ at which $f(\mathbf{x}^*) = 0$. This text uses $a = 1$ and $b = 5$. The Rosenbrock function is implemented in algorithm B.6.

```
rosenbrock(x; a=1, b=5) = (a-x[1])^2 + b*(x[2] - x[1]^2)^2
```

Algorithm B.6. The Rosenbrock function with two-dimensional input vector x and two optional parameters.

B.7   Wheeler's Ridge

Wheeler's ridge (figure B.7) is a two-dimensional function with a single global minimum in a deep curved peak. The function has two ridges, one along the positive and one along the negative first coordinate axis. A gradient descent method will diverge along the negative axis ridge. The function is very flat away from the optimum and the ridge.

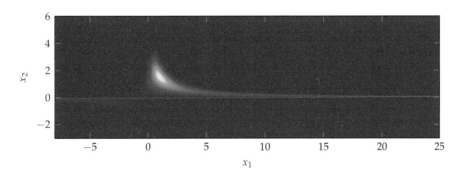

Figure B.7. Wheeler's ridge showing the two ridges and the peak containing the global minimum.

The function is given by

$$f(\mathbf{x}) = -\exp(-(x_1 x_2 - a)^2 - (x_2 - a)^2) \qquad (B.8)$$

with a typically equal to 1.5, for which the global optimum of -1 is at $[1, 3/2]$.

Figure B.8. A contour plot of the minimal region of Wheeler's ridge.

```
wheeler(x, a=1.5) = -exp(-(x[1]*x[2] - a)^2 -(x[2]-a)^2)
```

Algorithm B.7. Wheeler's ridge, which takes in a two-dimensional design point x and an optional scalar parameter a.

Wheeler's ridge has a smooth contour plot (figure B.8) when evaluated over $x_1 \in [0, 3]$ and $x_2 \in [0, 3]$. It is implemented in algorithm B.7.

B.8   Circle Function

The *circle function* (algorithm B.8) is a simple multiobjective test function given by

$$\mathbf{f}(\mathbf{x}) = \begin{bmatrix} 1 - r\cos(\theta) \\ 1 - r\sin(\theta) \end{bmatrix} \tag{B.9}$$

where $\theta = x_1$ and r is obtained by passing x_2 through

$$r = \frac{1}{2} + \frac{1}{2}\left(\frac{2x_2}{1 + x_2^2} \right) \tag{B.10}$$

The Pareto frontier has $r = 1$ and $\mod(\theta, 2\pi) \in [0, \pi/2]$ or $r = -1$ and $\mod(\theta, 2\pi) \in [\pi, 3\pi/2]$.

```
function circle(x)
    θ = x[1]
    r = 0.5 + 0.5*(2x[2]/(1+x[2]^2))
    y1 = 1 - r*cos(θ)
    y2 = 1 - r*sin(θ)
    return [y1, y2]
end
```

Algorithm B.8. The circle function, which takes in a two-dimensional design point x and produces a two-dimensional objective value.

C Mathematical Concepts

This appendix covers mathematical concepts used in the derivation and analysis of optimization methods. These concepts are used throughout this book.

C.1 Asymptotic Notation

Asymptotic notation is often used to characterize the growth of a function. This notation is sometimes called *big-Oh notation*, since the letter O is used because the growth rate of a function is often called its *order*. This notation can be used to describe the error associated with a numerical method or the time or space complexity of an algorithm. This notation provides an upper bound on a function as its argument approaches a certain value.

Mathematically, if $f(x) = O(g(x))$ as $x \rightarrow a$ then the absolute value of $f(x)$ is bounded by the absolute value of $g(x)$ times some positive and finite c for values of x sufficiently close to a:

$$|f(x)| \leq c|g(x)| \quad \text{for } x \rightarrow a \qquad (C.1)$$

Writing $f(x) = O(g(x))$ is a common abuse of the equal sign. For example, $x^2 = O(x^2)$ and $2x^2 = O(x^2)$, but, of course, $x^2 \neq 2x^2$. In some mathematical texts, $O(g(x))$ represents the set of all functions that do not grow faster than $g(x)$. One might write, for example, $5x^2 \in O(x^2)$. An example of asymptotic notation is given in example C.1.

If $f(x)$ is a *linear combination*[1] of terms, then $O(f)$ corresponds to the order of the fastest growing term. Example C.2 compares the orders of several terms.

[1] A linear combination is a weighted sum of terms. If the terms are in a vector $\mathbf{x}$, then the linear combination is $w_1 x_1 + w_2 x_2 + \cdots = \mathbf{w}^\top \mathbf{x}$.

Consider $f(x) = 10^6 e^x$ as $x \to \infty$. Here, f is a product of constant 10^6 and e^x. The constant can simply be incorporated into the bounding constant c:

$$|f(x)| \le c|g(x)|$$
$$10^6|e^x| \le c|g(x)|$$
$$|e^x| \le c|g(x)|$$

Thus, $f = O(e^x)$ as $x \to \infty$.

Example C.1. Asymptotic notation for a constant times a function.

Consider $f(x) = \cos(x) + x + 10x^{3/2} + 3x^2$. Here, f is a linear combination of terms. The terms $\cos(x)$, x, $x^{3/2}$, x^2 are arranged in order of increasing value as x approaches infinity. We plot $f(x)$ along with $c|g(x)|$, where c has been chosen for each term such that $c|g(x = 2)|$ exceeds $f(x = 2)$.

Example C.2. An illustration of finding the order of a linear combination of terms.

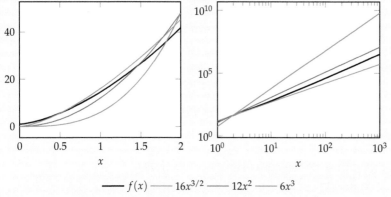

$$\text{—} f(x) \quad \text{——} 16x^{3/2} \quad \text{——} 12x^2 \quad \text{——} 6x^3$$

There is no constant c such that $f(x)$ is always less than $c|x^{3/2}|$ for sufficiently large values of x. The same is true for $\cos(x)$ and x.

We find that $f(x) = O(x^3)$, and in general $f(x) = O(x^m)$ for $m \ge 2$, along with other function classes like $f(x) = e^x$. We typically discuss the order that provides the tightest upper bound. Thus, $f = O(x^2)$ as $x \to \infty$.

C.2 Taylor Expansion

The *Taylor expansion*, also called the *Taylor series*, of a function is critical to understanding many of the optimization methods covered in this book, so we derive it here.

From the *first fundamental theorem of calculus*,[2] we know that

$$f(x + h) = f(x) + \int_0^h f'(x + a)\, da \qquad \text{(C.2)}$$

Nesting this definition produces the Taylor expansion of f about x:

$$f(x + h) = f(x) + \int_0^h \left(f'(x) + \int_0^a f''(x + b)\, db \right) da \qquad \text{(C.3)}$$

$$= f(x) + f'(x)h + \int_0^h \int_0^a f''(x + b)\, db\, da \qquad \text{(C.4)}$$

$$= f(x) + f'(x)h + \int_0^h \int_0^a \left(f''(x) + \int_0^b f'''(x + c)\, dc \right) db\, da \qquad \text{(C.5)}$$

$$= f(x) + f'(x)h + \frac{f''(x)}{2!}h^2 + \int_0^h \int_0^a \int_0^b f'''(x + c)\, dc\, db\, da \qquad \text{(C.6)}$$

$$\vdots \qquad \text{(C.7)}$$

$$= f(x) + \frac{f'(x)}{1!}h + \frac{f''(x)}{2!}h^2 + \frac{f'''(x)}{3!}h^3 + \dots \qquad \text{(C.8)}$$

$$= \sum_{n=0}^{\infty} \frac{f^{(n)}(x)}{n!}h^n \qquad \text{(C.9)}$$

In the formulation above, x is typically fixed and the function is evaluated in terms of h. It is often more convenient to write the Taylor expansion of $f(x)$ about a point a such that it remains a function of x:

$$f(x) = \sum_{n=0}^{\infty} \frac{f^{(n)}(a)}{n!}(x - a)^n \qquad \text{(C.10)}$$

The Taylor expansion represents a function as an infinite sum of polynomial terms based on repeated derivatives at a single point. Any analytic function can be represented by its Taylor expansion within a local neighborhood.

A function can be locally approximated by using the first few terms of the Taylor expansion. Figure C.1 shows increasingly better approximations for $\cos(x)$ about $x = 1$. Including more terms increases the accuracy of the local approximation, but error still accumulates as one moves away from the expansion point.

[2] The first fundamental theorem of calculus relates a function to the integral of its derivative:

$$f(b) - f(a) = \int_a^b f'(x)\, dx$$

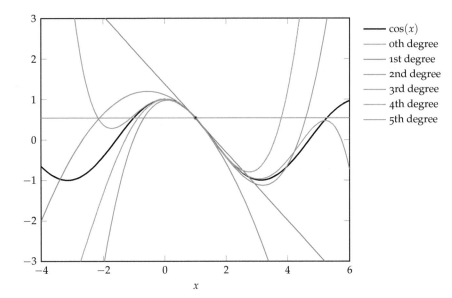

Figure C.1. Successive approximations of $\cos(x)$ about 1 based on the first n terms of the Taylor expansion.

A linear *Taylor approximation* uses the first two terms of the Taylor expansion:

$$f(x) \approx f(a) + f'(a)(x - a) \tag{C.11}$$

A quadratic Taylor approximation uses the first three terms:

$$f(x) \approx f(a) + f'(a)(x - a) + \frac{1}{2}f''(a)(x - a)^2 \tag{C.12}$$

and so on.

In multiple dimensions, the Taylor expansion about $\mathbf{a}$ generalizes to

$$f(\mathbf{x}) = f(\mathbf{a}) + \nabla f(\mathbf{a})^\top (\mathbf{x} - \mathbf{a}) + \frac{1}{2}(\mathbf{x} - \mathbf{a})^\top \nabla^2 f(\mathbf{a})(\mathbf{x} - \mathbf{a}) + \ldots \tag{C.13}$$

The first two terms form the tangent plane at $\mathbf{a}$. The third term incorporates local curvature. This text will use only the first three terms shown here.

C.3   Convexity

A *convex combination* of two vectors $\mathbf{x}$ and $\mathbf{y}$ is the result of

$$\alpha\mathbf{x} + (1 - \alpha)\mathbf{y} \tag{C.14}$$

for some $\alpha \in [0,1]$. Convex combinations can be made from m vectors,

$$w_1 \mathbf{v}^{(1)} + w_2 \mathbf{v}^{(2)} + \cdots + w_m \mathbf{v}^{(m)} \qquad \text{(C.15)}$$

with nonnegative weights $\mathbf{w}$ that sum to one.

A *convex set* is a set for which a line drawn between any two points in the set is entirely within the set. Mathematically, a set $\mathcal{S}$ is convex if we have

$$\alpha \mathbf{x} + (1 - \alpha)\mathbf{y} \in \mathcal{S}. \qquad \text{(C.16)}$$

for all $\mathbf{x}, \mathbf{y}$ in $\mathcal{S}$ and for all α in $[0,1]$. A convex and a nonconvex set are shown in figure C.2.

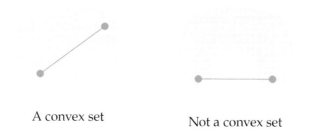

A convex set   Not a convex set

Figure C.2.   Convex and non-convex sets.

A *convex function* is a *bowl-shaped* function whose domain is a convex set. By bowl-shaped, we mean it is a function such that any line drawn between two points in its domain does not lie below the function. A function f is convex over a convex set $\mathcal{S}$ if, for all $\mathbf{x}, \mathbf{y}$ in $\mathcal{S}$ and for all α in $[0,1]$,

$$f(\alpha \mathbf{x} + (1 - \alpha)\mathbf{y}) \leq \alpha f(\mathbf{x}) + (1 - \alpha)f(\mathbf{y}) \qquad \text{(C.17)}$$

Convex and concave regions of a function are shown in figure C.3.

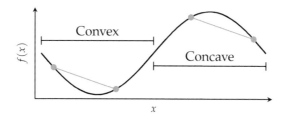

Figure C.3. Convex and nonconvex portions of a function.

A function f is *strictly convex* over a convex set $\mathcal{S}$ if, for all $\mathbf{x}, \mathbf{y}$ in $\mathcal{S}$ and α in $(0,1)$,

$$f(\alpha \mathbf{x} + (1 - \alpha)\mathbf{y}) < \alpha f(\mathbf{x}) + (1 - \alpha)f(\mathbf{y}) \qquad \text{(C.18)}$$

Strictly convex functions have at most one minimum, whereas a convex function can have flat regions.[3] Examples of strict and nonstrict convexity are shown in figure C.4.

[3] Optimization of convex functions is the subject of the textbook by S. Boyd and L. Vandenberghe, *Convex Optimization*. Cambridge University Press, 2004.

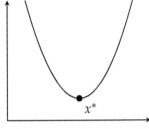

Strictly convex function with one global minimum.

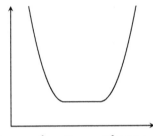

Convex function without a unique global minimum.

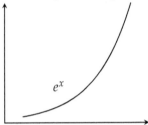

Strictly convex function without a global minimum.

Figure C.4. Not all convex functions have single global minima.

A function f is *concave* if $-f$ is convex. Furthermore, f is *strictly concave* if $-f$ is strictly convex.

Not all convex functions are unimodal and not all unimodal functions are convex, as shown in figure C.5.

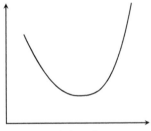

Unimodal and convex

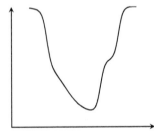

Unimodal but nonconvex

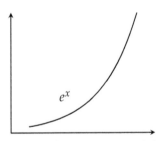

Convex but nonunimodal

Figure C.5. Convexity and unimodality are not the same thing.

C.4   Norms

A *norm* is a function that assigns a length to a vector. To compute the distance between two vectors, we evaluate the norm of the difference between those two vectors. For example, the distance between points $\mathbf{a}$ and $\mathbf{b}$ using the *Euclidean norm* is

$$\|\mathbf{a} - \mathbf{b}\|_2 = \sqrt{(a_1 - b_1)^2 + (a_2 - b_2)^2 + \cdots + (a_n - b_n)^2} \tag{C.19}$$

A function f is a norm if[4]

1. $f(\mathbf{x}) = 0$ if and only if $\mathbf{a}$ is the zero vector.

2. $f(a\mathbf{x}) = |a| f(\mathbf{x})$, such that lengths scale.

3. $f(\mathbf{a} + \mathbf{b}) \leq f(\mathbf{a}) + f(\mathbf{b})$, also known as the *triangle inequality*.

The L_p norms are a commonly used set of norms parameterized by a scalar $p \geq 1$. The Euclidean norm in equation (C.19) is the L_2 norm. Several L_p norms are shown in table C.1.

The L_p norms are defined according to:

$$\|\mathbf{x}\|_p = \lim_{\rho \to p} \left(|x_1|^\rho + |x_2|^\rho + \cdots + |x_n|^\rho \right)^{\frac{1}{\rho}} \tag{C.20}$$

where the limit is necessary for defining the infinity norm, L_∞.[5]

C.5   Matrix Calculus

This section derives two common gradients: $\nabla_\mathbf{x} \mathbf{b}^\top \mathbf{x}$ and $\nabla_\mathbf{x} \mathbf{x}^\top \mathbf{A} \mathbf{x}$.

To obtain $\nabla_\mathbf{x} \mathbf{b}^\top \mathbf{x}$, we first expand the dot product:

$$\mathbf{b}^\top \mathbf{x} = [b_1 x_1 + b_2 x_2 + \cdots + b_n x_n] \tag{C.21}$$

The partial derivative with respect to a single coordinate is:

$$\frac{\partial}{\partial x_i} \mathbf{b}^\top \mathbf{x} = b_i \tag{C.22}$$

Thus, the gradient is:

$$\nabla_\mathbf{x} \mathbf{b}^\top \mathbf{x} = \nabla_\mathbf{x} \mathbf{x}^\top \mathbf{b} = \mathbf{b} \tag{C.23}$$

[4] Some properties that follow from these axioms include:

$$f(-\mathbf{x}) = f(\mathbf{x})$$
$$f(\mathbf{x}) \geq 0$$

[5] The L_∞ norm is also referred to as the *max norm*, *Chebyschev distance*, or *chessboard distance*. The latter name comes from the minimum number of moves a chess king needs to move between two chess squares.

L_1: $\|\mathbf{x}\|_1 = |x_1| + |x_2| + \cdots + |x_n|$

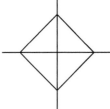

Table C.1. Common L_p norms. The illustrations show the shape of the norm contours in two dimensions. All points on the contour are equidistant from the origin under that norm.

L_2: $\|\mathbf{x}\|_2 = \sqrt{x_1^2 + x_2^2 + \cdots + x_n^2}$

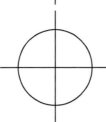

L_∞: $\|\mathbf{x}\|_\infty = \max(|x_1|, |x_2|, \cdots, |x_n|)$

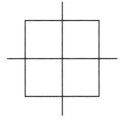

To obtain $\nabla_{\mathbf{x}}\mathbf{x}^\top\mathbf{A}\mathbf{x}$ for a square matrix $\mathbf{A}$, we first expand $\mathbf{x}^\top\mathbf{A}\mathbf{x}$:

$$\mathbf{x}^\top\mathbf{A}\mathbf{x} = \begin{bmatrix} x_1 \\ x_2 \\ \cdots \\ x_n \end{bmatrix}^\top \begin{bmatrix} a_{11} & a_{12} & \cdots & a_{1n} \\ a_{21} & a_{22} & \cdots & a_{2n} \\ \vdots & \vdots & \ddots & \vdots \\ a_{n1} & a_{n2} & \cdots & a_{nn} \end{bmatrix} \begin{bmatrix} x_1 \\ x_2 \\ \cdots \\ x_n \end{bmatrix} \tag{C.24}$$

$$= \begin{bmatrix} x_1 \\ x_2 \\ \cdots \\ x_n \end{bmatrix}^\top \begin{bmatrix} x_1 a_{11} + x_2 a_{12} + \cdots + x_n a_{1n} \\ x_1 a_{21} + x_2 a_{22} + \cdots + x_n a_{2n} \\ \vdots \\ x_1 a_{n1} + x_2 a_{n2} + \cdots + x_n a_{nn} \end{bmatrix} \tag{C.25}$$

$$= \begin{matrix} x_1^2 a_{11} + x_1 x_2 a_{12} + \cdots + x_1 x_n a_{1n} + \\ x_1 x_2 a_{21} + x_2^2 a_{22} + \cdots + x_2 x_n a_{2n} + \\ \vdots \\ x_1 x_n a_{n1} + x_2 x_n a_{n2} + \cdots + x_n^2 a_{nn} \end{matrix} \tag{C.26}$$

The partial derivative with respect to the ith component is

$$\frac{\partial}{\partial x_i}\mathbf{x}^\top\mathbf{A}\mathbf{x} = \sum_{j=1}^{n} x_j\left(a_{ij} + a_{ji}\right) \tag{C.27}$$

The gradient is thus:

$$\nabla_{\mathbf{x}}\mathbf{x}^\top\mathbf{A}\mathbf{x} = \begin{bmatrix} \sum_{j=1}^{n} x_j\left(a_{1j} + a_{j1}\right) \\ \sum_{j=1}^{n} x_j\left(a_{2j} + a_{j2}\right) \\ \vdots \\ \sum_{j=1}^{n} x_j\left(a_{nj} + a_{jn}\right) \end{bmatrix} \tag{C.28}$$

$$= \begin{bmatrix} a_{11} + a_{11} & a_{12} + a_{21} & \cdots & a_{1n} + a_{n1} \\ a_{21} + a_{12} & a_{22} + a_{22} & \cdots & a_{2n} + a_{n2} \\ \vdots & \vdots & \ddots & \vdots \\ a_{n1} + a_{1n} & a_{n2} + a_{2n} & \cdots & a_{nn} + a_{nn} \end{bmatrix} \begin{bmatrix} x_1 \\ x_2 \\ \vdots \\ x_n \end{bmatrix} \tag{C.29}$$

$$= \left(\mathbf{A} + \mathbf{A}^\top\right)\mathbf{x} \tag{C.30}$$

C.6 Positive Definiteness

The notion of a matrix being *positive definite* or *positive semidefinite* often arises in linear algebra and optimization for a variety of reasons. For example, if the matrix $\mathbf{A}$ is positive definite in the function $f(\mathbf{x}) = \mathbf{x}^\top \mathbf{A}\mathbf{x}$, then f has a unique global minimum.

Recall that the quadratic approximation of a twice-differentiable function f at $\mathbf{x}_0$ is

$$f(\mathbf{x}) \approx f(\mathbf{x}_0) + \nabla f(\mathbf{x}_0)^\top (\mathbf{x} - \mathbf{x}_0) + \frac{1}{2}(\mathbf{x} - \mathbf{x}_0)^\top \mathbf{H}_0(\mathbf{x} - \mathbf{x}_0) \qquad \text{(C.31)}$$

where $\mathbf{H}_0$ is the Hessian of f evaluated at $\mathbf{x}_0$. Knowing that $(\mathbf{x} - \mathbf{x}_0)^\top \mathbf{H}_0(\mathbf{x} - \mathbf{x}_0)$ has a unique global minimum is sufficient to determine whether the overall quadratic approximation has a unique global minimum.[6] .

A symmetric matrix $\mathbf{A}$ is positive definite if $\mathbf{x}^\top \mathbf{A}\mathbf{x}$ is positive for all points other than the origin: $\mathbf{x}^\top \mathbf{A}\mathbf{x} > 0$ for all $\mathbf{x} \neq \mathbf{0}$.

A symmetric matrix $\mathbf{A}$ is positive semidefinite if $\mathbf{x}^\top \mathbf{A}\mathbf{x}$ is always non-negative: $\mathbf{x}^\top \mathbf{A}\mathbf{x} \geq 0$ for all $\mathbf{x}$.

[6] The component $f(\mathbf{x}_0)$ merely shifts the function vertically. The component $\nabla f(\mathbf{x}_0)^\top (\mathbf{x} - \mathbf{x}_0)$ is a linear term which is dominated by the quadratic term.

C.7 Gaussian Distribution

The probability density function for a *univariate Gaussian*,[7] also called the *normal distribution*, is:

$$\mathcal{N}(x \mid \mu, \nu) = \frac{1}{\sqrt{2\pi\nu}} e^{-\frac{(x-\mu)^2}{2\nu}} \qquad \text{(C.32)}$$

where μ is the mean and ν is the variance.[8] This distribution is plotted in figure C.6.

The *cumulative distribution function* of a distribution maps x to the probability that drawing a value from that distribution will produce a value less than or equal to x. For a univariate Gaussian, the cumulative distribution function is given by

$$\Phi(x) \equiv \frac{1}{2} + \frac{1}{2}\text{erf}\left(\frac{x - \mu}{\sigma\sqrt{2}}\right) \qquad \text{(C.33)}$$

where erf is the *error function*:

$$\text{erf}(x) \equiv \frac{2}{\sqrt{\pi}} \int_0^x e^{-\tau^2} d\tau \qquad \text{(C.34)}$$

[7] The multivariate Gaussian is discussed in chapter 8 and chapter 15. The univariate Gaussian is used throughout.

[8] The variance is the standard deviation squared.

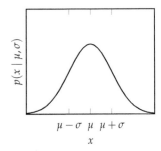

Figure C.6. A univariate Gaussian distribution, $\mathcal{N}(\mu, \nu)$.

C.8   Gaussian Quadrature

Gaussian quadrature is a technique for approximating integrals using a weighted sum of function evaluations.[9] The general form of the approximation is

$$\int_a^b p(x)f(x)\,dx \approx \sum_{i=1}^m w_i f(x_i)$$

(C.35)

[9] For a detailed overview of Gaussian quadrature, see J. Stoer and R. Bulirsch, *Introduction to Numerical Analysis*, 3rd ed. Springer, 2002.

where $p(x)$ is a known nonnegative weight function[10] over the finite or infinite interval $[a, b]$.

[10] The weight function is often a probability density function in practice.

An m-point *quadrature rule* is a unique choice of points $x_i \in (a, b)$ and weights $w_i > 0$ for $i \in \{1, \dots, m\}$ that define a Gaussian quadrature approximation such that any polynomial of degree $2m - 1$ or less is integrated exactly over $[a, b]$ with the given weight function.

Given a domain and a weight function, we can compute a class of orthogonal polynomials. We will use $b_i(x)$ to denote an orthogonal polynomial[11] of degree i. Any polynomial of degree m can be represented as a linear combination of the orthogonal polynomials up to degree m. We form a quadrature rule by selecting m points x_i to be the zeros of the orthogonal polynomial p_m and obtain the weights by solving the system of equations:

[11] Orthogonal polynomials are covered in chapter 18.

$$\sum_{i=1}^m b_k(x_i)w_i = \begin{cases} \int_a^b p(x)b_0(x)^2\,dx & \text{for } k = 0 \\ 0 & \text{for } k = 1, \dots, m-1 \end{cases}$$

(C.36)

Gauss solved equation (C.36) for the interval $[-1, 1]$ and the weighting function $p(x) = 1$. The orthogonal polynomials for this case are the *Legendre polynomials*. Algorithm C.1 implements Gaussian quadrature for Legendre polynomials and example C.3 works out a quadrature rule for integration over $[-1, 1]$.

We can transform any integral over the bounded interval $[a, b]$ to an integral over $[-1, 1]$ using the transformation

$$\int_a^b f(x)\,dx = \frac{b-a}{2} \int_{-1}^1 f\left(\frac{b-a}{2}x + \frac{a+b}{2}\right)dx$$

(C.37)

Quadrature rules can thus be precalculated for the Legendre polynomials and then applied to integration over any finite interval.[12] Example C.4 applies such a transformation and algorithm C.2 implements integral transformations in a Gaussian quadrature method for finite domains.

[12] Similar techniques can be applied to integration over infinite intervals, such as $[0, \infty)$ using the Laguerre polynomials and $(-\infty, \infty)$ using the Hermite polynomials.

Consider the Legendre polynomials for integration over $[-1, 1]$ with the weight function $p(x) = 1$. Suppose our function of interest is well approximated by a fifth degree polynomial. We construct a 3-point quadrature rule, which produces exact results for polynomials up to degree 5.

The Legendre polynomial of degree 3 is $\text{Le}_3(x) = \frac{5}{2}x^3 - \frac{3}{2}x$, which has roots at $x_1 = -\sqrt{3/5}$, $x_2 = 0$, and $x_3 = \sqrt{3/5}$. The Legendre polynomials of lesser degree are $\text{Le}_0(x) = 1$, $\text{Le}_1(x) = x$, and $\text{Le}_2(x) = \frac{3}{2}x^2 - \frac{1}{2}$. The weights are obtained by solving the system of equations:

$$\begin{bmatrix} \text{Le}_0(-\sqrt{3/5}) & \text{Le}_0(0) & \text{Le}_0(\sqrt{3/5}) \\ \text{Le}_1(-\sqrt{3/5}) & \text{Le}_1(0) & \text{Le}_1(\sqrt{3/5}) \\ \text{Le}_2(-\sqrt{3/5}) & \text{Le}_2(0) & \text{Le}_2(\sqrt{3/5}) \end{bmatrix} \begin{bmatrix} w_1 \\ w_2 \\ w_3 \end{bmatrix} = \begin{bmatrix} \int_{-1}^{1} \text{Le}_0(x)^2\, dx \\ 0 \\ 0 \end{bmatrix}$$

$$\begin{bmatrix} 1 & 1 & 1 \\ -\sqrt{3/5} & 0 & \sqrt{3/5} \\ 4/10 & -1/2 & 4/10 \end{bmatrix} \begin{bmatrix} w_1 \\ w_2 \\ w_3 \end{bmatrix} = \begin{bmatrix} 2 \\ 0 \\ 0 \end{bmatrix}$$

which yields $w_1 = w_3 = 5/9$ and $w_2 = 8/9$.

Consider integrating the 5th degree polynomial $f(x) = x^5 - 2x^4 + 3x^3 + 5x^2 - x + 4$. The exact value is $\int_{-1}^{1} p(x)f(x)\, dx = 158/15 \approx 10.533$. The quadrature rule produces the same value:

$$\sum_{i=1}^{3} w_i f(x_i) = \frac{5}{9}f\left(-\sqrt{\frac{3}{5}}\right) + \frac{8}{9}f(0) + \frac{5}{9}f\left(\sqrt{\frac{3}{5}}\right) \approx 10.533.$$

Example C.3. Obtaining a 3-term quadrature rule for exactly integrating polynomials up to degree 5.

Consider integrating $f(x) = x^5 - 2x^4 + 3x^3 + 5x^2 - x + 4$ over $[-3, 5]$. We can transform this into an integration over $[-1, 1]$ using equation (C.37):

$$\int_{-3}^{5} f(x)\, dx = \frac{5+3}{2} \int_{-1}^{1} f\left(\frac{5+3}{2}x + \frac{5-3}{2}\right) dx = 4 \int_{-1}^{1} f(4x + 1)\, dx$$

We use the 3-term quadrature rule obtained in example C.3 to integrate $g(x) = 4f(4x+1) = 4096x^5 + 3072x^4 + 1280y^3 + 768y^2 + 240y + 40$ over $[-1, 1]$:

$$\int_{-1}^{1} p(x)g(x)\, dx = \frac{5}{9}g\left(-\sqrt{3/5}\right) + \frac{8}{9}g(0) + \frac{5}{9}g\left(\sqrt{3/5}\right) = 1820.8$$

Example C.4. Integrals over finite regions can be transformed into integrals over $[-1, 1]$ and solved with quadrature rules for Legendre polynomials.

```
struct Quadrule
    ws
    xs
end
function quadrule_legendre(m)
    bs = [legendre(i) for i in 1 : m+1]
    xs = roots(bs[end])
    A = [bs[k](xs[i]) for k in 1 : m, i in 1 : m]
    b = zeros(m)
    b[1] = 2
    ws = A\b
    return Quadrule(ws, xs)
end
```

Algorithm C.1. A method for constructing m-point Legendre quadrature rules over $[-1, 1]$. The resulting type contains both the nodes xs and the weights ws.

```
quadint(f, quadrule) =
    sum(w*f(x) for (w,x) in zip(quadrule.ws, quadrule.xs))
function quadint(f, quadrule, a, b)
    α = (b-a)/2
    β = (a+b)/2
    g = x -> α*f(α*x+β)
    return quadint(g, quadrule)
end
```

Algorithm C.2. The function quadint for integrating a univariate function f with a given quadrature rule quadrule over the finite domain $[a, b]$.

D   Solutions

Exercise 1.1: $f(x) = x^3/3 - x$ at $x = 1$.

Exercise 1.2: It does not have a minimum, that is, the function is said to be unbounded below.

Exercise 1.3: No. Consider minimizing $f(x) = x$, subject to $x \geq 1$.

Exercise 1.4: The function f can be broken into two separate functions that depend only on their specific coordinate:

$$f(x,y) = g(x) + h(y)$$

where $g(x) = x^2$ and $h(y) = y$. Both g and h strictly increase for $x, y \geq 1$. While h is minimized for $y = 1$, we can merely approach $x \to 1$ due to the strict inequality $x > y$. Thus, f has no minima.

Exercise 1.5: An inflection point is a point on a curve where the sign of the curvature changes. A necessary condition for x to be an inflection point is that the second derivative is zero. The second derivative is $f''(x) = 6x$, which is only zero at $x = 0$.

   A sufficient condition for x to be an inflection point is that the second derivative changes sign around x. That is, $f''(x + \epsilon)$ and $f''(x - \epsilon)$ for $\epsilon \ll 1$ have opposite signs. This holds for $x = 0$, so it is an inflection point.

   There is thus only one inflection point on $x^3 - 10$.

Exercise 2.1: An entry of the Hessian can be computed using the forward difference method:

$$H_{ij} = \frac{\partial^2 f(\mathbf{x})}{\partial x_i \partial x_j} \approx \frac{\nabla f(\mathbf{x} + h\mathbf{e}_j)_i - \nabla f(\mathbf{x})_i}{h}$$

where $\mathbf{e}_i$ is the ith basis vector with $\mathbf{e}_i = 1$ and all other entries are zero.

We can thus approximate the jth column of the Hessian using:

$$\mathbf{H}_{\cdot j} \approx \frac{\nabla f(\mathbf{x} + h\mathbf{e}_j) - \nabla f(\mathbf{x})}{h}$$

This procedure can be repeated for each column of the Hessian.

Exercise 2.2: It requires two evaluations of the objective function.

Exercise 2.3: $f'(x) = \frac{1}{x} + e^x - \frac{1}{x^2}$. When x is close to zero, we find that $x < 1$. Hence $\frac{1}{x} > 1$, and finally $\frac{1}{x^2} > \frac{1}{x} > 0$, so $-\frac{1}{x^2}$ dominates.

Exercise 2.4: From the complex step method, we have $f'(x) \approx \text{Im}(2 + 4ih)/h = 4h/h = 4$.

Exercise 2.5: See the picture below:

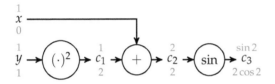

Exercise 2.6: The second-order derivative can be approximated using the central difference on the first-order derivative:

$$f''(x) \approx \frac{f'(x + h/2) - f'(x - h/2)}{h}$$

for small values of h.

Substituting in the forward and backwards different estimates of $f'(x + h/2)$ and $f'(x - h/2)$ yields:

$$f''(x) \approx \frac{\frac{f(x+h/2+h/2)-f(x+h/2-h/2)}{h} - \frac{f(x-h/2+h/2)-f(x-h/2-h/2)}{h}}{h}$$
$$= \frac{f(x + h) - f(x)}{h^2} - \frac{f(x) - f(x - h)}{h^2}$$
$$= \frac{f(x + h) - 2f(x) + f(x - h)}{h^2}$$

Exercise 3.1: Fibonacci search is preferred when derivatives are not available.

Exercise 3.2: The Shubert-Piyavskii method needs the Lipschitz constant, which may not be known.

Exercise 3.3: $f(x) = x^2$. Since the function is quadratic, after three evaluations, the quadratic model will represent this function exactly.

Exercise 3.4: We can use the bisection method to find the roots of $f'(x) = x - 1$. After the first update, we have $[0, 500]$. Then, $[0, 250]$. Finally, $[0, 125]$.

Exercise 3.5: No, the Lipschitz constant must bound the derivative everywhere on the interval, and $f'(1) = 2(1 + 2) = 6$.

Exercise 3.6: No. The best you can do is use Fibonacci Search and shrink the uncertainty by a factor of 3; that is, to $(32 - 1)/3 = 10\frac{1}{3}$.

Exercise 4.1: Consider running a descent method on $f(x) = 1/x$ for $x > 0$. The minimum does not exist and the descent method will forever proceed in the positive x direction with ever-increasing step sizes. Thus, only relying on a step-size termination condition would cause the method to run forever. Also terminating based on gradient magnitude would cause it to terminate.

A descent method applied to $f(x) = -x$ will also forever proceed in the positive x direction. The function is unbounded below, so neither a step-size termination condition nor a gradient magnitude termination condition would trigger. It is common to include an additional termination condition to limit the number of iterations.

Exercise 4.2: Applying the first Wolfe condition to our objective function yields $6 + (-1 + \alpha)^2 \leq 7 - 2\alpha \cdot 10^{-4}$, which can be simplified to $\alpha^2 - 2\alpha + 2 \cdot 10^{-4}\alpha \leq 0$. This equation can be solved to obtain $\alpha \leq 2(1 - 10^{-4})$. Thus, the maximum step length is $\alpha = 1.9998$.

Exercise 5.1: $\nabla f(\mathbf{x}) = 2\mathbf{A}\mathbf{x} + \mathbf{b}$.

Exercise 5.2: The derivative is $f'(x) = 4x^3$. Starting from $x^{(1)} = 1$:

$$f'(1) = 4 \qquad\qquad \rightarrow x^{(2)} = 1 - 4 = -3 \qquad (\text{D.1})$$
$$f'(-3) = 4 \cdot (-27) = -108 \qquad \rightarrow x^{(3)} = -3 + 108 = 105 \qquad (\text{D.2})$$

Exercise 5.3: We have $f'(x) = e^x - e^{-x} \approx e^x$ for large x. Thus $f'(x^{(1)}) \approx e^{10}$ and $x^{(2)} \approx -e^{10}$. If we apply an exact line search, $x^{(2)} = 0$. Thus, without a line search we are not guaranteed to reduce the value of the objective function.

Exercise 5.4: The Hessian is $2\mathbf{H}$, and

$$\nabla q(\mathbf{d}) = \mathbf{d}^\top \left(\mathbf{H} + \mathbf{H}^\top \right) + \mathbf{b} = \mathbf{d}^\top (2\mathbf{H}) + \mathbf{b}.$$

The gradient is $\mathbf{b}$ when $\mathbf{d} = \mathbf{0}$. The conjugate gradient method may diverge because $\mathbf{H}$ is not guaranteed to be positive definite.

Exercise 5.5: Nesterov momentum looks at the point where you will be after the update to compute the update itself.

Exercise 5.6: The conjugate gradient method implicitly reuses previous information about the function and thus may enjoy better convergence in practice.

Exercise 5.7: The conjugate gradient method initially follows the steepest descent direction. The gradient is

$$\nabla f(x,y) = [2x + y, 2y + x] \tag{D.3}$$

which for $(x,y) = (1,1)$ is $[1/\sqrt{2}, 1/\sqrt{2}]$. The direction of steepest descent is opposite the gradient, $\mathbf{d}^{(1)} = [-1/\sqrt{2}, -1/\sqrt{2}]$.

The Hessian is

$$\begin{bmatrix} 2 & 1 \\ 1 & 2 \end{bmatrix} \tag{D.4}$$

Since the function is quadratic and the Hessian is positive definite, the conjugate gradient method converges in at most two steps. Thus, the resulting point after two steps is the optimum, $(x,y) = (0,0)$, where the gradient is zero.

Exercise 5.8: No. If exact minimization is performed, then the descent directions between steps are orthogonal, but $[1,2,3]^\top [0,0,-3] \neq 0$.

Exercise 6.1: Second-order information can guarantee that one is at a local minimum, whereas a gradient of zero is necessary but insufficient to guarantee local optimality.

Exercise 6.2: We would prefer Newton's method if we start sufficiently close to the root and can compute derivatives analytically. Newton's method enjoys a better rate of convergence.

Exercise 6.3: $f'(x) = 2x$, $f''(x) = 2$. Thus, $x^{(2)} = x^{(1)} - 2x^{(1)}/2 = 0$; that is, you converge in one step from any starting point.

Exercise 6.4: Since $\nabla f(\mathbf{x}) = \mathbf{Hx}$, $\nabla^2 f(\mathbf{x}) = \mathbf{H}$, and $\mathbf{H}$ is nonsingular, it follows that $\mathbf{x}^{(2)} = \mathbf{x}^{(1)} - \mathbf{H}^{-1}\mathbf{Hx}^{(1)} = \mathbf{0}$. That is, Newton's method converges in one step.

Gradient descent diverges:

$$\mathbf{x}^{(2)} = [1,1] - [1,1000] = [0,-999] \tag{D.5}$$

$$\mathbf{x}^{(3)} = [0,-999] - [0,-1000 \cdot 999] = [0,998001] \tag{D.6}$$

Conjugate gradient descent uses the same initial search direction as gradient descent and converges to the minimum in the second step because the optimization objective is quadratic.

Exercise 6.5: The left plot shows convergence for Newton's method approaching floating-point resolution within nine iterations. The secant method is slower to converge because it can merely approximate the derivative.

The right plot shows the projection of the exact and approximate tangent lines with respect to f' for each method. The secant method's tangent lines have a higher slope, and thus intersect the x-axis prematurely.

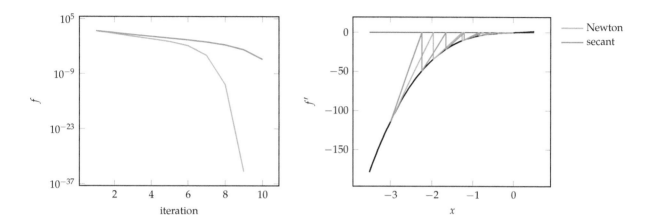

Exercise 6.6: Consider the sequence $x^{(k+1)} = x^{(k)}/2$ starting from $x^{(1)} = -1$ on the function $f(x) = x^2 - x$. Clearly the sequence converges to $x = 0$, the values for $f(x)$ are decreasing, and yet the sequence does not converge to a minimizer.

Exercise 6.7: It does not need computation or knowledge of the entries of the Hessian, and hence does not require solving a linear system at each iteration.

Exercise 6.8: The BFGS update does not exist when $\delta^\top \gamma \approx 0$. In that case, simply skip the update.

Exercise 6.9: The objective function is quadratic and can thus be minimized in one step. The gradient is $\nabla f = [2(x_1 + 1), 2(x_2 + 3)]$, which is zero at $\mathbf{x}^* = [-1, -3]$. The Hessian is positive definite, so $\mathbf{x}^*$ is the minimum.

Exercise 6.10: The new approximation has the form

$$f^{(k+1)}(\mathbf{x}) = y^{(k+1)} + \left(\mathbf{g}^{(k+1)}\right)^\top \left(\mathbf{x} - \mathbf{x}^{(k+1)}\right) + \frac{1}{2}\left(\mathbf{x} - \mathbf{x}^{(k+1)}\right)^\top \mathbf{H}^{(k+1)} \left(\mathbf{x} - \mathbf{x}^{(k+1)}\right)$$

using the true function value and gradient at $\mathbf{x}^{(k+1)}$ but requires an updated Hessian $\mathbf{H}^{(k+1)}$. This form automatically satisfies $f^{(k+1)}(\mathbf{x}^{(k+1)}) = y^{(k+1)}$ and $\nabla f^{(k+1)}(\mathbf{x}^{(k+1)}) = \mathbf{g}^{(k+1)}$. We must select the new Hessian to satisfy the third condition:

$$\nabla f^{(k+1)}\left(\mathbf{x}^{(k)}\right) = \mathbf{g}^{(k+1)} + \mathbf{H}^{(k+1)}\left(\mathbf{x}^{(k)} - \mathbf{x}^{(k+1)}\right)$$
$$= \mathbf{g}^{(k+1)} - \mathbf{H}^{(k+1)}\left(\mathbf{x}^{(k+1)} - \mathbf{x}^{(k)}\right)$$
$$= \mathbf{g}^{(k+1)} - \mathbf{H}^{(k+1)}\boldsymbol{\delta}^{(k+1)}$$
$$= \mathbf{g}^{(k)}$$

We can rearrange and substitute to obtain:

$$\mathbf{H}^{(k+1)}\boldsymbol{\delta}^{(k+1)} = \boldsymbol{\gamma}^{(k+1)}$$

Recall that a matrix $\mathbf{A}$ is positive definite if for every nonzero vector $\mathbf{x}^\top \mathbf{A} \mathbf{x} > 0$. If we multiply the secant equation by $\boldsymbol{\delta}^{(k+1)}$ we obtain the curvature condition:

$$\left(\boldsymbol{\delta}^{(k+1)}\right)^\top \mathbf{H}^{(k+1)}\boldsymbol{\delta}^{(k+1)} = \left(\boldsymbol{\delta}^{(k+1)}\right)^\top \boldsymbol{\gamma}^{(k+1)} > 0$$

We seek a new positive definite matrix $\mathbf{H}^{(k+1)}$. All positive definite matrices are symmetric, so calculating a new positive definite matrix requires specifying $n(n+1)/2$ variables. The secant equation imposes n conditions on these variables, leading to an infinite number of solutions. In order to have a unique solution, we choose the positive definite matrix closest to $\mathbf{H}^{(k)}$. This objective leads to the desired optimization problem.

Exercise 7.1: The derivative has n terms whereas the Hessian has n^2 terms. Each derivative term requires two evaluations when using finite difference methods: $f(\mathbf{x})$ and $f(\mathbf{x} + h\mathbf{e}^{(i)})$. Each Hessian term requires an additional evaluation when using finite difference methods:

$$\frac{\partial^2 f}{\partial x_i \partial x_j} \approx \frac{f(\mathbf{x} + h\mathbf{e}^{(i)} + h\mathbf{e}^{(j)}) - f(\mathbf{x} + h\mathbf{e}^{(i)}) - f(\mathbf{x} + h\mathbf{e}^{(j)}) + f(\mathbf{x})}{h^2}$$

Thus, to approximate the gradient, you need $n+1$ evaluations, and to approximate the Hessian you need on the order of n^2 evaluations.

Approximating the Hessian is prohibitively expensive for large n. Direct methods can take comparatively more steps using n^2 function evaluations, as direct methods need not estimate the derivative or Hessian at each step.

Exercise 7.2: Consider minimizing $f(x) = xy$ and $x_0 = [0, 0]$. Proceeding in either canonical direction will not reduce the objective function, but x_0 is clearly not a minimizer.

Exercise 7.3: At each iteration, the Hooke-Jeeves method samples $2n$ points along the coordinate directions with a step-size a. It stops when none of the points provides an improvement and the step size is no more than a given tolerance ϵ. While this often causes the Hooke-Jeeves method to stop when it has converged to within ϵ of a local minimum, that need not be the case. For example, a valley can descend between two coordinate directions farther than ϵ before arriving at a local minimum, and the Hooke-Jeeves method would not detect it.

Exercise 7.4: Minimizing the drag of an airfoil subject to a minimum thickness (to preserve structural integrity). Evaluating the performance of the airfoil using computational fluid dynamics involves solving partial differential equations. Because the function is not known analytically, we are unlikely to have an analytical expression for the derivative.

Exercise 7.5: The divided rectangles method samples at the center of the intervals and not where the bound derived from a known Lipschitz constant is lowest.

Exercise 7.6: It cannot be cyclic coordinate search since more than one component is changing. It can be Powell's method.

Exercise 8.1: The cross-entropy method must fit distribution parameters with every iteration. Unfortunately, no known analytic solutions for fitting multivariate mixture distributions exist. Instead, one commonly uses the iterative expectation maximization algorithm to converge on an answer.

Exercise 8.2: If the number of elite samples is close to the total number of samples, then the resulting distribution will closely match the population. There will not be a significant bias toward the best locations for a minimizer, and so convergence will be slow.

Exercise 8.3: The derivative of the log-likelihood with respect to ν is:

$$\frac{\partial}{\partial \nu} \ell(x \mid \mu, \nu) = \frac{\partial}{\partial \nu}\left(-\frac{1}{2}\ln 2\pi - \frac{1}{2}\ln \nu - \frac{(x - \mu)^2}{2\nu}\right)$$

$$= -\frac{1}{2\nu} + \frac{(x - \mu)^2}{2\nu^2}$$

The second term will be zero if the mean is already optimal. Thus, the derivative is $-1/2\nu$ and decreasing ν will increase the likelihood of drawing elite samples. Unfortunately, ν is optimized by approaching arbitrarily close to zero. The asymptote near zero in the gradient update will lead to large step sizes, which cannot be taken as ν must remain positive.

Exercise 8.4: The probability density of a design $\mathbf{x}$ under a multivariate normal distribution with mean $\boldsymbol{\mu}$ and covariance $\boldsymbol{\Sigma}$ is

$$p(\mathbf{x} \mid \boldsymbol{\mu}, \boldsymbol{\Sigma}) = \frac{1}{(2\pi|\boldsymbol{\Sigma}|)^{1/2}} \exp\left(-\frac{1}{2}(\mathbf{x}-\boldsymbol{\mu})^\top \boldsymbol{\Sigma}^{-1}(\mathbf{x}-\boldsymbol{\mu})\right)$$

We can simplify the problem by maximizing the log-likelihood instead.[1] The log-likelihood is:

[1] The log function is concave for positive inputs, so maximizing $\log f(x)$ also maximizes a strictly positive $f(x)$.

$$\ln p(\mathbf{x} \mid \boldsymbol{\mu}, \boldsymbol{\Sigma}) = -\frac{1}{2}\ln(2\pi|\boldsymbol{\Sigma}|) - \frac{1}{2}(\mathbf{x}-\boldsymbol{\mu})^\top \boldsymbol{\Sigma}^{-1}(\mathbf{x}-\boldsymbol{\mu})$$

$$= -\frac{1}{2}\ln(2\pi|\boldsymbol{\Sigma}|) - \frac{1}{2}\left(\mathbf{x}^\top \boldsymbol{\Sigma}^{-1}\mathbf{x} - 2\mathbf{x}^\top \boldsymbol{\Sigma}^{-1}\boldsymbol{\mu} + \boldsymbol{\mu}^\top \boldsymbol{\Sigma}^{-1}\boldsymbol{\mu}\right)$$

We begin by maximizing the log-likelihood of the m individuals with respect to the mean:

$$\ell\left(\boldsymbol{\mu} \mid \mathbf{x}^{(1)}, \cdots, \mathbf{x}^{(m)}\right) = \sum_{i=1}^{m} \ln p(\mathbf{x}^{(i)} \mid \boldsymbol{\mu}, \boldsymbol{\Sigma})$$

$$= \sum_{i=1}^{m} -\frac{1}{2}\ln(2\pi|\boldsymbol{\Sigma}|) - \frac{1}{2}\left(\left(\mathbf{x}^{(i)}\right)^\top \boldsymbol{\Sigma}^{-1}\mathbf{x}^{(i)} - 2\left(\mathbf{x}^{(i)}\right)^\top \boldsymbol{\Sigma}^{-1}\boldsymbol{\mu} + \boldsymbol{\mu}^\top \boldsymbol{\Sigma}^{-1}\boldsymbol{\mu}\right)$$

We compute the gradient using the facts that $\nabla_{\mathbf{z}}\mathbf{z}^\top \mathbf{A}\mathbf{z} = (\mathbf{A}+\mathbf{A}^\top)\mathbf{z}$, that $\nabla_{\mathbf{z}}\mathbf{a}^\top \mathbf{z} = \mathbf{a}$, and that $\boldsymbol{\Sigma}$ is symmetric and positive definite, and thus $\boldsymbol{\Sigma}^{-1}$ is symmetric:

$$\nabla_{\boldsymbol{\mu}}\ell\left(\boldsymbol{\mu} \mid \mathbf{x}^{(1)}, \cdots, \mathbf{x}^{(m)}\right) = \sum_{i=1}^{m} -\frac{1}{2}\left(\nabla_{\boldsymbol{\mu}}\left(-2\left(\mathbf{x}^{(i)}\right)^\top \boldsymbol{\Sigma}^{-1}\boldsymbol{\mu}\right) + \nabla_{\boldsymbol{\mu}}\left(\boldsymbol{\mu}^\top \boldsymbol{\Sigma}^{-1}\boldsymbol{\mu}\right)\right)$$

$$= \sum_{i=1}^{m} \left(\nabla_{\boldsymbol{\mu}}\left(\left(\mathbf{x}^{(i)}\right)^\top \boldsymbol{\Sigma}^{-1}\boldsymbol{\mu}\right) - \frac{1}{2}\nabla_{\boldsymbol{\mu}}\left(\boldsymbol{\mu}^\top \boldsymbol{\Sigma}^{-1}\boldsymbol{\mu}\right)\right)$$

$$= \sum_{i=1}^{m} \boldsymbol{\Sigma}^{-1}\mathbf{x}^{(i)} - \boldsymbol{\Sigma}^{-1}\boldsymbol{\mu}$$

We set the gradient to zero:

$$0 = \sum_{i=1}^{m} \Sigma^{-1} \mathbf{x}^{(i)} - \Sigma^{-1} \mu$$

$$\sum_{i=1}^{m} \mu = \sum_{i=1}^{m} \mathbf{x}^{(i)}$$

$$m\mu = \sum_{i=1}^{m} \mathbf{x}^{(i)}$$

$$\mu = \frac{1}{m} \sum_{i=1}^{m} \mathbf{x}^{(i)}$$

Next we maximize with respect to the inverse covariance, $\Lambda = \Sigma^{-1}$, using the fact that $|\mathbf{A}^{-1}| = 1/|\mathbf{A}|$ with $\mathbf{b}^{(i)} = \mathbf{x}^{(i)} - \mu$:

$$\ell\left(\Lambda \mid \mu, \mathbf{x}^{(1)}, \cdots, \mathbf{x}^{(m)}\right) = \sum_{i=1}^{m} -\frac{1}{2} \ln\left(2\pi |\Lambda|^{-1}\right) - \frac{1}{2} \left(\left(\mathbf{x}^{(i)} - \mu\right)^{\top} \Lambda \left(\mathbf{x}^{(i)} - \mu\right)\right)$$

$$= \sum_{i=1}^{m} \frac{1}{2} \ln(|\Lambda|) - \frac{1}{2} \left(\mathbf{b}^{(i)}\right)^{\top} \Lambda \mathbf{b}^{(i)}$$

We compute the gradient using the facts that $\nabla_{\mathbf{A}} |\mathbf{A}| = |\mathbf{A}| \mathbf{A}^{-\top}$ and $\nabla_{\mathbf{A}} \mathbf{z}^{\top} \mathbf{A} \mathbf{z} = \mathbf{z}\mathbf{z}^{\top}$:

$$\nabla_{\Lambda} \ell\left(\Lambda \mid \mu, \mathbf{x}^{(1)}, \cdots, \mathbf{x}^{(m)}\right) = \sum_{i=1}^{m} \nabla_{\Lambda} \left(\frac{1}{2} \ln(|\Lambda|) - \frac{1}{2} \left(\mathbf{b}^{(i)}\right)^{\top} \Lambda \mathbf{b}^{(i)}\right)$$

$$= \sum_{i=1}^{m} \frac{1}{2|\Lambda|} \nabla_{\Lambda} |\Lambda| - \frac{1}{2} \mathbf{b}^{(i)} \left(\mathbf{b}^{(i)}\right)^{\top}$$

$$= \sum_{i=1}^{m} \frac{1}{2|\Lambda|} |\Lambda| \Lambda^{-\top} - \frac{1}{2} \mathbf{b}^{(i)} \left(\mathbf{b}^{(i)}\right)^{\top}$$

$$= \frac{1}{2} \sum_{i=1}^{m} \Lambda^{-\top} - \mathbf{b}^{(i)} \left(\mathbf{b}^{(i)}\right)^{\top}$$

$$= \frac{1}{2} \sum_{i=1}^{m} \Sigma - \mathbf{b}^{(i)} \left(\mathbf{b}^{(i)}\right)^{\top}$$

and set the gradient to zero:

$$0 = \frac{1}{2} \sum_{i=1}^{m} \Sigma - \mathbf{b}^{(i)} \left(\mathbf{b}^{(i)} \right)^{\top}$$

$$\sum_{i=1}^{m} \Sigma = \sum_{i=1}^{m} \mathbf{b}^{(i)} \left(\mathbf{b}^{(i)} \right)^{\top}$$

$$\Sigma = \frac{1}{m} \sum_{i=1}^{m} \left(\mathbf{x}^{(i)} - \mu \right) \left(\mathbf{x}^{(i)} - \mu \right)^{\top}$$

Exercise 9.1: To bias survival to the fittest by biasing the selection toward the individuals with better objective function values.

Exercise 9.2: Mutation drives exploration using randomness. It is therefore essential in order to avoid local minima. If we suspect there is a better solution, we would need to increase the mutation rate and let the algorithm have time to discover it.

Exercise 9.3: Increase the population size or the coefficient that biases the search toward individual minima.

Exercise 10.1: First reformulate the problem as $f(x) = x + \rho \max(-x, 0)^2$ for which the derivative is

$$f'(x) = \begin{cases} 1 + 2\rho x & \text{if } x < 0 \\ 1 & \text{otherwise} \end{cases} \tag{D.7}$$

This unconstrained objective function can be solved by setting $f'(x) = 0$, which yields the solution $x^* = -\frac{1}{2\rho}$. Thus, as $\rho \to \infty$ we have that $x^* \to 0$.

Exercise 10.2: The problem is reformulated to $f(x) = x + \rho(x < 0)$. The unconstrained objective function is unbounded from below as long as ρ is finite and x approaches negative infinity. The correct solution is not found, whereas the quadratic penalty method was able to approach the correct solution.

Exercise 10.3: You might try to increase the penalty parameter ρ. It is possible that ρ is too small and the penalty term is ineffective. In such cases, the iterates may be reaching an infeasible region where the function decreases faster than the penalty terms causing the method to converge on an infeasible solution.

Exercise 10.4: Let x_p^* be the solution to the unconstrained problem. Notice that $x_p^* \not> 0$. Otherwise the penalty would be $\left(\min(x_p^*, 0) \right)^2 = 0$, which would imply that x_p^* is a solution to the original problem. Now, suppose $x_p^* = 0$. The first-order optimality conditions state that $f'(x_p^*) + \mu x_p^* = 0$, which implies $f'(x_p^*) = 0$, again a contradiction. Thus, if a minimizer exists, it must be infeasible.

Exercise 10.5: It does not require a large penalty ρ to produce an accurate solution.

Exercise 10.6: When iterates should remain feasible.

Exercise 10.7: Consider the following:

$$\underset{x}{\text{minimize}} \quad x^3 \tag{D.8}$$
$$\text{subject to} \quad x \geq 0$$

which is minimized for $x^* = 0$. Using the penalty method we can recast it as

$$\underset{x}{\text{minimize}} \, x^3 + \rho(\min(x,0))^2 \tag{D.9}$$

For any finite ρ, the function remains unbounded from below as x becomes infinitely negative. Furthermore, as x becomes infinitely negative the function becomes infinitely steep. In other words, if we start the steepest descent method too far to the left, we have $x^3 + \rho x^2 \approx x^3$, and the penalty would be ineffective, and the steepest descent method will diverge.

Exercise 10.8: You can frame finding a feasible point as an optimization problem with a constant objective function and a constraint that forces feasibility:

$$\underset{\mathbf{x}}{\text{minimize}} \quad 0$$
$$\text{subject to} \quad \mathbf{h}(\mathbf{x}) = \mathbf{0} \tag{D.10}$$
$$\mathbf{g}(\mathbf{x}) \leq \mathbf{0}$$

Such a problem can often be solved using penalty methods. Quadratic penalties are a common choice because they decrease in the direction of feasibility.

Exercise 10.9: The problem is minimized at $x^* = 1$, which is at the constraint boundary. Solving with the t-transform yields the unconstrained objective function:

$$f_t(\hat{x}) = \sin\left(\frac{4}{5.5 + 4.5\frac{2\hat{x}}{1+\hat{x}^2}}\right) \tag{D.11}$$

which has a single global minimum at $\hat{x} = -1$, correctly corresponding to x^*.

The sigmoid transform has an unconstrained objective function:

$$f_s(\hat{x}) = \sin\left(\frac{4}{1 + \frac{9}{1+e^{-\hat{x}}}}\right) \tag{D.12}$$

Unfortunately, the lower-bound a, in this case $x = 1$, is reached only as $\hat{x}$ approaches minus infinity. The unconstrained optimization problem obtained using the sigmoid transform does not have a solution, and the method fails to properly identify the solution of the original problem.

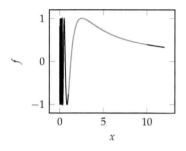 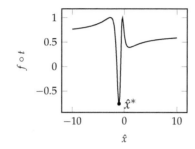 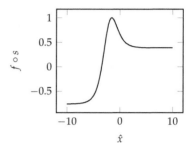

Exercise 10.10: Minimize $x_1^2 + x_2^2$ subject to $x_1 \geq 1$.

Exercise 10.11: We can rearrange the constraint in terms of x_1:

$$x_1 = 6 - 2x_2 - 3x_3 \tag{D.13}$$

and substitute the relation into the objective:

$$\underset{x_2, x_3}{\text{minimize}}\; x_2^2 + x_3 - (2x_2 + 3x_3 - 6)^3 \tag{D.14}$$

Exercise 10.12: The constraint must be aligned with the objective function. The orientation of the objective function is $[-1, -2]$. The orientation of the constraint is $[a, 1]$. The only value for a that aligns them is $a = 0.5$.

Exercise 10.13: The transformed objective function is $f(x) = 1 - x^2 + \rho p(x)$, where p is either a count penalty or a quadratic penalty:

$$p_{\text{count}}(x) = (|x| > 2) \qquad p_{\text{quadratic}}(x) = \max(|x| - 2, 0)^2 \tag{D.15}$$

The count penalty method does not provide any gradient information to the optimization process. An optimization algorithm initialized outside of the feasible set will be drawn away from the feasible region because $1 - x^2$ is minimized by moving infinitely far to the left or right from the origin. The large magnitude of the count penalty is not the primary issue; small penalties can lead to similar problems.

The quadratic penalty method does provide gradient information to the optimization process, guiding searches toward the feasible region. For very large penalties, the quadratic penalty method will produce large gradient values in the infeasible region. In this problem, the partial derivative is:

$$\frac{\partial f}{\partial x} = -2x + \rho \begin{cases} 2(x-2) & \text{if } x > 2 \\ 2(x+2) & \text{if } x < -2 \\ 0 & \text{otherwise} \end{cases} \tag{D.16}$$

For very large values of ρ, the partial derivative in the infeasible region is also large, which can cause problems for optimization methods. If ρ is not large, then infeasible points may not be sufficiently penalized, resulting in infeasible solutions.

Exercise 11.1: We have chosen to minimize a linear program by evaluating every vertex in the convex polytope formed by the constraints. Every vertex is thus a potential minimizer. Vertices are defined by intersections of active constraints. As every inequality constraint can either be active or inactive, and assuming there are n inequality constraints, we do not need to examine more than 2^n combinations of constraints.

This method does not correctly report unbounded linear constrained optimization problems as unbounded.

Exercise 11.2: The simplex method is guaranteed either to improve with respect to the objective function with each step or to preserve the current value of the objective function. Any linear program will have a finite number of vertices. So long as a heuristic, such as Bland's rule, is employed such that cycling does not occur, the simplex method must converge on a solution.

Exercise 11.3: We can add a slack variable x_3 and split x_1 and x_2 into x_1^+, x_1^-, x_2^+, and x_2^-. We minimize $6x_1^+ - 6x_1^- + 5x_2^+ - 5x_2^-$ subject to the constraints $-3x_1 + 2x_2 + x_3 = -5$ and $x_1^+, x_1^-, x_2^+, x_2^-, x_3 \geq 0$.

Exercise 11.4: If the current iterate $\mathbf{x}$ is feasible, then $\mathbf{w}^\top \mathbf{x} = b \geq 0$. We want the next point to maintain feasibility, and thus we require $\mathbf{w}^\top (\mathbf{x} + \alpha \mathbf{d}) \geq 0$. If the obtained value for α is positive, that α is an upper bound on the step length. If the obtained value for α is negative, it can be ignored.

Exercise 11.5: We can rewrite the problem:

$$\underset{\mathbf{x}}{\text{minimize}} \ \mathbf{c}^\top \mathbf{x} - \mu \sum_i \ln\left(\mathbf{A}_{\{i\}}^\top \mathbf{x}\right) \tag{D.17}$$

Exercise 12.1: The weighted sum method cannot find Pareto-optimal points in nonconvex regions of the Pareto frontier.

Exercise 12.2: Nonpopulation methods will identify only a single point in the Pareto frontier. The Pareto frontier is very valuable in informing the designer of the tradeoffs among a set of very good solutions. Population methods can spread out over the Pareto frontier and be used as an approximation of the Pareto frontier.

Exercise 12.3:

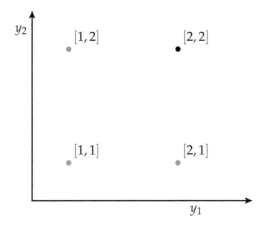

The only Pareto-optimal point is $[1,1]$. Both $[1,2]$ and $[2,1]$ are weakly Pareto optimal.

Exercise 12.4: The "gradient" of a vector is a matrix. Second-order derivatives would require using tensors and solving a tensor equation for a search direction is often computationally burdensome.

Exercise 12.5: The only Pareto-optimal point is $\mathbf{y} = [0,0]$. The rest of the points on the bottom-left border are weakly Pareto-optimal.

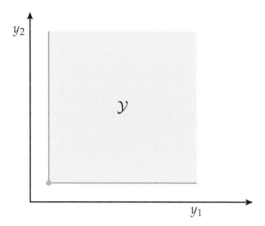

Exercise 12.6: Consider the square criterion space from the previous question. Using $\mathbf{w} = [0, 1]$ assigns zero value to the first objective, causing the entire bottom edge of the criterion space to have equal value. As discussed above, only $\mathbf{y} = [0, 0]$ is Pareto optimal, the rest are weakly Pareto optimal.

Exercise 12.7: For example, if $\mathbf{y}^{\text{goal}}$ is in the criterion set, the goal programming objective will be minimized by $\mathbf{y}^{\text{goal}}$. If $\mathbf{y}^{\text{goal}}$ is also not Pareto optimal, the solution will not be Pareto optimal either.

Exercise 12.8: The constraint method constrains all but one objective. A Pareto curve can be generated by varying the constraints. If we constrain the first objective, each optimization problem has the form:

$$\underset{x}{\text{minimize}} \qquad (x - 2)^2 \qquad \text{(D.18)}$$

$$\text{subject to} \qquad x^2 \leq c \qquad \text{(D.19)}$$

The constraint can be satisfied only for $c \geq 0$. This allows x to vary between $\pm\sqrt{c}$. The first objective is optimized by minimizing the deviation of x from 2. Thus, for a given value of c, we obtain:

$$x^* = \begin{cases} 2 & \text{if } c \geq 4 \\ \sqrt{c} & \text{if } c \in [0, 4) \\ \text{undefined} & \text{otherwise} \end{cases} \qquad \text{(D.20)}$$

The resulting Pareto curve is:

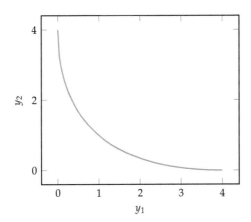

Exercise 12.9: The criterion space is the space of objective function values. The resulting plot is:

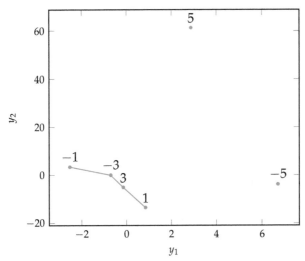

We find that four points are on the approximate Pareto frontier corresponding to our six sample points. The corresponding design points are $x = \{-1, -3, 3, 1\}$.

Exercise 13.1: The one-dimensional unit hypercube is $x \in [0, 1]$, and its volume is 1. In this case the required side length ℓ is 0.5. The two-dimensional unit hypercube is the unit square $x_i \in [0, 1]$ for $i \in \{1, 2\}$, which has a 2-dimensional volume, or area, of 1. The area of a square with side length ℓ is ℓ^2, so we solve:

$$\ell^2 = \frac{1}{2} \quad \Longrightarrow \quad \ell = \frac{\sqrt{2}}{2} \approx 0.707 \tag{D.21}$$

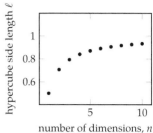

An n-dimensional hypercube has volume ℓ^n. We thus solve:

$$\ell^n = \frac{1}{2} \quad \Longrightarrow \quad \ell = 2^{-1/n} \tag{D.22}$$

The side length approaches one.

Exercise 13.2: The probability that a randomly sampled point is within ϵ-distance from the surface is just the ratio of the volumes. Thus:

$$P(\|x\|_2 > 1 - \epsilon) = 1 - P(\|x\|_2 < 1 - \epsilon) = 1 - (1 - \epsilon)^n \to 1 \tag{D.23}$$

as $n \to \infty$.

Exercise 13.3:

```
function pairwise_distances(X, p=2)
    m = length(X)
    [norm(X[i]-X[j], p) for i in 1:(m-1) for j in (i+1):m]
end
function phiq(X, q=1, p=2)
    dists = pairwise_distances(X, p)
    return sum(dists.^(-q))^(1/q)
end
X = [[cos(2π*i/10), sin(2π*i/10)] for i in 1 : 10]
@show phiq(X, 2)

phiq(X, 2) = 6.422616289332565
```

No. The Morris-Mitchell criterion is based entirely on pairwise distances. Shifting all of the points by the same amount does not change the pairwise distances and thus will not change $\Phi_2(X)$.

Exercise 13.4: A rational number can be written as a fraction of two integers a/b. It follows that the sequence repeats every b iterations:

$$x^{(k+1)} = x^{(k)} + \frac{a}{b} \pmod{1}$$

$$x^{(k)} = x^{(0)} + k\frac{a}{b} \pmod{1}$$

$$= x^{(0)} + k\frac{a}{b} + a \pmod{1}$$

$$= x^{(0)} + (k+b)\frac{a}{b} \pmod{1}$$

$$= x^{(k+b)}$$

Exercise 14.1: The linear regression objective function is

$$\|\mathbf{y} - \mathbf{X}\boldsymbol{\theta}\|_2^2$$

We take the gradient and set it to zero:

$$\nabla (\mathbf{y} - \mathbf{X}\boldsymbol{\theta})^\top (\mathbf{y} - \mathbf{X}\boldsymbol{\theta}) = -2\mathbf{X}^\top (\mathbf{y} - \mathbf{X}\boldsymbol{\theta}) = \mathbf{0}$$

which yields the normal equation

$$\mathbf{X}^\top \mathbf{X}\boldsymbol{\theta} = \mathbf{X}^\top \mathbf{y}$$

Exercise 14.2: As a general rule, more descriptive models should be used when more data are available. If only few samples are available such models are prone to overfitting, and a simpler model (with fewer degrees of freedom) should be used.

Exercise 14.3: The model at hand may have a very large number of parameters. In such case, the resulting linear system will be too large and will require memory that grows quadratically with the parameter space. Iterative procedures like stochastic gradient descent require memory linear in the size of the parameter space and are sometimes the only viable solution.

Exercise 14.4: The leave-one-out cross-validation estimate is obtained by running k-fold cross validation with k equal to the number of samples in X. This means we must run 4-fold cross validation for each polynomial degree.

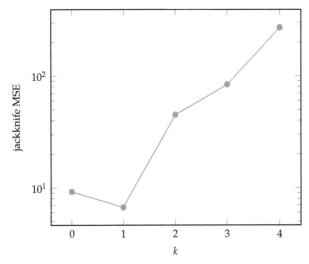

The lowest mean squared error is obtained for a linear model, $k = 1$. We fit a new linear model on the complete dataset to obtain our parameters:

```
X = [[1],[2],[3],[4]]
y = [0,5,4,6]
bases = polynomial_bases(1, 1)
B = [b(x) for x in X, b in bases]
θ = pinv(B)*y
@show θ
```

θ = [-0.49999999999999956, 1.6999999999999997]

Exercise 15.1: Gaussian processes are *nonparametric*, whereas linear regression models are *parametric*. This means that the number of degrees of freedom of the model grows with the amount of data, allowing the Gaussian process to maintain a balance between bias and variance during the optimization process.

Exercise 15.2: Obtaining the conditional distribution of a Gaussian process requires solving equation (15.13). The most expensive operation is inverting the $m \times m$ matrix $\mathbf{K}(X, X)$, which is $O(m^3)$.

Exercise 15.3: The derivative of f is

$$\frac{(x^2 + 1)\cos(x) - 2x\sin(x)}{(x^2 + 1)^2}$$

Below we plot the predictive distribution for Gaussian processes with and without derivative information. The maximum standard deviation in the predicted distribution over $[-5, 5]$ for the Gaussian process with derivative information is approximately 0.377 at $x \approx \pm 3.8$.

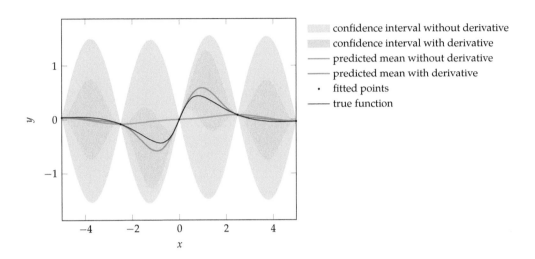

Incorporating derivative information significantly decreases the confidence interval because more information is available to inform the prediction. Below we plot the maximum standard deviation in the predicted distribution over $[-5, 5]$ for Gaussian processes without derivative information with a varying number of evenly-spaced evaluations. At least eight points are needed in order to outperform the Gaussian process with derivative information.

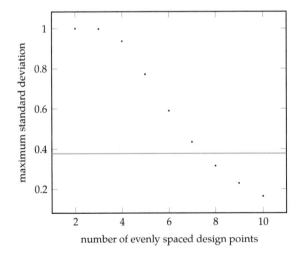

Exercise 15.4: This can be derived according to:

$$k_{f\nabla}(\mathbf{x}, \mathbf{x}')_i = \text{cov}\left(f(\mathbf{x}), \frac{\partial}{\partial x_i'} f(\mathbf{x}')\right)$$

$$= \mathbb{E}\left[(f(\mathbf{x}) - \mathbb{E}[f(\mathbf{x})])\left(\frac{\partial}{\partial x_i'} f(\mathbf{x}') - \mathbb{E}\left[\frac{\partial}{\partial x_i'} f(\mathbf{x}')\right]\right)\right]$$

$$= \mathbb{E}\left[(f(\mathbf{x}) - \mathbb{E}[f(\mathbf{x})])\left(\frac{\partial}{\partial x_i'} f(\mathbf{x}') - \frac{\partial}{\partial x_i'}\mathbb{E}[f(\mathbf{x}')]\right)\right]$$

$$= \mathbb{E}\left[(f(\mathbf{x}) - \mathbb{E}[f(\mathbf{x})])\frac{\partial}{\partial x_i'}(f(\mathbf{x}') - \mathbb{E}[f(\mathbf{x}')])\right]$$

$$= \frac{\partial}{\partial x_i'}\mathbb{E}\left[(f(\mathbf{x}) - \mathbb{E}[f(\mathbf{x})])(f(\mathbf{x}') - \mathbb{E}[f(\mathbf{x}')])\right]$$

$$= \frac{\partial}{\partial x_i'}\text{cov}(f(\mathbf{x}), f(\mathbf{x}'))$$

$$= \frac{\partial}{\partial x_i'}k_{ff}(\mathbf{x}, \mathbf{x}')$$

where we have used $\mathbb{E}\left[\frac{\partial}{\partial x} f\right] = \frac{\partial}{\partial x}\mathbb{E}[f]$. We can convince ourselves that this is true:

$$\mathbb{E}\left[\frac{\partial}{\partial x} f\right] = \mathbb{E}\left[\lim_{h \to 0} \frac{f(x+h) - f(x)}{h}\right]$$

$$= \lim_{h \to 0} \mathbb{E}\left[\frac{f(x+h) - f(x)}{h}\right]$$

$$= \lim_{h \to 0} \frac{1}{h}(\mathbb{E}[f(x+h)] - \mathbb{E}[f(x)])$$

$$= \frac{\partial}{\partial x}\mathbb{E}[f(x)]$$

provided that the objective function is differentiable.

Exercise 15.5: Let us write the joint Gaussian distribution as:

$$\begin{bmatrix} a \\ b \end{bmatrix} \sim \mathcal{N}\left(\begin{bmatrix} \mu_a \\ \mu_b \end{bmatrix}, \begin{bmatrix} v_a & v_c \\ v_c & v_b \end{bmatrix}\right) \tag{D.24}$$

The marginal distribution over a is $\mathcal{N}(\mu_a, v_a)$, which has variance v_a. The conditional distribution for a has variance $v_a - v_c^2/v_b$. We know v_b must be positive in order for the original covariance matrix to be positive definite. Thus, v_c^2/v_b is positive and $v_a - v_c^2/v_b \leq v_a$.

It is intuitive that the conditional distribution has no greater variance than the marginal distribution because the conditional distribution incorporates more information about a. If a and b are correlated, then knowing the value of b informs us about the value of a and decreases our uncertainty.

Exercise 15.6: We can tune the parameters to our kernel function or switch kernel functions using generalization error estimation or by maximizing the likelihood of the observed data.

Exercise 15.7: Maximizing the product of the likelihoods is equivalent to maximizing the sum of the log likelihoods. Here are the log likelihoods of the third point given the other points using each kernel:

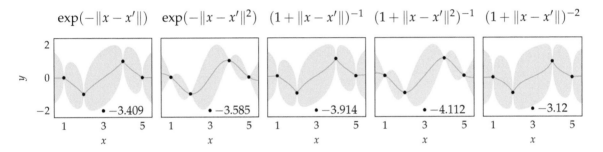

Computing these values over all five folds yields the total log likelihoods:

$$\exp(-\|x - x'\|) \to -8.688$$
$$\exp(-\|x - x'\|^2) \to -9.010$$
$$(1 + \|x - x'\|)^{-1} \to -9.579$$
$$(1 + \|x - x'\|^2)^{-1} \to -10.195$$
$$(1 + \|x - x'\|)^{-2} \to -8.088$$

It follows that the kernel that maximizes the leave-one-out cross-validated likelihood is the rational quadratic kernel $(1 + \|x - x'\|)^{-2}$.

Exercise 16.1: Prediction-based optimization with Gaussian processes can repeatedly sample the same point. Suppose we have a zero-mean Gaussian process and we start with a single point $\mathbf{x}^{(1)}$, which gives us some $y^{(1)}$. The predicted mean has a single global minimizer at $\mathbf{x}^{(1)}$. Prediction-based optimization will continue to sample at $\mathbf{x}^{(1)}$.

Exercise 16.2: Error-based exploration wastes effort in reducing the variance and does not actively seek to minimize the function.

Exercise 16.3: The Gaussian process looks like this:

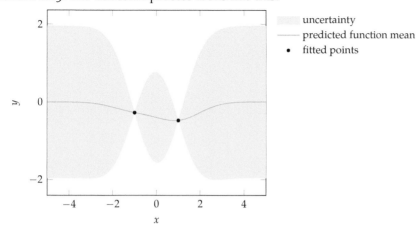

The probability of improvement and expected improvement look like:

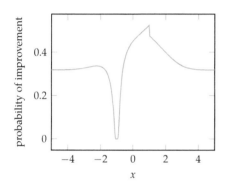 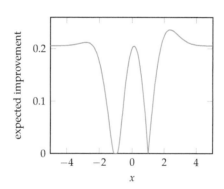

The maximum probability of improvement is at $x = 0.98$ for $P = 0.523$.

The maximum expected improvement is at $x = 2.36$ for $E = 0.236$.

Exercise 17.1: The objective is to minimize $\mathbb{E}_{z \sim \mathcal{N}}[f(x + z)] - \sqrt{\mathrm{Var}_{z \sim \mathcal{N}}[f(x + z)]}$. The first term, corresponding to the mean, is minimized at design point a. The second term, corresponding to the standard deviation, is also maximized at design point a because perturbations to the design at that location cause large variations in the output. The optimal design is thus $x^* = a$.

Exercise 17.2: The deterministic optimization problem is:

$$\begin{aligned} \underset{x}{\text{minimize}} \quad & x^2 \\ \text{subject to} \quad & \gamma x^{-2} \leq 1 \end{aligned}$$

The optimal cross-section length as a function of the factor of safety is $x = \sqrt{\gamma}$. We can thus substitute $\sqrt{\gamma}$ for the cross-section length in each uncertainty formulation and evaluate the probability that the design does not fail. Note that all designs have a 50% chance of failure when the factor of safety is one, due to the symmetry of the normal distribution.

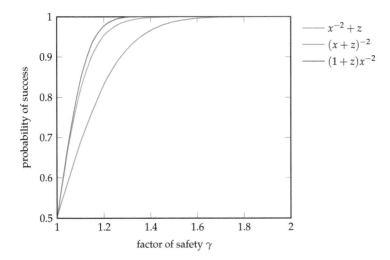

Exercise 17.3: Figure D.1 shows the noise-free feasible region. Without noise, the optimum lies with x_1 infinitely negative. We have noise and have chosen not to accept any outliers with magnitude greater than 6. Such outliers occur approximately $1.973 \times 10^{-7}\%$ of the time.

The feasible region for x_2 lies between e^{x_1} and $2e^{x_1}$. The noise is symmetric, so the most robust choice for x_2 is $1.5e^{x_1}$.

The width of the feasible region for x_2 is e^{x_1}, which increases with x_1. The objective function increases with x_1 as well, so the optimal x_1 is the lowest such that the width of the feasible region is at least 12. This results in $x_1 = \ln 12 \approx 2.485$ and $x_2 = 18$.

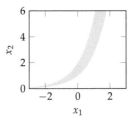

Figure D.1. The noise-free feasible region.

Exercise 18.1: The cumulative distribution function can be used to calculate these values:

```julia
julia> using Distributions
julia> N = Normal(0,1);
julia> cdf(N, 1) - cdf(N, -1)
Error: UndefVarError: Grisu not defined
```

```
julia> cdf(N, 1)
Error: UndefVarError: Grisu not defined
```

Thus, our sample falls within one standard deviation of the mean about 68.3% of the time and is less than one standard deviation above the mean 84.1% of the time.

Exercise 18.2: We begin by substituting in the definition of the sample mean:

$$\text{Var}(\hat{\mu}) = \text{Var}\left(\frac{x^{(1)} + x^{(2)} + \cdots + x^{(m)}}{m}\right)$$

$$= \text{Var}\left(\frac{1}{m}x^{(1)} + \frac{1}{m}x^{(2)} + \cdots + \frac{1}{m}x^{(m)}\right)$$

The variance of the sum of two independent variables is the sum of the variances of the two variables. It follows that:

$$\text{Var}(\hat{\mu}) = \text{Var}\left(\frac{1}{m}x^{(1)}\right) + \text{Var}\left(\frac{1}{m}x^{(2)}\right) + \cdots + \text{Var}\left(\frac{1}{m}x^{(m)}\right)$$

$$= \frac{1}{m^2}\text{Var}\left(x^{(1)}\right) + \frac{1}{m^2}\text{Var}\left(x^{(2)}\right) + \cdots + \frac{1}{m^2}\text{Var}\left(x^{(m)}\right)$$

$$= \frac{1}{m^2}(v + v + \cdots + v)$$

$$= \frac{1}{m^2}(mv)$$

$$= \frac{v}{m}$$

Exercise 18.3: The three-term recurrence relation for orthogonal polynomials is central to their construction and use. A key to the derivation is noticing that a multiple of z can be shifted from one basis to the other:

$$\int_{\mathcal{Z}}(zb_i(z))b_j(z)p(z)\,dz = \int_{\mathcal{Z}}b_i(z)(zb_j(z))p(z)\,dz$$

We must show that

$$b_{i+1}(z) = \begin{cases} (z - \alpha_i)b_i(z) & \text{for } i = 1 \\ (z - \alpha_i)b_i(z) - \beta_i b_{i-1}(z) & \text{for } i > 1 \end{cases}$$

produces orthogonal polynomials.

We notice that $b_{i+1} - zb_i$ is a polynomial of degree at most i, so one can write it as a linear combination of the first i orthogonal polynomials:

$$b_{i+1}(z) - zb_i(z) = -\alpha_i b_i(z) - \beta_i b_{i-1}(z) + \sum_{j=0}^{i-2} \gamma_{ij} b_j(z)$$

for constants α_i, β_i, and γ_{ij}.

Multiplying both sides by b_i and p and then integrating yields:

$$\int_{\mathcal{Z}} (b_{i+1}(z) - zb_i(z)) b_i(z) p(z) \, dz = \int_{\mathcal{Z}} \left(-\alpha_i b_i(z) - \beta_i b_{i-1}(z) + \sum_{j=0}^{i-2} \gamma_{ij} b_j(z) \right) b_i(z) p(z) \, dz$$

$$\int_{\mathcal{Z}} b_{i+1}(z) b_i(z) p(z) \, dz - \int_{\mathcal{Z}} zb_i(z) b_i(z) p(z) \, dz = -\int_{\mathcal{Z}} \alpha_i b_i(z) b_i(z) p(z) \, dz -$$

$$-\int_{\mathcal{Z}} \beta_i b_{i-1}(z) b_i(z) p(z) \, dz +$$

$$+ \sum_{j=0}^{i-2} \int_{\mathcal{Z}} \gamma_{ij} b_j(z) b_i(z) p(z) \, dz$$

$$-\int_{\mathcal{Z}} zb_i^2(z) p(z) \, dz = -\alpha_i \int_{\mathcal{Z}} b_i^2(z) p(z) \, dz$$

producing our expression for α_i:

$$\alpha_i = \frac{\int_{\mathcal{Z}} zb_i^2(z) p(z) \, dz}{\int_{\mathcal{Z}} b_i^2(z) p(z) \, dz}$$

The expression for β_i with $i \geq 1$ is obtained instead by multiplying both sides by b_{i-1} and p and then integrating.

Multiplying both sides by b_k, with $k < i - 1$, similarly produces:

$$-\int_{\mathcal{Z}} zb_i(z) b_k(z) p(z) \, dz = \gamma_{ik} \int_{\mathcal{Z}} b_k^2(z) p(z) \, dz$$

The shift property can be applied to yield:

$$\int_{\mathcal{Z}} zb_i(z) b_k(z) p(z) \, dz = \int_{\mathcal{Z}} b_i(z) (zb_k(z)) p(z) \, dz = 0$$

as $zb_k(z)$ is a polynomial of at most order $i - 1$, and, by orthogonality, the integral is zero. It follows that all γ_{ik} are zero, and the three term recurrence relation is established.

Exercise 18.4: We can derive a gradient approximation using the partial derivative of f with respect to a design component x_i:

$$\frac{\partial}{\partial x_i} f(\mathbf{x}, \mathbf{z}) \approx b_1(\mathbf{z}) \frac{\partial}{\partial x_i} \theta_1(\mathbf{x}) + \cdots + b_k(\mathbf{z}) \frac{\partial}{\partial x_i} \theta_k(\mathbf{x})$$

If we have m samples, we can write these partial derivatives in matrix form:

$$\begin{bmatrix} \frac{\partial}{\partial x_i} f(\mathbf{x}, \mathbf{z}^{(1)}) \\ \vdots \\ \frac{\partial}{\partial x_i} f(\mathbf{x}, \mathbf{z}^{(m)}) \end{bmatrix} \approx \begin{bmatrix} b_1(\mathbf{z}^{(1)}) & \cdots & b_k(\mathbf{z}^{(1)}) \\ \vdots & \ddots & \vdots \\ b_1(\mathbf{z}^{(m)}) & \cdots & b_k(\mathbf{z}^{(m)}) \end{bmatrix} \begin{bmatrix} \frac{\partial}{\partial x_i} \theta_1(\mathbf{x}) \\ \vdots \\ \frac{\partial}{\partial x_i} \theta_k(\mathbf{x}) \end{bmatrix}$$

We can solve for approximations of $\frac{\partial}{\partial x_i} \theta_1(\mathbf{x}), \dots, \frac{\partial}{\partial x_i} \theta_k(\mathbf{x})$ using the pseudoinverse:

$$\begin{bmatrix} \frac{\partial}{\partial x_i} \theta_1(\mathbf{x}) \\ \vdots \\ \frac{\partial}{\partial x_i} \theta_k(\mathbf{x}) \end{bmatrix} \approx \begin{bmatrix} b_1(\mathbf{z}^{(1)}) & \cdots & b_k(\mathbf{z}^{(1)}) \\ \vdots & \ddots & \vdots \\ b_1(\mathbf{z}^{(m)}) & \cdots & b_k(\mathbf{z}^{(m)}) \end{bmatrix}^{+} \begin{bmatrix} \frac{\partial}{\partial x_i} f(\mathbf{x}, \mathbf{z}^{(1)}) \\ \vdots \\ \frac{\partial}{\partial x_i} f(\mathbf{x}, \mathbf{z}^{(m)}) \end{bmatrix}$$

Exercise 18.5: The estimated mean and variance have coefficients which depend on the design variables:

$$\hat{\mu}(\mathbf{x}) = \theta_1(\mathbf{x})$$
$$\hat{v}(\mathbf{x}) = \sum_{i=2}^{k} \theta_i^2(\mathbf{x}) \int_{\mathcal{Z}} b_i(\mathbf{z})^2 p(\mathbf{z}) \, d\mathbf{z} \tag{D.25}$$

The partial derivative of f_{mod} with respect to the ith design component is

$$\frac{\partial}{\partial x_i} f_{\text{mod}}(\mathbf{x}) = \alpha \frac{\partial \theta_1(\mathbf{x})}{\partial x_i} + 2(1 - \alpha) \sum_{i=2}^{k} \theta_i(\mathbf{x}) \frac{\partial \theta_i(\mathbf{x})}{x_i} \int_{\mathcal{Z}} b_i(\mathbf{z})^2 p(\mathbf{z}) \, d\mathbf{z} \tag{D.26}$$

Computing equation (D.26) requires the gradient of the coefficients with respect to $\mathbf{x}$, which is estimated in exercise 18.4.

Exercise 19.1: Enumeration tries all designs. Each component can either be true or false, thus resulting in 2^n possible designs in the worst case. This problem has $2^3 = 8$ possible designs.

```
f(x) = (!x[1] || x[3]) && (x[2] || !x[3]) && (!x[1] || !x[2])
using IterTools
for x in Iterators.product([true,false], [true,false], [true,false])
    if f(x)
        @show(x)
        break
    end
end
```

```
x = (false, true, true)
```

Exercise 19.2: The Boolean satisfiability problem merely seeks a valid solution. As such, we set **c** to zero.

The constraints are more interesting. As with all integer linear programs, **x** is constrained to be nonnegative and integral. Furthermore, we let 1 correspond to true and 0 correspond to false and introduce the constraint **x** $\leq$ **1**.

Next, we look at the objective function and observe that the $\wedge$ "and" statements divide f into separate Boolean expressions, each of which must be true. We convert the expressions to linear constraints:

$$x_1 \implies x_1 \geq 1$$
$$x_2 \vee \neg x_3 \implies x_2 + (1 - x_3) \geq 1$$
$$\neg x_1 \vee \neg x_2 \implies (1 - x_1) + (1 - x_2) \geq 1$$

where each expression must be satisfied (≥ 1) and a negated variable $\neg x_i$ is simply $1 - x_i$.

The resulting integer linear program is:

$$\begin{array}{ll} \underset{\mathbf{x}}{\text{minimize}} & 0 \\ \text{subject to} & x_1 \geq 1 \\ & x_2 - x_3 \geq 0 \\ & -x_1 - x_2 \geq -1 \\ & \mathbf{x} \in \mathbb{N}^3 \end{array}$$

This approach is general and can be used to transform any Boolean satisfiability problem into an integer linear program.

Exercise 19.3: Totally unimodular matrices have inverses that are also integer matrices. Integer programs for which **A** is totally unimodular and **b** is integral can be solved exactly using the simplex method.

A matrix is totally unimodular if every square nonsingular submatrix is unimodular. A single matrix entry is a square submatrix. The determinant of a 1×1 matrix is the absolute value of its single entry. A single-entry submatrix is only unimodular if it has a determinant of ± 1, which occurs only for entries of ± 1. The single-entry submatrix can also be nonsingular, which allows for 0. No other entries are permitted, and thus every totally unimodular matrix contains only entries that are 0, 1, or -1.

Exercise 19.4: The branch and bound method requires that we can perform the branching and bounding operations on our design.[2] The decisions being made in 0-1 knapsack are whether or not to include each item. Each item therefore represents a branch; either the item is included or it is excluded.

A tree is constructed for every such enumeration according to:

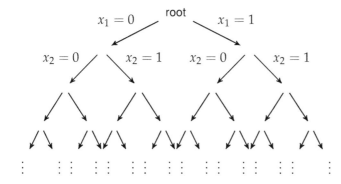

[2] P. J. Kolesar, "A Branch and Bound Algorithm for the Knapsack Problem," *Management Science*, vol. 13, no. 9, pp. 723–735, 1967.

Each node represents a subproblem in which certain items have already been included or excluded. The subproblem has the associated decision variables, values, and weights removed, and the capacity is effectively reduced by the total weight of the included items.

The branch and bound method avoids constructing the entire tree using information from the bounding operation. A bound can be constructed by solving a relaxed version of the knapsack subproblem. This *fractional knapsack* problem allows fractional values of items to be allocated, $0 \leq x_i \leq 1$.

The relaxed knapsack problem can be efficiently solved with a greedy approach. Items are added one at a time by selecting the next item with the greatest ratio of value to weight. If there is enough remaining capacity, the item is fully assigned with $x_i = 1$. If not, a fractional value is assigned such that the remaining capacity is saturated and all remaining items have $x_i = 0$.

We begin by branching on the first item. The subtree with $x_1 = 0$ has the subproblem:

$$\underset{x_{2:6}}{\text{minimize}} \quad -\sum_{i=2}^{6} v_i x_i$$

$$\text{subject to} \quad \sum_{i=2}^{6} w_i x_i \leq 20$$

whereas the subtree with $x_1 = 1$ has the subproblem:

$$\underset{x_{2:6}}{\text{minimize}} \quad -9 - \sum_{i=2}^{6} v_i x_i$$

$$\text{subject to} \quad \sum_{i=2}^{6} w_i x_i \leq 13$$

We can construct a lower bound for both subtrees using the greedy approach. We sort the remaining items by value to weight:

item:	6	4	5	3	2
ratio:	3/4	3/5	5/9	2/4	4/8
	0.75	0.6	0.556	0.5	0.5

For the subtree with $x_1 = 0$, we fully allocate items 6, 4, and 5. We then partially allocate item 3 because we have remaining capacity 2, and thus set $x_3 = 2/4 = 0.5$. The lower bound is thus $-(3 + 5 + 3 + 0.5 \cdot 2) = -12$.

For the subtree with $x_1 = 1$, we allocate items 6 and 4 and partially allocate item 5 to $x_5 = 4/9$. The lower bound is thus $-(3 + 5 + (4/9) \cdot 3) \approx -18.333$.

The subtree with $x_1 = 1$ has the better lower bound, so the algorithm continues by splitting that subproblem. The final solution is $\mathbf{x} = [1, 0, 0, 0, 1, 1]$.

Exercise 20.1: Six expression trees can be generated:

$\mathbb{I}$	$\mathbb{I}$	$\mathbb{F}$	$\mathbb{R}$	$\mathbb{R}$	$\mathbb{R}$
↓	↓	↓	↓	↓	↓
1	2	π	$\mathbb{I}$	$\mathbb{I}$	$\mathbb{F}$
			↓	↓	↓
			1	2	π

Exercise 20.2: Only one expression of height 0 exists and that is the empty expression. Let us denote this as $a_0 = 1$. Similarly, only one expressions of height 1 exists and that is the expression $\{\}$. Let us denote this as $a_1 = 1$. Three expressions exist for depth 2, 21 for depth 3, and so on.

Suppose we have constructed all expressions up to height h. All expressions of height $h+1$ can be constructed using a root node with left and right subexpressions selected according to:

1. A left expression of height h and a right expression of height less than h

2. A right expression of height h and a left expression of height less than h

3. Left and right expressions of height h

It follows that the number of expressions of height $h+1$ are:[3]

$$a_{h+1} = 2a_h(a_0 + \cdots + a_{h-1}) + a_h^2$$

[3] This corresponds to OEIS sequence A001699.

Exercise 20.3: One can use the following grammar and the starting symbol $\mathbb{I}$:

$$\mathbb{I} \mapsto \mathbb{D} + 10 \times \mathbb{I}$$
$$\mathbb{D} \mapsto 0 \mid 1 \mid 2 \mid 3 \mid 4 \mid 5 \mid 6 \mid 7 \mid 8 \mid 9$$

Exercise 20.4: Constructing exception-free grammars can be challenging. Such issues can be avoided by catching the exceptions in the objective function and suitably penalizing them.

Exercise 20.5: There are many reasons why one must constrain the types of the variables manipulated during expression optimization. Many operators are valid only on certain inputs[4] and matrix multiplication requires that the dimensions of the inputs be compatible. Physical dimensionality of the variables is another concern. The grammar must reason about the units of the input values and of the valid operations that can be performed on them.

[4] One typically does not take the square root of a negative number.

For instance, $x \times y$, with x having units $\mathrm{kg}^a \mathrm{m}^b \mathrm{s}^c$, and y having units $\mathrm{kg}^d \mathrm{m}^e \mathrm{s}^f$, will produce a value with units $\mathrm{kg}^{a+d} \mathrm{m}^{b+e} \mathrm{s}^{c+f}$. Taking the square root of x will produce a value with units $\mathrm{kg}^{a/2} \mathrm{m}^{b/2} \mathrm{s}^{c/2}$. Furthermore, operations such as sin can be applied only to unitless inputs.

One approach for handling physical units is to associate an n-tuple with each node in the expression tree. The tuple records the exponent with respect to the allowable elementary units, which are specified by the user. If the elementary units involved are mass, length, and time, then each node would have a 3-tuple (a, b, c) to represent units $kg^a m^b s^c$. The associated grammar must take these units into account when assigning production rules.[5]

Exercise 20.6: The grammar can be encoded using `ExprRules.jl` using string composition:

```
grammar = @grammar begin
    S = NP * " " * VP
    NP = ADJ * " " * NP
    NP = ADJ * " " * N
    VP = V  * " " * ADV
    ADJ = |(["a", "the", "big", "little", "blue", "red"])
    N = |(["mouse", "cat", "dog", "pony"])
    V = |(["ran", "sat", "slept", "ate"])
    ADV = |(["quietly", "quickly", "soundly", "happily"])
end
```

We can use our phenotype method to obtain the solution.

```
eval(phenotype([2,10,19,0,6], grammar, :S)[1], grammar)
```

The phenotype is "little dog ate quickly".

Exercise 20.7: We define a grammar for the clock problem. Let $\mathbb{G}_r$ be the symbol for a gear of radius r, let $\mathbb{A}$ be an axle, let $\mathbb{R}$ be a rim, and let $\mathbb{H}$ be a hand. Our grammar is:

$$\mathbb{G}_r \mapsto \mathbb{R}\,\mathbb{A} \mid \epsilon$$
$$\mathbb{R} \mapsto \mathbb{R}\,\mathbb{R} \mid \mathbb{G}_r \mid \epsilon$$
$$\mathbb{A} \mapsto \mathbb{A}\,\mathbb{A} \mid \mathbb{G}_r \mid \mathbb{H} \mid \epsilon$$

which allows each gear to have any number of rim and axle children. The expression ϵ is an empty terminal.

A clock with a single second hand can be constructed according to:

[5] For a more thorough overview, see, for example, A. Ratle and M. Sebag, "Genetic Programming and Domain Knowledge: Beyond the Limitations of Grammar-Guided Machine Discovery," in *International Conference on Parallel Problem Solving from Nature*, 2000.

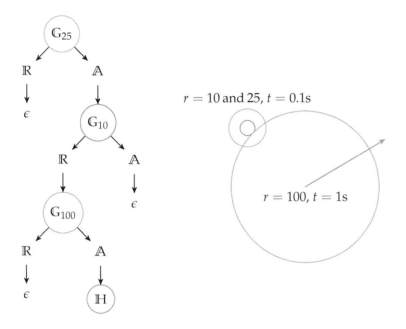

$r = 10 \text{ and } 25, t = 0.1\text{s}$

$r = 100, t = 1\text{s}$

Note that the grammar does not compute the rotation period of the gears. This is handled by the objective function. A recursive procedure can be written to return the rotation periods of all hands. The list of rotation periods is subsequently used to compute the objective function value.

Exercise 20.8: Any of the methods covered in this chapter can be used to complete the four 4s puzzle. A simple approach is to use Monte Carlo sampling on a suitable grammar. Sampled expressions with exactly four 4s are evaluated and, if suitable, are recorded. This procedure is repeated until an expression has been found for each integer.

One such suitable grammar is:[6]

$$\mathbb{R} \mapsto 4 \mid 44 \mid 444 \mid 4444 \mid \mathbb{R} + \mathbb{R} \mid \mathbb{R} - \mathbb{R} \mid \mathbb{R} \times \mathbb{R} \mid \mathbb{R}/\mathbb{R} \mid$$
$$\mathbb{R}^{\mathbb{R}} \mid \lfloor \mathbb{R} \rfloor \mid \lceil \mathbb{R} \rceil \mid \sqrt{\mathbb{R}} \mid \mathbb{R}! \mid \Gamma(\mathbb{R})$$

We round evaluated expressions to the nearest integer. An expression that is rounded up can be contained inside a ceiling operation, and an expression that is rounded down can be contained inside a floor operation, so all such expressions are valid.

[6] The gamma function $\Gamma(x)$ is an extension of the factorial function which accepts real and complex-valued inputs. For positive integers x it produces $(x - 1)!$.

Exercise 20.9: The expression is obtained by applying the production rules:

$$
\begin{aligned}
\mathbb{R} &\mapsto \mathbb{R} + \mathbb{R} & P &= 1/12 \\
\mathbb{R} &\mapsto \mathbb{F} & P &= 5/12 \\
\mathbb{F} &\mapsto 1.5 & P &= 4/7 \\
\mathbb{R} &\mapsto \mathbb{I} & P &= 5/12 \\
\mathbb{I} &\mapsto & P &= 1/3
\end{aligned}
$$

which has a probability of

$$
\frac{1}{12} \frac{5}{12} \frac{4}{7} \frac{5}{12} \frac{1}{3} = \frac{25}{9072} \approx 0.00276
$$

Exercise 20.10: The learning update clears all counts and then increments each production rule each time it is applied. The five applied rules are each incremented once, resulting in:

$$
\begin{aligned}
\mathbb{R} &\mapsto \mathbb{R} + \mathbb{R} \mid \mathbb{R} \times \mathbb{R} \mid \mathbb{F} \mid \mathbb{I} & w_{\mathbb{R}} &= [1, 0, 1, 1] \\
\mathbb{F} &\mapsto 1.5 \mid \infty & p_{\mathbb{F}} &= [1, 0] \\
\mathbb{I} &\mapsto 1 \mid 2 \mid 3 & p_{\mathbb{I}} &= [0, 1, 0]
\end{aligned}
$$

Exercise 21.1: Maximize the lift-to-drag ratio of an airfoil shape subject to a constraint on the structural stability of the airfoil.

Exercise 21.2: Consider a problem where the disciplinary dependency graph is a (directed) tree: if the optimization starts from the root and proceeds by following the topological order, then convergence occurs after one traversal of the tree.

Exercise 21.3: It can execute disciplinary analyses in parallel.

Exercise 21.4: The spring-pendulum problem under the multidisciplinary design feasible architecture is:

$$
\begin{aligned}
\underset{k}{\text{minimize}} \quad & f(\text{MDA}(k)) & \text{subject to} \quad & k > 0 \\
& \theta \leq \theta_{\max} \\
& \text{MDA}(k) \text{ converged}
\end{aligned}
$$

where MDA(k) performs a multidisciplinary analysis on the two disciplinary analyses: a loads analysis that computes $\mathcal{A}[M] = mg\ell \cos(\mathcal{A}[\theta])$ and a displacement analysis that computes $\mathcal{A}[\theta] = \mathcal{A}[M]/\mathcal{A}[k]$. Whether the multidisciplinary design analysis converged is included as an additional response variable in order to enforce convergence.

Solving the optimization problem produces $k \approx 55.353$ /N.

Exercise 21.5: The spring-pendulum problem under the individual design feasible architecture is:

$$\underset{k,\theta_c,M_c}{\text{minimize}} \quad k$$

$$\text{subject to} \quad k > 0$$

$$\theta_c = F_{\text{displacement}}(k, M_c)$$

$$M_c = F_{\text{loads}}(\theta_c)$$

$$\theta \leq \theta_{\max}$$

where θ_c and M_c are the additional coupling variables under control of the optimizer. The two disciplinary analyses can be executed in parallel.

Exercise 21.6: The two disciplinary optimization problems for the spring-pendulum problem under the collaborative optimization architecture are:

$$\underset{k,M}{\text{minimize}} \quad J_{\text{displacement}} \quad = (k_g - \mathcal{A}[k_g])^2 + (M_g - \mathcal{A}[M_g])^2 + (F_{\text{displacement}}(k_g, M_g) - \theta)^2$$

$$\text{subject to} \quad \theta_g \leq \theta_{\max}$$

$$k > 0$$

and

$$\underset{\theta_g}{\text{minimize}} \quad J_{\text{loads}} = (\theta_g - \theta)^2 + (F_{\text{loads}}(\theta_g) - M)^2$$

where the subscript g indicates a global variable. The global variables are $\mathcal{A}_g = \{k_g, \theta_g, M_g\}$.

The system-level optimization problem is:

$$\underset{k_g,\theta_g,M_g}{\text{minimize}} \quad k_g$$

$$\text{subject to} \quad J_{\text{structures}} = 0$$

$$J_{\text{loads}} = 0$$

Bibliography

1. N. M. Alexandrov and M. Y. Hussaini, eds., *Multidisciplinary Design Optimization: State of the Art*. SIAM, 1997 (cit. on p. 387).

2. S. Amari, "Natural Gradient Works Efficiently in Learning," *Neural Computation*, vol. 10, no. 2, pp. 251–276, 1998 (cit. on p. 90).

3. Aristotle, *Metaphysics*, trans. by W. D. Ross. 350 BCE, Book I, Part 5 (cit. on p. 2).

4. L. Armijo, "Minimization of Functions Having Lipschitz Continuous First Partial Derivatives," *Pacific Journal of Mathematics*, vol. 16, no. 1, pp. 1–3, 1966 (cit. on p. 56).

5. J. Arora, *Introduction to Optimum Design*, 4th ed. Academic Press, 2016 (cit. on p. 4).

6. R. K. Arora, *Optimization: Algorithms and Applications*. Chapman and Hall/CRC, 2015 (cit. on p. 6).

7. T. W. Athan and P. Y. Papalambros, "A Note on Weighted Criteria Methods for Compromise Solutions in Multi-Objective Optimization," *Engineering Optimization*, vol. 27, no. 2, pp. 155–176, 1996 (cit. on p. 221).

8. C. Audet and J. E. Dennis Jr., "Mesh Adaptive Direct Search Algorithms for Constrained Optimization," *SIAM Journal on Optimization*, vol. 17, no. 1, pp. 188–217, 2006 (cit. on pp. 105, 126).

9. D. A. Bader, W. E. Hart, and C. A. Phillips, "Parallel Algorithm Design for Branch and Bound," in *Tutorials on Emerging Methodologies and Applications in Operations Research*, H. J. Greenberg, ed., Kluwer Academic Press, 2004 (cit. on p. 346).

10. W. W. R. Ball, *Mathematical Recreations and Essays*. Macmillan, 1892 (cit. on p. 385).

11. D. Barber, *Bayesian Reasoning and Machine Learning*. Cambridge University Press, 2012 (cit. on p. 377).

12. A. G. Baydin, R. Cornish, D. M. Rubio, M. Schmidt, and F. Wood, "Online Learning Rate Adaptation with Hypergradient Descent," in *International Conference on Learning Representations (ICLR)*, 2018 (cit. on p. 82).

13. A. D. Belegundu and T. R. Chandrupatla, *Optimization Concepts and Applications in Engineering*, 2nd ed. Cambridge University Press, 2011 (cit. on p. 6).

14.   R. Bellman, "On the Theory of Dynamic Programming," *Proceedings of the National Academy of Sciences of the United States of America*, vol. 38, no. 8, pp. 716–719, 1952 (cit. on p. 3).

15.   R. Bellman, *Eye of the Hurricane: An Autobiography*. World Scientific, 1984 (cit. on p. 351).

16.   H. Benaroya and S. M. Han, *Probability Models in Engineering and Science*. Taylor & Francis, 2005 (cit. on p. 322).

17.   F. Berkenkamp, A. P. Schoellig, and A. Krause, "Safe Controller Optimization for Quadrotors with Gaussian Processes," in *IEEE International Conference on Robotics and Automation (ICRA)*, 2016 (cit. on p. 297).

18.   H.-G. Beyer and B. Sendhoff, "Robust Optimization—A Comprehensive Survey," *Computer Methods in Applied Mechanics and Engineering*, vol. 196, no. 33, pp. 3190–3218, 2007 (cit. on p. 307).

19.   T. L. Booth and R. A. Thompson, "Applying Probability Measures to Abstract Languages," *IEEE Transactions on Computers*, vol. C-22, no. 5, pp. 442–450, 1973 (cit. on p. 375).

20.   C. Boutilier, R. Patrascu, P. Poupart, and D. Schuurmans, "Constraint-Based Optimization and Utility Elicitation Using the Minimax Decision Criterion," *Artificial Intelligence*, vol. 170, no. 8-9, pp. 686–713, 2006 (cit. on p. 231).

21.   G. E. P. Box, W. G. Hunter, and J. S. Hunter, *Statistics for Experimenters: An Introduction to Design, Data Analysis, and Model Building*, 2nd ed. Wiley, 2005 (cit. on pp. 235, 307).

22.   S. Boyd and L. Vandenberghe, *Convex Optimization*. Cambridge University Press, 2004 (cit. on pp. 6, 178, 438).

23.   D. Braziunas and C. Boutilier, "Minimax Regret-Based Elicitation of Generalized Additive Utilities," in *Conference on Uncertainty in Artificial Intelligence (UAI)*, 2007 (cit. on p. 231).

24.   D. Braziunas and C. Boutilier, "Elicitation of Factored Utilities," *AI Magazine*, vol. 29, no. 4, pp. 79–92, 2009 (cit. on p. 230).

25.   R. P. Brent, *Algorithms for Minimization Without Derivatives*. Prentice Hall, 1973 (cit. on p. 51).

26.   S. J. Colley, *Vector Calculus*, 4th ed. Pearson, 2011 (cit. on p. 19).

27.   V. Conitzer, "Eliciting Single-Peaked Preferences Using Comparison Queries," *Journal of Artificial Intelligence Research*, vol. 35, pp. 161–191, 2009 (cit. on p. 228).

28.   S. Cook, "The Complexity of Theorem-Proving Procedures," in *ACM Symposium on Theory of Computing*, 1971 (cit. on p. 358).

29.   W. Cook, A. M. Gerards, A. Schrijver, and É. Tardos, "Sensitivity Theorems in Integer Linear Programming," *Mathematical Programming*, vol. 34, no. 3, pp. 251–264, 1986 (cit. on p. 341).

30.   A. Corana, M. Marchesi, C. Martini, and S. Ridella, "Minimizing Multimodal Functions of Continuous Variables with the 'Simulated Annealing' Algorithm," *ACM Transactions on Mathematical Software*, vol. 13, no. 3, pp. 262–280, 1987 (cit. on pp. 130, 132).

31.   G. B. Dantzig, "Origins of the Simplex Method," in *A History of Scientific Computing*, S. G. Nash, ed., ACM, 1990, pp. 141–151 (cit. on p. 195).

32.   S. Das and P. N. Suganthan, "Differential Evolution: A Survey of the State-of-the-Art," *IEEE Transactions on Evolutionary Computation*, vol. 15, no. 1, pp. 4–31, 2011 (cit. on p. 157).

33.   W. C. Davidon, "Variable Metric Method for Minimization," Argonne National Laboratory, Tech. Rep. ANL-5990, 1959 (cit. on p. 92).

34.   W. C. Davidon, "Variable Metric Method for Minimization," *SIAM Journal on Optimization*, vol. 1, no. 1, pp. 1–17, 1991 (cit. on p. 92).

35.   A. Dean, D. Voss, and D. Draguljić, *Design and Analysis of Experiments*, 2nd ed. Springer, 2017 (cit. on p. 235).

36.   K. Deb, A. Pratap, S. Agarwal, and T. Meyarivan, "A Fast and Elitist Multiobjective Genetic Algorithm: NSGA-II," *IEEE Transactions on Evolutionary Computation*, vol. 6, no. 2, pp. 182–197, 2002 (cit. on p. 223).

37.   T. J. Dekker, "Finding a Zero by Means of Successive Linear Interpolation," in *Constructive Aspects of the Fundamental Theorem of Algebra*, B. Dejon and P. Henrici, eds., Interscience, 1969 (cit. on p. 51).

38.   R. Descartes, "La Géométrie," in *Discours de la Méthode*. 1637 (cit. on p. 2).

39.   E. D. Dolan, R. M. Lewis, and V. Torczon, "On the Local Convergence of Pattern Search," *SIAM Journal on Optimization*, vol. 14, no. 2, pp. 567–583, 2003 (cit. on p. 103).

40.   M. Dorigo, G. Di Caro, and L. M. Gambardella, "Ant Algorithms for Discrete Optimization," *Artificial Life*, vol. 5, no. 2, pp. 137–172, 1999 (cit. on p. 356).

41.   M. Dorigo, V. Maniezzo, and A. Colorni, "Ant System: Optimization by a Colony of Cooperating Agents," *IEEE Transactions on Systems, Man, and Cybernetics, Part B (Cybernetics)*, vol. 26, no. 1, pp. 29–41, 1996 (cit. on pp. 354, 355).

42.   J. Duchi, E. Hazan, and Y. Singer, "Adaptive Subgradient Methods for Online Learning and Stochastic Optimization," *Journal of Machine Learning Research*, vol. 12, pp. 2121–2159, 2011 (cit. on p. 77).

43.   R. Eberhart and J. Kennedy, "A New Optimizer Using Particle Swarm Theory," in *International Symposium on Micro Machine and Human Science*, 1995 (cit. on p. 159).

44.   B. Efron, "Bootstrap Methods: Another Look at the Jackknife," *The Annals of Statistics*, vol. 7, pp. 1–26, 1979 (cit. on p. 270).

45.   B. Efron, "Estimating the Error Rate of a Prediction Rule: Improvement on Cross-Validation," *Journal of the American Statistical Association*, vol. 78, no. 382, pp. 316–331, 1983 (cit. on p. 273).

46.   B. Efron and R. Tibshirani, "Improvements on Cross-Validation: The .632+ Bootstrap Method," *Journal of the American Statistical Association*, vol. 92, no. 438, pp. 548–560, 1997 (cit. on p. 273).

47.   Euclid, *The Elements*, trans. by D. E. Joyce. 300 BCE (cit. on p. 2).

48.   R. Fletcher, *Practical Methods of Optimization*, 2nd ed. Wiley, 1987 (cit. on p. 92).

49.   R. Fletcher and M. J. D. Powell, "A Rapidly Convergent Descent Method for Minimization," *The Computer Journal*, vol. 6, no. 2, pp. 163–168, 1963 (cit. on p. 92).

50.   R. Fletcher and C. M. Reeves, "Function Minimization by Conjugate Gradients," *The Computer Journal*, vol. 7, no. 2, pp. 149–154, 1964 (cit. on p. 73).

51.   J. J. Forrest and D. Goldfarb, "Steepest-Edge Simplex Algorithms for Linear Programming," *Mathematical Programming*, vol. 57, no. 1, pp. 341–374, 1992 (cit. on p. 201).

52.   A. Forrester, A. Sobester, and A. Keane, *Engineering Design via Surrogate Modelling: A Practical Guide*. Wiley, 2008 (cit. on p. 291).

53.   A. I. J. Forrester, A. Sóbester, and A. J. Keane, "Multi-Fidelity Optimization via Surrogate Modelling," *Proceedings of the Royal Society of London A: Mathematical, Physical and Engineering Sciences*, vol. 463, no. 2088, pp. 3251–3269, 2007 (cit. on p. 244).

54.   J. H. Friedman, "Exploratory Projection Pursuit," *Journal of the American Statistical Association*, vol. 82, no. 397, pp. 249–266, 1987 (cit. on p. 323).

55.   W. Gautschi, *Orthogonal Polynomials: Computation and Approximation*. Oxford University Press, 2004 (cit. on p. 328).

56.   A. Girard, C. E. Rasmussen, J. Q. Candela, and R. Murray-Smith, "Gaussian Process Priors with Uncertain Inputs—Application to Multiple-Step Ahead Time Series Forecasting," in *Advances in Neural Information Processing Systems (NIPS)*, 2003 (cit. on p. 335).

57.   D. E. Goldberg and J. Richardson, "Genetic Algorithms with Sharing for Multimodal Function Optimization," in *International Conference on Genetic Algorithms*, 1987 (cit. on p. 227).

58.   D. E. Goldberg, *Genetic Algorithms in Search, Optimization, and Machine Learning*. Addison-Wesley, 1989 (cit. on p. 148).

59. G. H. Golub and J. H. Welsch, "Calculation of Gauss Quadrature Rules," *Mathematics of Computation*, vol. 23, no. 106, pp. 221–230, 1969 (cit. on p. 331).

60. R. E. Gomory, "An Algorithm for Integer Solutions to Linear Programs," *Recent Advances in Mathematical Programming*, vol. 64, pp. 269–302, 1963 (cit. on p. 342).

61. I. Goodfellow, Y. Bengio, and A. Courville, *Deep Learning*. MIT Press, 2016 (cit. on p. 4).

62. A. Griewank and A. Walther, *Evaluating Derivatives: Principles and Techniques of Algorithmic Differentiation*, 2nd ed. SIAM, 2008 (cit. on p. 23).

63. S. Guo and S. Sanner, "Real-Time Multiattribute Bayesian Preference Elicitation with Pairwise Comparison Queries," in *International Conference on Artificial Intelligence and Statistics (AISTATS)*, 2010 (cit. on p. 228).

64. B. Hajek, "Cooling Schedules for Optimal Annealing," *Mathematics of Operations Research*, vol. 13, no. 2, pp. 311–329, 1988 (cit. on p. 130).

65. T. C. Hales, "The Honeycomb Conjecture," *Discrete & Computational Geometry*, vol. 25, pp. 1–22, 2001 (cit. on p. 2).

66. J. H. Halton, "Algorithm 247: Radical-Inverse Quasi-Random Point Sequence," *Communications of the ACM*, vol. 7, no. 12, pp. 701–702, 1964 (cit. on p. 248).

67. N. Hansen, "The CMA Evolution Strategy: A Tutorial," *ArXiv*, no. 1604.00772, 2016 (cit. on pp. 138, 143).

68. F. M. Hemez and Y. Ben-Haim, "Info-Gap Robustness for the Correlation of Tests and Simulations of a Non-Linear Transient," *Mechanical Systems and Signal Processing*, vol. 18, no. 6, pp. 1443–1467, 2004 (cit. on p. 312).

69. G. Hinton and S. Roweis, "Stochastic Neighbor Embedding," in *Advances in Neural Information Processing Systems (NIPS)*, 2003 (cit. on p. 125).

70. R. Hooke and T. A. Jeeves, "Direct Search Solution of Numerical and Statistical Problems," *Journal of the ACM (JACM)*, vol. 8, no. 2, pp. 212–229, 1961 (cit. on p. 102).

71. H. Ishibuchi and T. Murata, "A Multi-Objective Genetic Local Search Algorithm and Its Application to Flowshop Scheduling," *IEEE Transactions on Systems, Man, and Cybernetics*, vol. 28, no. 3, pp. 392–403, 1998 (cit. on p. 225).

72. V. S. Iyengar, J. Lee, and M. Campbell, "Q-EVAL: Evaluating Multiple Attribute Items Using Queries," in *ACM Conference on Electronic Commerce*, 2001 (cit. on p. 229).

73. D. Jones and M. Tamiz, *Practical Goal Programming*. Springer, 2010 (cit. on p. 219).

74. D. R. Jones, C. D. Perttunen, and B. E. Stuckman, "Lipschitzian Optimization Without the Lipschitz Constant," *Journal of Optimization Theory and Application*, vol. 79, no. 1, pp. 157–181, 1993 (cit. on p. 108).

75.   A. B. Kahn, "Topological Sorting of Large Networks," *Communications of the ACM*, vol. 5, no. 11, pp. 558–562, 1962 (cit. on p. 390).

76.   L. Kallmeyer, *Parsing Beyond Context-Free Grammars*. Springer, 2010 (cit. on p. 361).

77.   L. V. Kantorovich, "A New Method of Solving Some Classes of Extremal Problems," in *Proceedings of the USSR Academy of Sciences*, vol. 28, 1940 (cit. on p. 3).

78.   A. F. Kaupe Jr, "Algorithm 178: Direct Search," *Communications of the ACM*, vol. 6, no. 6, pp. 313–314, 1963 (cit. on p. 104).

79.   A. Keane and P. Nair, *Computational Approaches for Aerospace Design*. Wiley, 2005 (cit. on p. 6).

80.   J. Kennedy, R. C. Eberhart, and Y. Shi, *Swarm Intelligence*. Morgan Kaufmann, 2001 (cit. on p. 158).

81.   D. Kingma and J. Ba, "Adam: A Method for Stochastic Optimization," in *International Conference on Learning Representations (ICLR)*, 2015 (cit. on p. 79).

82.   S. Kiranyaz, T. Ince, and M. Gabbouj, *Multidimensional Particle Swarm Optimization for Machine Learning and Pattern Recognition*. Springer, 2014, Section 2.1 (cit. on p. 2).

83.   S. Kirkpatrick, C. D. Gelatt Jr., and M. P. Vecchi, "Optimization by Simulated Annealing," *Science*, vol. 220, no. 4598, pp. 671–680, 1983 (cit. on p. 128).

84.   T. H. Kjeldsen, "A Contextualized Historical Analysis of the Kuhn-Tucker Theorem in Nonlinear Programming: The Impact of World War II," *Historia Mathematica*, vol. 27, no. 4, pp. 331–361, 2000 (cit. on p. 176).

85.   L. Kocis and W. J. Whiten, "Computational Investigations of Low-Discrepancy Sequences," *ACM Transactions on Mathematical Software*, vol. 23, no. 2, pp. 266–294, 1997 (cit. on p. 249).

86.   P. J. Kolesar, "A Branch and Bound Algorithm for the Knapsack Problem," *Management Science*, vol. 13, no. 9, pp. 723–735, 1967 (cit. on p. 475).

87.   B. Korte and J. Vygen, *Combinatorial Optimization: Theory and Algorithms*, 5th ed. Springer, 2012 (cit. on p. 339).

88.   J. R. Koza, *Genetic Programming: On the Programming of Computers by Means of Natural Selection*. MIT Press, 1992 (cit. on p. 364).

89.   K. W. C. Ku and M.-W. Mak, "Exploring the Effects of Lamarckian and Baldwinian Learning in Evolving Recurrent Neural Networks," in *IEEE Congress on Evolutionary Computation (CEC)*, 1997 (cit. on p. 162).

90.   L. Kuipers and H. Niederreiter, *Uniform Distribution of Sequences*. Dover, 2012 (cit. on p. 239).

91.   J. C. Lagarias, J. A. Reeds, M. H. Wright, and P. E. Wright, "Convergence Properties of the Nelder–Mead Simplex Method in Low Dimensions," *SIAM Journal on Optimization*, vol. 9, no. 1, pp. 112–147, 1998 (cit. on p. 105).

92. R. Lam, K. Willcox, and D. H. Wolpert, "Bayesian Optimization with a Finite Budget: An Approximate Dynamic Programming Approach," in *Advances in Neural Information Processing Systems (NIPS)*, 2016 (cit. on p. 291).

93. A. H. Land and A. G. Doig, "An Automatic Method of Solving Discrete Programming Problems," *Econometrica*, vol. 28, no. 3, pp. 497–520, 1960 (cit. on p. 346).

94. C. Lemieux, *Monte Carlo and Quasi-Monte Carlo Sampling*. Springer, 2009 (cit. on p. 245).

95. J. R. Lepird, M. P. Owen, and M. J. Kochenderfer, "Bayesian Preference Elicitation for Multiobjective Engineering Design Optimization," *Journal of Aerospace Information Systems*, vol. 12, no. 10, pp. 634–645, 2015 (cit. on p. 228).

96. K. Levenberg, "A Method for the Solution of Certain Non-Linear Problems in Least Squares," *Quarterly of Applied Mathematics*, vol. 2, no. 2, pp. 164–168, 1944 (cit. on p. 61).

97. S. Linnainmaa, "The Representation of the Cumulative Rounding Error of an Algorithm as a Taylor Expansion of the Local Rounding Errors," M.S. thesis, University of Helsinki, 1970 (cit. on p. 30).

98. M. Manfrin, "Ant Colony Optimization for the Vehicle Routing Problem," Ph.D. dissertation, Université Libre de Bruxelles, 2004 (cit. on p. 355).

99. R. T. Marler and J. S. Arora, "Survey of Multi-Objective Optimization Methods for Engineering," *Structural and Multidisciplinary Optimization*, vol. 26, no. 6, pp. 369–395, 2004 (cit. on p. 211).

100. J. R. R. A. Martins and A. B. Lambe, "Multidisciplinary Design Optimization: A Survey of Architectures," *AIAA Journal*, vol. 51, no. 9, pp. 2049–2075, 2013 (cit. on p. 387).

101. J. R. R. A. Martins, P. Sturdza, and J. J. Alonso, "The Complex-Step Derivative Approximation," *ACM Transactions on Mathematical Software*, vol. 29, no. 3, pp. 245–262, 2003 (cit. on p. 25).

102. J. H. Mathews and K. D. Fink, *Numerical Methods Using MATLAB*, 4th ed. Pearson, 2004 (cit. on p. 24).

103. K. Miettinen, *Nonlinear Multiobjective Optimization*. Kluwer Academic Publishers, 1999 (cit. on p. 211).

104. D. J. Montana, "Strongly Typed Genetic Programming," *Evolutionary Computation*, vol. 3, no. 2, pp. 199–230, 1995 (cit. on p. 365).

105. D. C. Montgomery, *Design and Analysis of Experiments*. Wiley, 2017 (cit. on p. 235).

106. M. D. Morris and T. J. Mitchell, "Exploratory Designs for Computational Experiments," *Journal of Statistical Planning and Inference*, vol. 43, no. 3, pp. 381–402, 1995 (cit. on p. 242).

107.   K. P. Murphy, *Machine Learning: A Probabilistic Perspective*. MIT Press, 2012 (cit. on pp. 254, 265).

108.   S. Narayanan and S. Azarm, "On Improving Multiobjective Genetic Algorithms for Design Optimization," *Structural Optimization*, vol. 18, no. 2-3, pp. 146–155, 1999 (cit. on p. 227).

109.   S. Nash and A. Sofer, *Linear and Nonlinear Programming*. McGraw-Hill, 1996 (cit. on p. 178).

110.   J. A. Nelder and R. Mead, "A Simplex Method for Function Minimization," *The Computer Journal*, vol. 7, no. 4, pp. 308–313, 1965 (cit. on p. 105).

111.   Y. Nesterov, "A Method of Solving a Convex Programming Problem with Convergence Rate $O(1/k^2)$," *Soviet Mathematics Doklady*, vol. 27, no. 2, pp. 543–547, 1983 (cit. on p. 76).

112.   J. Nocedal, "Updating Quasi-Newton Matrices with Limited Storage," *Mathematics of Computation*, vol. 35, no. 151, pp. 773–782, 1980 (cit. on p. 94).

113.   J. Nocedal and S. J. Wright, *Numerical Optimization*, 2nd ed. Springer, 2006 (cit. on pp. 57, 189).

114.   J. Nocedal and S. J. Wright, "Trust-Region Methods," in *Numerical Optimization*. Springer, 2006, pp. 66–100 (cit. on p. 65).

115.   A. O'Hagan, "Some Bayesian Numerical Analysis," *Bayesian Statistics*, vol. 4, J. M. Bernardo, J. O. Berger, A. P. Dawid, and A. F. M. Smith, eds., pp. 345–363, 1992 (cit. on p. 282).

116.   M. Padberg and G. Rinaldi, "A Branch-and-Cut Algorithm for the Resolution of Large-Scale Symmetric Traveling Salesman Problems," *SIAM Review*, vol. 33, no. 1, pp. 60–100, 1991 (cit. on p. 342).

117.   P. Y. Papalambros and D. J. Wilde, *Principles of Optimal Design*. Cambridge University Press, 2017 (cit. on p. 6).

118.   G.-J. Park, T.-H. Lee, K. H. Lee, and K.-H. Hwang, "Robust Design: An Overview," *AIAA Journal*, vol. 44, no. 1, pp. 181–191, 2006 (cit. on p. 307).

119.   G. C. Pflug, "Some Remarks on the Value-at-Risk and the Conditional Value-at-Risk," in *Probabilistic Constrained Optimization: Methodology and Applications*, S. P. Uryasev, ed. Springer, 2000, pp. 272–281 (cit. on p. 318).

120.   S. Piyavskii, "An Algorithm for Finding the Absolute Extremum of a Function," *USSR Computational Mathematics and Mathematical Physics*, vol. 12, no. 4, pp. 57–67, 1972 (cit. on p. 45).

121.   E. Polak and G. Ribière, "Note sur la Convergence de Méthodes de Directions Conjuguées," *Revue Française d'informatique et de Recherche Opérationnelle, Série Rouge*, vol. 3, no. 1, pp. 35–43, 1969 (cit. on p. 73).

122. M. J. D. Powell, "An Efficient Method for Finding the Minimum of a Function of Several Variables Without Calculating Derivatives," *Computer Journal*, vol. 7, no. 2, pp. 155–162, 1964 (cit. on p. 100).

123. W. H. Press, S. A. Teukolsky, W. T. Vetterling, and B. P. Flannery, *Numerical Recipes in C: The Art of Scientific Computing*. Cambridge University Press, 1982, vol. 2 (cit. on p. 100).

124. C. E. Rasmussen and Z. Ghahramani, "Bayesian Monte Carlo," in *Advances in Neural Information Processing Systems (NIPS)*, 2003 (cit. on p. 335).

125. C. E. Rasmussen and C. K. I. Williams, *Gaussian Processes for Machine Learning*. MIT Press, 2006 (cit. on pp. 277, 278, 287).

126. A. Ratle and M. Sebag, "Genetic Programming and Domain Knowledge: Beyond the Limitations of Grammar-Guided Machine Discovery," in *International Conference on Parallel Problem Solving from Nature*, 2000 (cit. on p. 478).

127. I. Rechenberg, *Evolutionsstrategie Optimierung technischer Systeme nach Prinzipien der biologischen Evolution*. Frommann-Holzboog, 1973 (cit. on p. 137).

128. R. G. Regis, "On the Properties of Positive Spanning Sets and Positive Bases," *Optimization and Engineering*, vol. 17, no. 1, pp. 229–262, 2016 (cit. on p. 103).

129. A. M. Reynolds and M. A. Frye, "Free-Flight Odor Tracking in Drosophila is Consistent with an Optimal Intermittent Scale-Free Search," *PLoS ONE*, vol. 2, no. 4, e354, 2007 (cit. on p. 162).

130. R. T. Rockafellar and S. Uryasev, "Optimization of Conditional Value-at-Risk," *Journal of Risk*, vol. 2, pp. 21–42, 2000 (cit. on p. 316).

131. R. T. Rockafellar and S. Uryasev, "Conditional Value-at-Risk for General Loss Distributions," *Journal of Banking and Finance*, vol. 26, pp. 1443–1471, 2002 (cit. on p. 318).

132. H. H. Rosenbrock, "An Automatic Method for Finding the Greatest or Least Value of a Function," *The Computer Journal*, vol. 3, no. 3, pp. 175–184, 1960 (cit. on p. 430).

133. R. Y. Rubinstein and D. P. Kroese, *The Cross-Entropy Method: A Unified Approach to Combinatorial Optimization, Monte-Carlo Simulation, and Machine Learning*. Springer, 2004 (cit. on p. 133).

134. D. E. Rumelhart, G. E. Hinton, and R. J. Williams, "Learning Representations by Back-Propagating Errors," *Nature*, vol. 323, pp. 533–536, 1986 (cit. on p. 30).

135. C. Ryan, J. J. Collins, and M. O. Neill, "Grammatical Evolution: Evolving Programs for an Arbitrary Language," in *European Conference on Genetic Programming*, 1998 (cit. on p. 370).

136. T. Salimans, J. Ho, X. Chen, and I. Sutskever, "Evolution Strategies as a Scalable Alternative to Reinforcement Learning," *ArXiv*, no. 1703.03864, 2017 (cit. on p. 137).

137. R. Salustowicz and J. Schmidhuber, "Probabilistic Incremental Program Evolution," *Evolutionary Computation*, vol. 5, no. 2, pp. 123–141, 1997 (cit. on p. 377).

138. J. D. Schaffer, "Multiple Objective Optimization with Vector Evaluated Genetic Algorithms," in *International Conference on Genetic Algorithms and Their Applications*, 1985 (cit. on p. 221).

139. C. Schretter, L. Kobbelt, and P.-O. Dehaye, "Golden Ratio Sequences for Low-Discrepancy Sampling," *Journal of Graphics Tools*, vol. 16, no. 2, pp. 95–104, 2016 (cit. on p. 247).

140. A. Shapiro, D. Dentcheva, and A. Ruszczyński, *Lectures on Stochastic Programming: Modeling and Theory*, 2nd ed. SIAM, 2014 (cit. on p. 314).

141. A. Shmygelska, R. Aguirre-Hernández, and H. H. Hoos, "An Ant Colony Algorithm for the 2D HP Protein Folding Problem," in *International Workshop on Ant Algorithms (ANTS)*, 2002 (cit. on p. 355).

142. B. O. Shubert, "A Sequential Method Seeking the Global Maximum of a Function," *SIAM Journal on Numerical Analysis*, vol. 9, no. 3, pp. 379–388, 1972 (cit. on p. 45).

143. D. Simon, *Evolutionary Optimization Algorithms*. Wiley, 2013 (cit. on p. 162).

144. J. Sobieszczanski-Sobieski, A. Morris, and M. van Tooren, *Multidisciplinary Design Optimization Supported by Knowledge Based Engineering*. Wiley, 2015 (cit. on p. 387).

145. I. M. Sobol, "On the Distribution of Points in a Cube and the Approximate Evaluation of Integrals," *USSR Computational Mathematics and Mathematical Physics*, vol. 7, no. 4, pp. 86–112, 1967 (cit. on p. 249).

146. D. C. Sorensen, "Newton's Method with a Model Trust Region Modification," *SIAM Journal on Numerical Analysis*, vol. 19, no. 2, pp. 409–426, 1982 (cit. on p. 61).

147. K. Sörensen, "Metaheuristics—the Metaphor Exposed," *International Transactions in Operational Research*, vol. 22, no. 1, pp. 3–18, 2015 (cit. on p. 162).

148. T. J. Stieltjes, "Quelques Recherches sur la Théorie des Quadratures Dites Mécaniques," *Annales Scientifiques de l'École Normale Supérieure*, vol. 1, pp. 409–426, 1884 (cit. on p. 328).

149. J. Stoer and R. Bulirsch, *Introduction to Numerical Analysis*, 3rd ed. Springer, 2002 (cit. on pp. 89, 443).

150. M. Stone, "Cross-Validatory Choice and Assessment of Statistical Predictions," *Journal of the Royal Statistical Society*, vol. 36, no. 2, pp. 111–147, 1974 (cit. on p. 269).

151. T. Stützle, "MAX-MIN Ant System for Quadratic Assignment Problems," Technical University Darmstadt, Tech. Rep., 1997 (cit. on p. 355).

152. Y. Sui, A. Gotovos, J. Burdick, and A. Krause, "Safe Exploration for Optimization with Gaussian Processes," in *International Conference on Machine Learning (ICML)*, vol. 37, 2015 (cit. on p. 296).

153. H. Szu and R. Hartley, "Fast Simulated Annealing," *Physics Letters A*, vol. 122, no. 3-4, pp. 157–162, 1987 (cit. on p. 130).

154. G. B. Thomas, *Calculus and Analytic Geometry*, 9th ed. Addison-Wesley, 1968 (cit. on p. 22).

155. V. Torczon, "On the Convergence of Pattern Search Algorithms," *SIAM Journal of Optimization*, vol. 7, no. 1, pp. 1–25, 1997 (cit. on p. 103).

156. M. Toussaint, "The Bayesian Search Game," in *Theory and Principled Methods for the Design of Metaheuristics*, Y. Borenstein and A. Moraglio, eds. Springer, 2014, pp. 129–144 (cit. on p. 291).

157. R. J. Vanderbei, *Linear Programming: Foundations and Extensions*, 4th ed. Springer, 2014 (cit. on p. 189).

158. D. Wierstra, T. Schaul, T. Glasmachers, Y. Sun, and J. Schmidhuber, "Natural Evolution Strategies," *ArXiv*, no. 1106.4487, 2011 (cit. on p. 139).

159. H. P. Williams, *Model Building in Mathematical Programming*, 5th ed. Wiley, 2013 (cit. on p. 189).

160. D. H. Wolpert and W. G. Macready, "No Free Lunch Theorems for Optimization," *IEEE Transactions on Evolutionary Computation*, vol. 1, no. 1, pp. 67–82, 1997 (cit. on p. 6).

161. P. K. Wong, L. Y. Lo, M. L. Wong, and K. S. Leung, "Grammar-Based Genetic Programming with Bayesian Network," in *IEEE Congress on Evolutionary Computation (CEC)*, 2014 (cit. on p. 375).

162. X.-S. Yang, *Nature-Inspired Metaheuristic Algorithms*. Luniver Press, 2008 (cit. on pp. 159, 160).

163. X.-S. Yang, "A Brief History of Optimization," in *Engineering Optimization*. Wiley, 2010, pp. 1–13 (cit. on p. 2).

164. X.-S. Yang and S. Deb, "Cuckoo Search via Lévy Flights," in *World Congress on Nature & Biologically Inspired Computing (NaBIC)*, 2009 (cit. on p. 161).

165. P. L. Yu, "Cone Convexity, Cone Extreme Points, and Nondominated Solutions in Decision Problems with Multiobjectives," *Journal of Optimization Theory and Applications*, vol. 14, no. 3, pp. 319–377, 1974 (cit. on p. 219).

166. Y. X. Yuan, "Recent Advances in Trust Region Algorithms," *Mathematical Programming*, vol. 151, no. 1, pp. 249–281, 2015 (cit. on p. 61).

167. L. Zadeh, "Optimality and Non-Scalar-Valued Performance Criteria," *IEEE Transactions on Automatic Control*, vol. 8, no. 1, pp. 59–60, 1963 (cit. on p. 218).

168. M. D. Zeiler, "ADADELTA: An Adaptive Learning Rate Method," *ArXiv*, no. 1212.5701, 2012 (cit. on p. 78).

Index